RECOMBINANT DNA AND GENETIC EXPERIMENTATION

RECOMBINANT DNA AND GENETIC EXPERIMENTATION

Proceedings of a Conference on Recombinant DNA, jointly organised by the Committee on Genetic Experimentation (COGENE) and The Royal Society of London, held at Wye College, Kent, UK, 1-4 April, 1979

Editors

JOAN MORGAN

and

W. J. WHELAN

Published for
COGENE
by
PERGAMON PRESS

OXFORD · NEW YORK · TORONTO · SYDNEY · PARIS · FRANKFURT

U.K.	Pergamon Press Ltd., Headington Hill Hall, Oxford OX3 OBW, England
U.S.A.	Pergamon Press Inc., Maxwell House, Fairview Park, Elmsford, New York 10523, U.S.A.
CANADA	Pergamon of Canada, Suite 104, 150 Consumers Road, Willowdale, Ontario M2J 1P9, Canada
AUSTRALIA	Pergamon Press (Aust.) Pty. Ltd., P.O. Box 544, Potts Point, N.S.W. 2011, Australia
FRANCE	Pergamon Press SARL, 24 rue des Ecoles, 75240 Paris, Cedex 05, France
FEDERAL REPUBLIC OF GERMANY	Pergamon Press GmbH, 6242 Kronberg-Taunus, Pferdstrasse 1, Federal Republic of Germany

First edition 1979

British Library Cataloguing in Publication Data

Recombinant DNAand genetic experimentation.
1. Recombinant DNA - Congresses
I. Morgan, Joan II. Whelan, William Joseph
III. Committee on Genetic Experimentation
IV. Royal Society
574.8'732 QH442 79-40962

ISBN 0-08-024427-0

In order to make this volume available as economically and as rapidly as possible the authors' typescripts have been reproduced in their original forms. This method has its typographical limitations but it is hoped that they in no way distract the reader.

Printed in Great Britain by A. Wheaton & Co., Ltd., Exeter

CONTENTS

ACHIEVEMENTS OF RECOMBINANT DNA RESEARCH

PRACTICAL BENEFITS OF RECOMBINANT DNA RESEARCH

GUIDELINES AND LEGISLATION

RE-EXAMINATION OF BASIC ASSUMPTIONS

PREFACE

The Royal Society and COGENE (The Committee on Genetic Experimentation) were the joint sponsors of an international Conference on recombinant DNA which was held at Wye College, Kent, UK between 1-4 April, 1979. In the belief that the papers presented, and the ensuing discussions, will be of interest to a much wider audience than could be at Wye College, the proceedings are being published on behalf of COGENE.

The Conference was called to review the current status of recombinant DNA research and to discuss matters such as the scientific advances in, and practical applications of, recombinant DNA research, the conjectural hazards and the regulation of the research through guidelines and legislation. The Conference was attended by 150 scientists and other participants from more than 30 nations.

All the sessions were recorded and the master tapes have been given to the Recombinant DNA History Project at the Massachusetts Institute of Technology, Cambridge, Mass., USA, to which there will be free access. During the meeting the discussions following the lectures were transcribed and made available for the participants to check and clarify where necessary. Where substantial changes or additions were made, the fact has been noted on the text.

These proceedings are grouped into the sessions as they occurred at the meeting and comprise, in the main, a chairman's introduction, a summary of the lecture, and the entire discussion following a lecture or lectures.

Our arrangement with the publishers is to achieve a rapid publication and we thank the speakers for prompt submission of their manuscripts and those who participated in the discussions for their cooperation in quickly checking and returning their comments.

We would also like to thank Wye College and in particular their Audio Visual Unit for recording the sessions. Special gratitude is due to Carolyn Ison and Linda Reading for transcribing the tapes at the meeting and again to Carolyn Ison for putting these into their final manuscript form and assembling the typescripts.

Joan Morgan W.J. Whelan

PARTICIPANTS

MARIJA ALACEVIC, Faculty of Technology, University of Zagreb, Laboratory of Industrial Microbiology, 41000 Zagreb, Yugoslavia

I. ALTOSAAR, Biochemistry Department, University of Ottawa, Ottawa, Ontario, Canada

L.J. ARCHER, Molecular Genetics Group, Gulbenkian Institute of Science, Oeiras, Portugal

K. AUNSTRUP, NOVO Research Institute, NOVO ALLE, DK-2880, Bagsvaerd, Denmark

A.A. BAYEV (COGENE), Institute of Molecular Biology, Academy of Sciences, 32 Vavilov Street, Moscow 117312, USSR

G. BERNARDI (COGENE), I.R.B.M., CNRS, Universite Paris VII - Tour 43, 2 Place Jussieu, 75221 Paris Cedex 05, France

G. BERTANI, Institute of Microbial Genetics, Karolinska Institutet, S-104 01 Stockholm, Sweden

J.H. BIRKNER, 5008 Alta Loma Road, Colorado Springs, Colorado 80918, USA

J. BIRNBAUM, Merck Institute for Therapeutic Research, PO Box 2000, Rahway, New Jersey 07065, USA

F. BLASI, Centro di Endocrinologia e Oncologia, Sperimentale del CNR, Via S. Pansini 5, 80131 Naples, Italy

W.F. BODMER, Genetics Laboratory, Department of Biochemistry, University of Oxford, South Parks Road, Oxford, UK

R. BREATHNACH, Laboratoire de Genetique Moleculaire des Eukaryotes du CNRS, Faculte de Medecine, 11 Rue Humann, 67085 Strasbourg Cedex, France

F. BROWN, Animal Virus Research Institute, Pirbright, Woking, Surrey, UK

A. CAMPBELL, Department of Biological Sciences, Stanford University, Stanford, California 94305, USA

R. CAPE, Cetus Corporation, 600 Bancroft Way, Berkeley, California 94710, USA

A.R. CASHMORE, Royal Society of New Zealand, c/o Applied Biochemistry Division, DSIR Private Bag, Palmerston North, New Zealand

S. CHANG, Cetus Corporation, Recombinant Molecular Research, 600 Bancroft Way, Berkeley, California 94710, USA

T-Y CHOW, Institute of Botany, Academia Sinica, Taipei, Taiwan 115

PATRICIA CLARKE, Department of Biochemistry, University College London, Gower Street, London WC1E 6BT, UK

S.N. COHEN (COGENE), Department of Genetics, Stanford University, Room S-337, Stanford, California 94305, USA

J.D. COOMBES, Hoechst Pharmaceutical Research Laboratories, Walton Manor, Milton Keynes, Bucks. MK17 9PJ, UK

P.E. CORNELIS, Universite Catholique de Louvain, Laboratoire de Cytogenetique, Institute Carnoy, 4 Place Croix du Sud, 1348 Louvain-la-Neuve, Belgium

R.A. COX, National Institute for Medical Research, The Ridgeway, Mill Hill, London NW7 1AA, UK

R.K. CRAIG, Courtauld Institute of Biochemistry, Middlesex Hospital Medical School, Mortimer Street, London W1P 7PN, UK

J. DAVISON, Unit of Molecular Biology, International Institute of Cellular and Molecular Pathology, Avenue Hippocrate 75, B-1200 Brussels, Belgium

V. DEBABOV, Institute of Genetics & Selection of Industrial Microorganisms, Dorozhnaya 8, Moscow 113545, USSR

S.D. EHRLICH, Institut de Recherche en Biologie Moleculaire, Universite Paris VII, 2 Place Jussieu, 75221 Paris, France

I. FEDORCSAK (COGENE), Division of Scientific Research and Higher Education, UNESCO House, 7 Place de Fontenoy, 75700 Paris, France

W. FIERS, Laboratory of Molecular Biology, Ledeganekstraat 35, B-9000 Ghent, Belgium

MAGDALENA FIKUS, Institute of Biochemistry & Biophysics, Polish Academy of Sciences, Rakowiecka 36, 02-532 Warsaw, Poland

J.R.S. FINCHAM, Department of Genetics, University of Edinburgh, King's Buildings, West Mains Road, Edinburgh, Scotland, UK

S.W. FISHER, Imperial Cancer Research Fund, 44 Lincoln's Inn Fields, London WC2A 3PX, UK

R.B. FLAVELL, Plant Breeding Institute, Trumpington, Cambridge CB2 2LQ, UK

I.I. FODOR, Institute of Biochemistry & Physiology of Microorganisms, Pushchino 142292, Moscow, USSR

M. FRIED, Imperial Cancer Research Fund Laboratories, PO Box 123, Lincoln's Inn Fields, London WC2A 3PX, UK

A. FURUYA, Tokyo Research Laboratory, Kyowa Hakko Kogyo Co. Limited, 3-6-6 Asahicho, Machidashi, Tokyo 194, Japan

W.J. GARTLAND, Director, Office of Recombinant DNA Activities, NIGMS, Building 31, Room 4A52, Bethesda, Maryland 20014, USA

T. GEFFEN, Department of Health, Alexander Fleming House, Elephant and Castle, London SE1 6BY, UK

E. GEISSLER, Akademie der Wissenschaften der DDR, Zentralinstitut fur Molekularbiologie, Lindenberger Weg 70, DDR-1115, Berlin-Buch, DDR

MARY-JANE GETHING, Imperial Cancer Research Fund, PO Box 123, Lincoln's Inn Fields, London WC2A 3PX, UK

K.I. GIBSON, Department of Education and Science, Medical Research Council, 20 Park Crescent, London W.1., UK

JACQUELINE GITS, Smith Kline-Rit S.A., 89 Rue de l'Institut, B-1330 Rixensart, Belgium

W. GOEBEL, Institut fur Genetik und Mikrobiologie, Rontgenring 11, 87 Wurzburg, FRG

DONNA HARBER, Association of Scientific Technical and Managerial Staffs, 10-26A Jamestown Road, London NW1 7DT, UK

D.O. HAINES, Health & Safety Executive, 25 Chapel Street, London NW1 5DT, UK

T.J.R. HARRIS, Animal Virus Research Institute, Pirbright, Woking, Surrey, U.K.

M. HASHMI, United Nations Environment Programme, PO Box 30552, Nairobi, Kenya, East Africa

L. HATCH, National Institute for Occupational Safety and Health, 4676 Columbia Parkway (C-17), Cincinnati, Ohio 45226, USA

G. HOLT, Commission of the European Communities, 14 Adams Square, Crook Log, Bexleyheath, Kent DA6 8BX, UK

T. IINO, Laboratory of Genetics, Faculty of Science, University of Tokyo, Hongo, Tokyo, Japan

J. INGLE, Agricultural Research Council, 160 Great Portland Street, London W1N 6DT, UK

I.S. JOHNSON, Eli Lilly & Company, Lilly Research Laboratories, 307 East McCarty Street, Indianapolis, Indiana 46206, USA

L. KAARIAINEN, Department of Virology, University of Helsinki, Haartmaninkatu 3, 00290 Helsinki 29, Finland

SIR J. KENDREW (COGENE), EMBL, Postfach 10.2209, 6900 Heidelberg, FRG

N.O. KIELDGAARD, Institute of Molecular Biology, Aarhus University, DK-8000 Aarhus C, Denmark

H.F. KLEFENZ, Schiller Strasse 10, D-6701 Hochdorf-Assenheim 2, FRG

M.A. KOCH, Robert-Koch Institut, Nordufer 20, D-1000 Berlin, FRG

M. KOLOSOV, Institute of Bio-organic Chemistry, USSR Academy of Sciences, 117312 Moscow, Vavilov Street 32, USSR

SIR HANS KORNBERG, Department of Biochemistry, University of Cambridge, Tennis Court Road, Cambridge CB2 1QW, UK

E. LAING, Department of Botany, University of Ghana, Legon, Accra, Ghana

B.W. LANGLEY, ICI Corporate Laboratory, PO Box 11, The Heath, Runcorn, Cheshire, UK

A.F. LANGLYKKE (COGENE), Frederick Cancer Research Center, PO Box B, Frederick, Maryland 21701, USA

ELEANOR LAWRENCE, Association of British Science Writers, Nature, 4 Little Essex Street, London WC2, UK

R. LEWIN, Association of British Science Writers, New Scientist, Kings Reach Tower, Stamford Street, London SE1, UK

H.W. LEWIS, Division of PCM Biology, National Science Foundation, Washington, DC 20550, USA

B-C LIN, Institute of Botany, Academia Sinica, Taipei, Taiwan

A. LINDBERG, Swedish Natural Science Research Council, Statens Bakteriologiska Laboratorium, Fack, S-105 21 Stockholm, Sweden

R.G. LLOYD, Genetics Department, University of Nottingham, Nottingham NG7 2RD, UK

EBBA LUND, Department of Veterinary Virology and Immunology, Royal Veterinary & Agricultural University of Copenhagen, 13 Bulowsvej, DK-1870, Copenhagen, Denmark

A.P. MACLENNAN, WHO Special Programme on Safety Measures in Microbiology, World Health Organisation, 1211 Geneva 27, Switzerland

K. MARKOV, Institute of Molecular Biology, Sofia 1113, Acad. Bontschev Street, Block 6, Bulgaria

M. MARTIN, Department of Health, Education and Welfare, Public Health Service, NIH, Bethesda, Maryland 20014, USA

B.J. MIFLIN, Biochemistry Department, Rothamsted Experimental Station, Harpenden, Herts. AL5 2JQ, UK

DIANE J. MITCHELL, Department of Plant Biology and Microbiology, Queen Mary College, University of London, Mile End Road, London E1 4NS, UK

J.H. MORRIS, Medical Research Council, 20 Park Crescent, London W1N 4AL, UK

K. MURRAY, Department of Molecular Biology, Edinburgh University, Mayfield Road, Edinburgh, Scotland, UK

V.R. OVIATT, WHO Special Programme on Safety Measures in Microbiology, World Health Organisation, Geneva, Switzerland

H. PANNEKOEK, Biochemical Laboratory, State University, Leiden, Wassenaarseweg 64, Netherlands

W.J. PEACOCK, CSIRO Division of Plant Industry, PO Box 1600, Canberra City, ACT 2601, Australia

G.B. PETERSEN, Department of Biochemistry, University of Otago, PO Box 56, Dunedin, New Zealand

V. PIRROTTA, EMBL, Postfach 102209, Heidelberg, FRG

R.H. PRITCHARD, Department of Genetics, University of Leicester, Adrian Building, University Road, Leicester LE1 7RH, UK

O.W. PROZESKY, Foreign Relations Division CSIR Pretoria, National Institute for Virology, Private Bag X4, Sandringham 2131, Republic of South Africa

H.A. RAUE, Department of Biochemistry, Free University, De Boelelaan 1085, Amsterdam, Netherlands

M. RICHARDS, Beecham Pharmaceuticals Research Division, Brockham Park, Betchworth, Surrey RH3 7AJ, UK

M.H. RICHMOND, Department of Bacteriology, University of Bristol, University Walk, Bristol BS9 4AN, UK

R. RILEY (COGENE), Director, Agricultural Research Council, 160 Great Portland Street, London W1N 6DT, UK

GILLIAN RITCHIE, Elsevier Publishing Company, Amsterdam, Netherlands

F.R. ROLLESTON, Medical Research Council, Ottawa, Ontario K1A 0W9, Canada

W.C. RUSSELL, Division of Virology, National Institute for Medical Research, Mill Hill, London NW7 1AA, UK

W.J. RUTTER, Department of Biochemistry & Biophysics, University of California School of Medicine, San Francisco, California 94143, USA

J. SAMBROOK, Cold Spring Harbor Laboratory, PO Box 100, Cold Spring Harbor, New York 11724, USA

G.F. SAUNDERS, University of Texas System Cancer Center, M.D. Anderson Hospital, Department of Biochemistry, 6723 Bertner Avenue, Houston, Texas 77030, USA

J.S. SCHELL (COGENE), Max-Planck Institute fur Zuchtingsforschung, Egelspfad, D-5000 Koln 30, FRG

JANE K. SETLOW, Biology Department, Brookhaven National Laboratory, Upton Long Island, New York 11973, USA

V. SGARAMELLA (COGENE), Consiglio Nazionale delle Ricerche. Lab. CNR Genetica, Biochimica Evoluzionistica, Via S. Spifanio 14, 27100 Pavia, Italy

G. SHARPE, ICI Corporate Laboratory, PO Box 11, The Heath, Runcorn, Cheshire WA7 4QE, UK

G. SIEWERT, Schering AG, Mikrobiologische Biochemie, Postfach 65 03 11, D-1000 Berlin 65, FRG

MAXINE SINGER, Building 37, Room 4A-01, National Institutes of Health, Bethesda, Maryland 20014, USA

ANNA MARIE SKALKA (COGENE), The Roche Institute of Molecular Biology, Department of Cell Biology, Nutley, New Jersey 07110, USA

K.G. SKRYABIN, USSR Academy of Sciences, Moscow B 312, Vavilov Street 32, USSR

D.A. SOUTHGATE (COGENE), Food Research Institute, Colney Lane, Norwich NR4 7UA, Norfolk, UK

E. SPROTT, Unilever Limited, URL Colworth, Sharnbrook, Bedford, Beds. MK44 1LQ, UK

P. STARLINGER, Institut fur Genetik, Weyertal 121, 5 Koln 41, FRG

H.D. STETTEN JR. AND MARJORIE STETTEN, Building 1, Room 122, National Institutes of Health, Bethesda, Maryland 20014, USA

M.G.P. STOKER (COGENE, Foreign Secretary Royal Society), Imperial Cancer Research Fund Laboratories, Lincoln's Inn Fields, London WC2A 3PX, UK

J. SUBAK-SHARPE, Institute of Virology, University of Glasgow, Church Street, Glasgow G11 5JR, UK

Y. SVERDLOV, Institute of Bio-organic Chemistry, USSR Academy of Sciences, 117312 Moscow, Vavilov Street 32, USSR

W. SZYBALSKI, McArdle Laboratory, University of Wisconsin, Madison, Wisconsin 53706, USA

A. TALAVERA, Consejo Superior de Investigaciones Cientificas, Instituto Biologia del Desarrollo, Centro de Biologia Molecular, Universidad Autonoma de Madrid, Canto Blanco, Madrid 34, Spain

S-T TAN, Kumpulan Guthrie SDN.BHD Malaysia, Chemara Research Station, Jalan Labu, Seremban, Negri Sembilan, Malaysia

V.I. TANYASHIN, Institute of Biochemistry & Physiology of Microorganisms, USSR Academy of Sciences, Pushchino on Oka, Moscow 142292, USSR

J.R. TATA, National Institute for Medical Research, Mill Hill, London NW7 1AA, UK

NATALIE M. TEICH, Imperial Cancer Research Fund Laboratories, Lincoln's Inn Fields, London WC2A 3PX, UK

R. THOMAS, Academie Royale de Belgique, 2 Rue du Plagniau, B-1330 Rixensart, Belgium

CONGRESSMAN R. THORNTON, PO Box 521, Sheridan, Arkansas 72150, USA

G. THRELFALL, Searle Research Laboratories, Lane End Road, High Wycombe, Bucks., UK

C.F. THURSTON, Microbiology Department, Queen Elizabeth College, Campden Hill Road, London W8 7AH, UK

J. TOOZE (COGENE), EMBO, Postfach 1022.40, Heidelberg 69, FRG

A. VAN KAMMEN, Royal Dutch Academy of Sciences, Department of Molecular Biology, Agricultural University, De Dreijen 11, 6703 BC Wageningen, Netherlands

M.C. VAN MONTAGU, Rijksuniversiteit Gent, Lab. Histologie en Genetika, Ledeganckstraat 35, B-9000 Gent, Belgium

J.K. VASS, Beatson Institute for Cancer Research, Garscube Estate, Switchback Road, Bearsden, Glasgow, Scotland, UK

P. VENETIANER (COGENE), Biol. Research Centre, Hungarian Academy of Sciences, 6701 PO Box 521, Szeged, Hungary

M.V. VIOLA, The Medical School, University of Connecticut Health Centre, Room L. 2025, Farmington, Connecticut 06032, USA

L.P. VISENTIN, Division of Biological Science, N.R.C. Canada, 100 Sussex Drive, Ottawa, Canada KiA OR6

R. WALLGATE, Association of British Science Writers, Nature, 4 Little Essex Street, London WC2, UK

SIR F. WARNER, 140 Buckingham Palace Road, London SW1, UK

I. WATANABE (COGENE), Department of Molecular Biology, School of Medicine, Keio University, 35 Shinanomachi, Shinjuku-ku, Tokyo 160, Japan

J.D. WATSON, Cold Spring Harbor Laboratory, Cold Spring Harbor, New York 11724, USA

C. WEINER, Program in Science Technology & Society, Massachusetts Institute of Technology 20B-231, Cambridge, Mass. 02139, USA

D.L. WEISS, National Research Council, National Academy of Sciences, 2101 Constitution Avenue, N.W. Washington, DC 20418, USA

EDITH BROWN WEISS, American Society of International Law, Georgetown University Law Center, 600 New Jersey Avenue, N.W. Washington, DC 20001, USA

R.A. WEISS, Imperial Cancer Research Fund Laboratories, PO Box 123, Lincoln's Inn Fields, London WC2A 3PX, UK

C. WEISSMANN, Institut fur Molekularbiologie I, Universitat Zurich, Honggerberg, 8093 Zurich, Switzerland

W.J. WHELAN (COGENE), Biochemistry - UMED, PO Box 520875, Miami, Florida 33152, USA

R.O. WILLIAMS, International Laboratory for Research on Animal Diseases, PO Box 30709, Nairobi, Kenya, East Africa

W.A. WOLF, Bochringer Monnheim GmbH, Biochemica Werk Tutzing, 8132 Tutzing, Postfach, FRG

E.L. WOLLMAN (COGENE), Institut Pasteur, 28 Rue du Docteur Roux, Paris 75724, France

SIR G. WOLSTENHOLME, 41 Portland Place, London W1N 4BN, UK

C. WRAY, Ministry of Agriculture, Fisheries and Food, Central Veterinary Laboratory, New Haw, Weybridge, Surrey KT15 3NB, UK

E.J. YOXEN, Department of Liberal Studies in Science, University of Manchester, Manchester, UK

S. ZADRAZIL, Institute of Molecular Genetics, Czechoslovak Academy of Sciences, Flemingovo 2, 166 10 Praha 6, Czechoslovakia

N.D. ZINDER, The Rockefeller University, 1230 York Avenue, New York, N.Y. 10021, USA

S T A F F

SUE DARELL-BROWN, 47 Bradbourne Street, London SW6

CAROLYN ISON, 3 Hurstwood Road, London NW11

JOAN MORGAN, 58 Muswell Hill Road, London N10

NANCY MORRIS, Bruins, 30 Longdown Road, Lower Bourne, Farnham, Surrey

LINDA READING, Department of Biochemistry, Charing Cross Hospital Medical School, Fulham Palace Road, London W6

INTRODUCTION AND WELCOME

M. G. P. Stoker

Imperial Cancer Research Fund Laboratories, Lincoln's Inn Fields, London WC2A 3PX, U.K.

My task is to start this meeting on Recombinant DNA and to welcome you to Wye College on behalf of COGENE and The Royal Society the co-sponsors. Since there is much to do in the next few days the welcome is brief but nonetheless sincere. So - welcome! Not just as scientists skilled in the art of the soluble, nor as politicians dealing in the possible. The art which is our immediate concern is the art of the improbable.

The particular importance of our meeting, among many on the same subject, lies in its international authorisation. To be sure there has been an international component in several meetings before this, starting with Asilomar. But ICSU, the International Council of Scientific Unions, is the only non-governmental body representative of world science which has tackled the problem. It has done so by setting up COGENE, the Committee on Genetic Experimentation in 1977, under the Chairmanship of Bill Whelan, who will be shortly expounding on the events leading up to this conference. COGENE has already had meetings to discuss particular aspects of recombinant DNA research, but this is its first attempt at a global assessment of the situation which has arisen since the initiative taken in 1974 by our colleagues in the United States.

I stress that COGENE and its parent body, ICSU, are non-governmental and represent the world's scientists and no one else. As such COGENE is certainly qualified to give an expert view, but we must be careful not to claim any representation beyond this point. I need hardly list the groups that COGENE does *not* represent but who are deeply interested in the subject of this meeting, governments included.

COGENE's intention then, has been to organise a widely based international conference of scientists knowledgeable and concerned about recombinant DNA research risks and implications. But the organisers were also concerned that discussions could take place in a cool, uncharged atmosphere, at least until the conference had assessed the situation and come to some conclusions. COGENE, therefore, looked around for a small, quiet country where a display of emotion is, to say the least, embarrassing. Britain was the obvious choice. The Royal Society was happy to co-sponsor such an important event, the Wye College, which many of us knew through Harden conferences, seemed suitably remote. (Remoter than you may think - because the river can rise in a timely way and cut us all off completely.)

During the meeting we shall no doubt hear a good deal about the possibility or impossibility of risk quantitation. Indeed there has now emerged a popular statistical game of "safer than thou". In this context I particularly like Pochin's (1978) figures for a one in a million risk of dying. These include one month of background radiation, one and a half minutes of severe rock climbing, 400 miles by air, 60 miles by car, three quarters of a cigarette, and, of personal interest to me, being 60 years old for 20 minutes. No doubt many many years would be needed in a P3 laboratory, unless of course you happen to be 60. Unfortunately, the figures for real risk, where they can be measured, are quite different from those of perceived risk, and it is the latter which usually determines the rules.

Talking of rules, British scientists have a particular reason to welcome the conference in the United Kingdom. Despite an orderly start, based on pre-existing safety legislation which gave Britain some initial advantage over other countries, we are now hoist with our own petard. Those responsible are trying to devise a means of escape from the safety net or should I say prison bars, which prevent us from reacting rapidly to changing ideas evident in the rest of the world. So we hope that the final conclusions of this conference, whatever they are, will be noticed by those with effective authority over our experiments (this last phrase being our greatest worry).

The motto of The Royal Society is "Nullius in Verba", adopted over 300 hundred years ago. It is extracted in a curious way from some lines of Horace and unfortunately no one knows how to translate it. However, since the Society was founded by pioneers of experiment and observation, in an age of scholasticism and ancient dogma, the motto is taken as a determination to withstand the authority of the given word, and to appeal to facts. Roughly "Nothing by word alone", or more loosely "Find out for yourself". There can surely be no field in which this advice is more needed than in assessment of risks and regulations relating to the new recombinant DNA research.

Fortunately a good deal of finding out by experiment has been achieved in risk assessment in the last few years, as you can see from the recent COGENE report. But it is too much to expect a rational response and the dogma persists. Let us hope that it does not last, like medieval dogma, for hundreds of years.

REFERENCE

Pochin, E. (1978). Estimates of industrial and other risks. *J.R. Coll. Physns. Lond.*, 12, 210-218.

THE PURPOSE OF THE MEETING

W. J. Whelan

Department of Biochemistry, University of Miami School of Medicine, P.O. Box 016129, Miami, Florida 33101, U.S.A.

I am grateful to have Dr. Stoker as my co-Chairman because it permits me as the Chairman of COGENE to convey in a direct manner our thanks to The Royal Society for co-sponsoring this meeting, for providing us with a home base and for providing the able help of Dr. Ronald Keay, the Executive Secretary of The Royal Society, and Miss Nancy Morris, who arranged many of the details of the meeting. Within COGENE itself, I want to thank Dr. John Tooze and his colleagues on the Organizing Committee for the Conference and again Dr. Stoker, who was a member of this Committee.

I would like to expand on the brief introduction you have already had to ICSU, the parent body of COGENE. This organization, celebrating its 60th birthday next July, is a unique non-governmental, non-political consortium of 18 independent scientific unions - astronomy, geography, chemistry, biophysics and the like, together with more than 60 national members who are academies, like The Royal Society, or research councils.

ICSU's main purpose is "to encourage international scientific activity for the benefit of mankind". The primary means by which ICSU fulfils this objective is to initiate, design and coordinate research programs, as for example the International Geophysical Year and the International Biological Program. In addition, ICSU acts as a focus for the communication of scientific information and the development of standards in methodology, nomenclature and units. The various members of the ICSU family organize international conferences, congresses, symposia and publish journals. As one example, a number of us here tonight will soon be meeting again in Toronto when our own union, Biochemistry, holds its triennial congress.

We cooperate closely with, and receive sizeable financial support from, intergovernmental organizations such as WHO, FAO and especially UNESCO.

When scientific activities of a wide-ranging nature arise and the scope of the project is clearly of interest to several unions, ICSU itself moves in to bring the interested unions together to form a Scientific Committee. There are many such committees already in existence, for example, the Committees on Space Research, Scientific Problems of the Environment and Science and Technology in Developing Countries.

Recombinant DNA technology was clearly a topic that would call for a Scientific Committee. That the perception of ICSU's role in the field came early rather than late was certainly the result of Sir John Kendrew becoming the Secretary General almost coincident with the publication of the "Berg" letter in 1974. Exploratory discussions led to a recommendation to ICSU to take such action and COGENE came into being in October 1976 with the support of 7 of the ICSU unions. The original recommendation to form a committee for recombinant DNA was widened to one on genetic experimentation in general. We have had the support of some of the ablest practitioners in the field such as Giorgio Bernardi, Stanley Cohen, Ken Murray, Anna Skalka and some of those with influence such as Alexander Bayev, Chairman of the USSR Recombinant DNA Committee. Also we operate in part with the financial help of UNESCO and WHO, whose representatives to COGENE are here today.

Our original goals were primarily expressed through two Working Groups, one on Recombinant DNA Guidelines and one on Risk Assessment. You have their first reports as part of the Conference materials.

More recently, a Working Group on Benefits and Applications, which I convene, was established. It was, we thought, a promising reflection of the changing mood that at COGENE's annual business meeting yesterday we heard that the Risk Assessment Group is lowering the pace of its activity, while the Benefits and Applications Group, plus a previously inactive Training and Education Group, will now move to initiate training courses in recombinant DNA methodology in areas of the world where such courses are not presently available. These decisions reflect the lowering of the concerns over conjectural hazards and the corresponding lowering of containment requirements for laboratory experiments, such that the methodology is now becoming easily usable. It is our wish to see that those who would want to use the techniques but who presently have no access to training, will now receive expert instruction.

In carrying out these tasks we have been fulfilling the first three of the four articles of our Terms of Reference. The fourth reads that COGENE shall act to provide a forum through which interested national, regional and other international bodies may communicate.

We have perceived our ability to act as the convenor of an international meeting of this type in a way that no other body could act and we have succeeded in bringing together scientists, research directors, legislators, lawyers and public health experts and an audience drawn from 30 nations. I hope our success in bringing you together will be matched by a successful meeting. I cannot, however, promise you that it will take on the elements of a religious revivalist meeting, a description that Dr. Stetten recently applied to Asilomar II.

The truly international aspect of the problems that have beset this branch of research hardly need dwelling on. Yet I have often felt in listening to some of the debates that there is far too much parochial national concern. If there are six-headed green-eyed monsters to be created by recombinant DNA methodology they will not be suppressed simply by banning the research in any one country. There should be international agreement. At the same time, if the technique is not going to give rise to these dangers, it is as important to be aware of this in Calcutta as it is in Chicago so that if someone here has spent money on risk assessment or has good reasons for not spending such money, someone elsewhere does not waste money repeating or initiating futile endeavours.

Given that Asilomar II was four years ago and given that that particular span of time, in terms of what has happened in the field, makes Asilomar seem more like fourteen years ago, we felt it was timely to call an international conference that would provide first an exposition of the current state of the science and technology - this constitutes the first half of the meeting, and secondly a study of the problems as they are reflected in guidelines, legislation and risk assessment, followed by a look at the future - the second half of the conference.

Credit for the idea for this meeting should go to one of the members of COGENE who would, however, rather remain anonymous. Where Michael Stoker borrowed from Horace, I will borrow from Greek mythology which records one of the earliest examples of genetic engineering, that of Zeus, who received a blow on the head, and from his head sprang Pallas Athene, fully armed. When from the head of our own Zeus in COGENE there sprang the proposal for this meeting, fully detailed, some eighteen months ago, the rest of us were immediately and equally seized with the idea and quickly assisted in its delivery. Mythology also records how Zeus suffered an immortal headache as a result of the traumatic experience, a condition that some of us have often reflected in recent months, especially over the matter of press reporting.

That then is the setting of the meeting, our attempt to bring people together to discuss international science and international concerns. You may recently have read or have heard of ulterior motives why COGENE is supposed to have chosen this particular time to hold this meeting in this particular country. There were no ulterior motives and there is no hidden agenda.

There is obviously and explicitly something in this new branch of science that stirs very basic emotions. Even the word chimera, as applied to the hybrid DNA plasmid, recalls the mythical chimera that the dictionary defines as a horrible, frightening and fantastic combination of incongruous parts. The notion of gene transplantation by chimeras conveys this same unnerving aspect to many people and Lewis Thomas has reminded us that the mixed beings of classical mythology, part man, part animal and part plant are mostly associated with tragedy.

New ideas that involve inter-species interaction have always been scorned and fought. Jenner's use of a cow vaccine against smallpox was ridiculed in this way. A Gilray cartoon depicted miniature cows erupting from the bodies of people so vaccinated.

The modern day counterpart was to be seen in the *Boston Globe*, whose cartoonist pictured the joy of a scientist who has learned that the city of Cambridge, Massachusetts, did not intend to ban recombinant DNA research. He conveyed the glad news to the monsters that peopled his laboratory with the cry: "Crack open the liquid nitrogen dumplings, we're on our way!".

Still others concentrate on their fears that the technique will be used for human genetic experimentation, as we were reminded at the U.S. National Academy of Sciences Forum on Recombinant DNA two years ago, when the demonstrators carried a banner to the platform bearing a quotation from Adolf Hitler: "We will create the perfect race". Another hotly debated aspect of the recombinant DNA scene is the fear of an invasion by money grubbers from industry who would patent and exploit the new discoveries. This also gave rise to its own banner at the same Academy meeting, namely: "Public debate before private profit".

Sometimes the comments are in lighter vein such as the comparison of the dangers of this research with Mark Twain's opinion of Wagner's music: "It's not as bad as it sounds", or the *New Yorker* cartoonist who depicts a worried gentleman getting out of bed and saying to his wife: "I have the weirdest feeling that someone was fiddling with my genes during the night".

COGENE did not come into being with the intent of presiding over the suppression of this new methodology, but of helping it. Nevertheless, we have no wish to become known as a partisan group, advocating unbridled and unprincipled exploitation of these new research tools, and I owe my own position as Chairman of COGENE to the fact that I have no personal research stake in this field. At the same time I consider it our duty and mission, backed by the authority and prestige of ICSU, to speak up forcefully on behalf of the science and of the scientists who practice it, to see that the pursuit of new knowledge proceeds optimally, with the minimum of restrictions, consonant with whatever are the real needs for safety precautions.

It is in this vein that I would like to conclude, like Michael Stoker, by borrowing a quotation from The Royal Society, namely from its history, as written by Thomas Sprat in 1667 where we read:

"It is strange that we are not able to inculcate into the minds of many men, the necessity of that distinction of my Lord *Bacons*, that there ought to be *Experiments* of *Light*, as well as of *Fruit*. It is their usual word, *What solid good will come from thence*? They are, indeed to be commended for being so severe *Exactors of goodness*. And it were to be wish'd, that they would not only exercise this vigour, about *Experiments*, but on their own *lives* and *actions*: that they would still question with themselves, in all that they do; what *solid good* will come from thence? But they are to know, that in so large, and so various an *Art* as this of *Experiments*, there are many degrees of usefulness: some may serve for real, and plain *benefit*, without much *delight*: some for *teaching* without apparent *profit*: some for *light* now, and for *use* hereafter; some only for *ornament*, and *curiosity*. If they will persist in contemning all *Experiments*, except those which bring with them immediate *gain* and a present *harvest*: they may as well cavil at the Providence of God, that he has not made all the seasons of the year, to be times of *mowing*, *reaping*, and *vintage*."

Or, as Michael Stoker said: "Find out for yourself!"

RECOMBINANT DNA TECHNIQUES PUT INTO BIOLOGICAL PERSPECTIVE

CHAIRMAN'S INTRODUCTION

Sir Hans Kornberg

Department of Biochemistry, University of Cambridge, Tennis Court Road, Cambridge CB2 1QW, U.K.

It will not have escaped your notice that distinguished scientists, whose utterances are normally characterised by a sober adherence to the contemporary scene, seem to be irresistibly drawn to Greek mythology when discussing work with recombinant DNA. Bill Whelan's reference last night to Pallas Athene's startling emergence from the head of Zeus is a case in point. It thus did not surprise me to observe, when I last had the privilege of standing here (which was at a Harden Conference some two years ago on the-then-new techniques of recombinant DNA technology) that some speakers who discussed the implications of that novel technique, would gloomily refer - and what is more would gloomily refer with great frequency - to the legend of Pandora and her box. Even I, whose task it was to sum up the Conference, could not resist the temptation of an historical allusion. I likened our endeavours then to those that confronted the medieval navigators who set out to explore the New World, whose hopes were high but mingled with anxiety, and whose maps carried great blank areas adorned with pictures of mythical beasts inscribed "Here be dragons".

It now appears to us in the comforting retrospect of more than two years of work in this area that those mythical beasts may have as much substance as the mythical beasts that, in the event, were not encountered by those navigators. It was perhaps apparent even then that much of the discussion and the anxieties emerging from that discussion took place in a framework that largely forgot or ignored the biological context of our work. There were few voices, but only a relatively few voices, that were raised to remind us of the realities of genetic exchange in the world outside the laboratory and of the factors that would affect the likelihood of an epidemic catastrophe.

If I may borrow again from the classics, though this time from a Greek author, it may be that at the end of our discussion this morning we will feel like the fly that sat on the axle-wheel of the chariot and said "Look what a dust I raised": man may be able, even were he to attempt deliberately to do so, to do very little to change the processes that go on around him. This morning's discussion places the accent on the biological perspective of recombinant DNA techniques. It is appropriate that our first speaker should be one of those whose voices were raised even then - Walter Bodmer of Oxford.

EVOLUTION AND STABILITY OF SPECIES

W. F. Bodmer

Genetics Laboratory, University of Oxford,
Oxford OX1 3QU, U.K.
from 1 July 1979
Imperial Cancer Research Fund, P.O. Box 123,
Lincoln's Inn Fields, London, WC2A 3PX, U.K.

This brief survey of evolution and the nature of species is meant to serve as a background to considering the possible implications of the heterologous combinations that can be synthesized by recombinant DNA techniques. The essence of life is the capacity for self reproduction, which is mediated by the genetic material, DNA (or RNA), and the machinery for its expression and duplication. Without faithful reproduction there can be no biological continuity and so it is an inevitable property of biological systems that genetic changes, or mutations, should be comparatively rare. Mutations do, however, occur and they provide the basic genetic variation which allows evolution to take place. Evolution is progressive genetic change and, as Darwin taught us, has natural selection as its main driving force. Mutation, because it is rare, is not itself a significant force for genetic change. Adaptive evolution thus depends on the differential reproduction and survival of different genotypes leading to increases in frequency of the advantageous at the expense of the less favoured genotypes. Evolution by natural selection may be thought of as Nature's way of counteracting the second law of thermodynamics.

MUTATION

No doubt the earliest, most primitive, organisms were poorly adapted to their environments and may have had to tolerate substantial and uncontrolled heritable variations imposed by a primitive and imperfect system for reproduction. However, as soon as there had been any substantial evolutionary adaptation, correlated presumably with increased organisational complexity, a premium must have been placed on the fidelity of the system of reproduction. The more complex the biological organisation, and the higher the level of adaptation, the less likely it becomes that an arbitrary genetic change can be advantageous. Direct observation clearly supports the *a priori* argument that most mutations should be deleterious and, indeed, the greater the extent of the genetic change, the more deleterious is its effect likely to be. Thus, while point mutations may still lead to functional proteins, most deletions will not do so and, clearly, the more extensive the deletion, the larger the number of genes affected. As pointed out by R.A. Fisher (1930), many years ago, mutation rates presumably reach an equilibrium maintained by the opposing evolutionary pressures, first to keep them down because of their overall deleterious effects and second to favour them in order to provide enough variation for the evolutionary process to take place.

Fisher also argued that the more highly adapted the organism, the less likely it would be that a single large change would be advantageous. Conversely, advantageous mutations are expected nearly always to have relatively small overall phenotypic effects.

Evolution by natural selection is not a process involving grand a priori designs. It leads to a somewhat messy continuing patching up of existing systems, taking advantage of duplications, which increase the amount of available genetic material, to develop new functions. If new proteins can be produced from duplicated nucleotide sequences that are translated in a different reading frame, as seems likely, these will have little, if any, relationship to their immediate evolutionary precursors and so their initial functions may be carried out extremely inefficiently. The genetic system has built into it extensive redundancy, which provides both resilience in the face of unwanted variation and a source of material for new developments, as needed. There are no opportunities for neat reconstructions of the genotype in order to clear away the accumulated messy patches.

SEX AND RECOMBINATION

Early primitive organisms must, of course, have reproduced asexually. In an asexual population two different advantageous mutations can only be incorporated into a single individual, and so into the population, if one of the mutations occurs in a descendant of an individual in which the other has occurred. In a sexual population, on the other hand, two mutations can be brought together in the same individual by recombination, interpreted in the broadest sense. Thus, as clearly pointed out by Fisher and Muller in the 1930s, sexual reproduction greatly accelerates the rate of evolution by facilitating the accumulation in a single individual of different independently occurring advantageous mutations. This feature of sexual reproduction is undoubtedly responsible for its evolution. As Muller (1958) aptly put it, the difference between asexual and sexual evolution is essentially a contrast between evolution in series as compared to evolution in parallel.

Sexual mechanisms, and genetic exchange, comprise two very different and quite separate processes. First, there is the mechanism by which different individuals or organisms, get together and in some way juxtapose their genetic information so that it can be reassorted, or exchanged. In classical genetic terms this includes all those aspects of the sexual process which lead up to synapsis, namely the pairing of chromosomes in meiosis. The second component which is much more specific, is the actual exchange of genetic information at the chromosomal level. Genetic recombination and reassortment can occur in the absence of recombination in this second more restricted sense. Thus, the independent segregation of whole chromosomes during meiosis is an important part of the process of genetic exchange and recombination from an evolutionary point of view. There are mechanisms of genetic exchange which do not even go as far as synapsis, but involve only whole genetic units or chromosomes. In bacteria, for example, this would include transfer of drug resistance factors. Somatic cell fusion is another process leading to genetic assortment which involves no recombination between homologous chromosomes. It is indeed tempting to speculate that cell fusion and hybridisation, accompanied by more or less erratic chromosomal segregation, may have been the precursor of an organised meiosis in primitive unicellular eukaryotic organisms. Such fusions could have led to evolutionarily effective recombination of genetic information by segregation of whole chromosomes (and also possibly chromosome fragments), before the evolution of precise pairing and recombination between homologous chromosomes (Bodmer, 1972). In the absence of precisely controlled mitosis, a unicellular organism, may tolerate extra chromosomes much more readily than a complex multicellular organism and this is observed to be the case in many cell cultures. Cell fusion, which could easily have happened at first by chance, might therefore have led to the advantageous production of increased genetic variability without an immediate need for refinement of the recombination process. Meiosis might thus have evolved as a regularisation of cell division following primitive cell fusion. The precise mechanism of genetic recombination between homologous chromosomes depends on the proper co-ordinated functioning of a number of different genes. As is true for most complex organs, there is no advantage to the system until it works in its entirety. The complexity of recombination precludes its evolution as a functioning system in one step, especially if population advantage was the main selective pressure favouring this development. Many aspects of the system must therefore originally have evolved as byproducts of some advantage conferred at the individual level, that have later been exploited for the evolution of genetic recombination. Thus, many of the steps of genetic recombination at the molecular level can be traced to functions required for DNA metabolism and replication. Recombination must be precise and reciprocal, otherwise it leads to gross mutations whose disadvantages would surely outweigh any possible advantages from new combinations. Genetic reassortment can only be an advantage if it can occur without disturbing functional organisation. The precision of homologous recombination depends on DNA base pairing between homologous chromosomes.

The evolution of multicellular sexual organisms depended on the development of a separate germ line. This produces the cells, eggs and sperm, which fuse to give rise to a new individual. The complexity of mechanisms for sexual reproduction and genetic recombination indicates the strength of the selective pressures which must have favoured their development, and so the importance of the increased rate of evolution that can be achieved by sexual as compared to asexual organisms.

SPECIES AND THEIR FORMATION

Originally the concept of a species was essentially typological. Organisms were classified into groups, or species, on the basis of their readily ascertainable similarities. Relationships between species were then analysed in terms of the differences between typical organisms representative of the species. If these measured similarities are taken to be measures of evolutionary relatedness, they can form the basis for construction of a phylogenetic tree. Since evolution is genetic change, measures of evolutionary differences must eventually be made at the genetic level. A typological approach to the definition of species poses a major problem as to where to draw the line that separates two species. When during a gradual evolutionary process can a new species be said to emerge? How many genetic differences between two groups of individuals must there be before they can be considered to form different species? With asexual organisms the answers to these questions must always be fairly arbitrary.

Sexuality introduces a new dimension to the concept of species, through the phenomenon of interbreeding. Now, a species must be defined at the population level to be the most inclusive group of individuals that can potentially interbreed and produce fertile offspring. The criterion for classifying individuals into different species is now much clearer. It is, basically, whether or not they can interbreed. This still does not solve the problem for the paleontologist, who is left with the difficulty of deciding when the morphological differences between two organisms are so great that it is unlikely that they could have interbred. Even for the student of contemporary organisms, however, difficulties will arise when groups of organisms are separated by geographical or other barriers which, in nature, prevent their interbreeding but where the organisms can, under experimental conditions, still produce fertile offspring. Indeed, if the natural barriers disappeared such organisms could, perhaps, once again form an interbreeding population. In such situations, no doubt, one is observing incipient speciation.

The "interbreeding population" concept of the species, as emphasised most clearly by Ernst Mayr, invokes reproductive barriers as the major basis for speciation. These barriers are often geographical but may also, for example, be due to seasonal or even diurnal separation, or may arise through the existence of different ecological niches within the same geographic area. The initial stages of reproductive isolation may often be largely due to behavioural factors and may depend on only a few genes. Thus, it is quite conceivable that speciation in its earliest stages involves only a relatively few genetic differences. Once, however, two populations are reproductively isolated from each other their separate subsequent evolutionary pathways will inevitably tend to diverge, leading eventually to physiological and other incompatibilities that give rise to hybrid infertility and inviability. Such hybrid incompatibility need not be, and indeed mostly probably is not, positively selected for. It arises simply as a natural consequence of reproductive isolation. Within the species, genetic changes will be selected for that are compatible with each other, as they must be. But no such opportunity for co-adaptation of genetic changes exists between species. Too much emphasis has, in my view, been placed in the past on the notion of positive selection for isolation by hybrid infertility or inviability. The pressure for this, especially in the early stages of speciation, is not strong. Only after two species have diverged significantly does it become a major disadvantage if the mechanisms of mate selection do not ensure the choice of a mate from one's own species.

RATES OF EVOLUTION AND POLYMORPHISM

The ability to study differences between proteins has provided a measure for variation and evolution, effectively at the genetic level. The number of amino acids by which homologous proteins in two species differ is a direct function of the corresponding genetic difference. Such observations first of all provide objective measures at the genetic level for the construction phylogenetic trees, and secondly provide proper estimates of evolutionary rates, again at the genetic level. Fortunately, but perhaps not unexpectedly, molecular phylogenies agree well with their more conventional counterparts.

The measurement of evolutionary rates, on the other hand, has led to a revival of the old controversy concerning the relative roles in evolution of natural selection and random genetic drift. Thus, while so far I have emphasised natural selection as the major driving force for genetic change, theoretical studies by Fisher, Wright and later Kimura and others clearly show that chance variation can lead to significant increases in frequency of genes which have neither a selective advantage nor disadvantage, namely are neutral. The probability that any given neutral mutation will establish itself in a population and will reach significant frequencies is very small, essentially 1/2N, where N is the population size. But given the large number of possible mutations that can be formed, if many of these are neutral, then appreciable genetic divergence will occur simply by chance. Following Haldane (1957), Kimura (1968 and later) has argued that observed rates of evolution are too fast to be compatible with natural selection being the agent responsible for most of the observed differences. Others, including myself, believe that this is not the case (see e.g. Bodmer and Cavalli-Sforza, 1972). The argument, essentially, becomes one of establishing what proportion of observed divergence is due to chance effects and what proportion to natural selection. All, however, are agreed that disadvantageous mutants have a negligible probability of achieving success and that the majority of new mutants will be disadvantageous.

Studies of protein differences between individuals, mainly using the technique of electrophoresis, have contributed dramatically to our knowledge of the extent of genetic variation in natural populations. Following the work of Lewontin and Hubby (1966) in *Drosophila* and Harris (1966) in man, and subsequently many others, it is now clear that nearly all genes may be polymorphic, namely have two or more alleles occuring in a population each with frequencies of more than 1%. These observations have raised the same argument about random drift effects versus natural selection as in the case of evolutionary rates. Now the question is, to what extent are the observed polymorphisms maintained by selection as opposed to having arisen purely by chance fluctuations in gene frequency, and so being neutral in their effects. Once again, Kimura and his colleagues argue for neutrality, while others point to a variety of arguments concerned with the general distributions of gene frequencies, as well as examination of specific polymorphisms, which they consider favour natural selection as the mechanism. I believe that the level of polymorphism observed in natural populations should be thought of as a dynamic equilibrium between the rate at which polymorphisms are formed and the rate at which they are lost. Formation of polymorphisms can occur either by chance increases in the frequencies of neutral genes or by periodic selection favouring a particular mutation which, once having reached a polymorphic frequency, will not rapidly disappear from the population even when the selection which might originally have favoured its increase disappears. Many genetic variants may thus remain in a population as relics of previous episodes of natural selection, perhaps eventually to disappear due to chance fluctuations or because of subsequent adverse selective pressures. So long as the rate at which polymorphisms are formed is greater than that at which they are lost, polymorphisms will accumulate indefinitely and so could easily reach presently observed levels (Bodmer, 1968). On this basis, most polymorphisms might originally have been established by selective effects but may now no longer be selectively maintained.

New approaches to the detection of polymorphisms at the DNA level, using restriction enzymes and cloned DNA probes (Kan and Dozy, 1978), may provide yet another quantum jump in the range of genetic variation that can be detected in natural populations and used as a basis for measurement of evolutionary rates and phylogenetic divergence. These variations will certainly pose new problems for the interpretation of the action of natural selection especially in relation to differences between changes occurring in a coding sequence and those occurring in intervening sequences or the extensive non-coding regions found between the coding regions. Such studies should also provide major new insights as to the nature of new variants, and the mechanisms by which mutations are produced. These may more often be by some form of intra-chromosomal unequal crossing over, than simple base pair substitution.

Studies at the molecular level clearly confirm that the basic nature of genetic variation within and between species is the same. There do not seem to be any obvious discrete changes which accompany speciation, though certainly some evolutionary changes are pivotal, in the sense that they allow rapid subsequent evolutionary divergence to take place. This must surely have been true, for example, following the evolution of sexual reproduction, even at the most primitive level. At the other end of the evolutionary scale, it must also have been true for the substantial increase in effective brain size that accompanied the evolutionary transition from higher primates to *Homo Sapiens*.

CONCLUSIONS AND IMPLICATIONS

In the light of this brief review of evolutionary mechanisms, what can be said about the possible effects of heterologous insertions, such as could be mediated by recombinant DNA technology? First, it must be emphasised that there is nothing to contradict the view that most mutants, and more especially those with more extensive effects, are almost certainly disadvantageous and that this will be true whether, for example, human genes are put into *E. coli* or *E. coli* genes into human cells. A second major point is that there is nothing different in nature about the genetic differences between species as compared to those within. There is no particular fundamental integrity of the genome which would be disrupted by a transspecific insertion. The fact, for example, that human-mouse hybrid cells function perfectly adequately in cell culture supports this view. Co-adaptation fails at the organismal rather than the cellular level. A third important point is that there must have existed ample opportunity for DNA exchange to occur between widely disparate species throughout evolution. Bacterial breakdown, whether in the gut, the spleen or elsewhere, must surely release DNA that can be taken up by adjacent cells. Organisms continuously ingest foreign DNA when they eat, and many infectious organisms, including especially viruses easily cross wide species barriers. Though DNA transformation in mammalian cells seems to be a very inefficient process, the cells do take up DNA (see e.g. Raspe, 1972) and, over an evolutionary time scale, the opportunities for taking advantage of such foreign material must have been legion. Yet, in higher organisms there is little evidence for homologies with, say, prokaryotic genes, which would indicate such transfers. This suggests that heterologous exchanges of this sort have hardly been significant throughout evolution, and neither are they particularly likely to have been selected against. It should be remembered that for an organism to cause an epidemic it must develop an extraordinary selective advantage. The evolutionary principles I have reviewed certainly would not lead one to expect such an advantage to develop from, for example, heterologous exchanges between prokaryotes and higher eukaryotes. A fourth point to make is that the discovery of insertion sequences, which can recombine into a chromosome without apparent need for homology (see e.g. Nevers and Saedler, 1977) suggests that promiscuous genetic exchanges may be more common, or at least more possible, than was until recently thought to be the case. The final point to make is that the one genetic change which heterologous insertion would be expected to lead to, is inactivation of the gene into which a foreign sequence was inserted. Whether such inactivating mutations could form the basis for tumorigenic potential is not clear, but in any case their overall effect would depend very much on the frequency with which such mutations were produced. It seems *a priori* unlikely that simple functional loss mutations could form a major basis for oncogenic transformation, and of course, any such mutagenic effect would have to be assessed in relation to background mutation rates. Several events, all with exceedingly small probabilities, are required before there could be successful heterologous insertions say from a recombinant plasmid carried in *E. coli* into a human cell.

I believe that evolutionary considerations make it almost impossible to conceive of a situation where a recombinant plasmid, or an infected host, acquires unwittingly a special advantage. I therefore do not see the case for significant concern, even about conjectured hazards, with recombinant DNA technology, unless work is carried out with known dangerous pathogens.

REFERENCES

Specific References (given the broad and general nature of this review, these have been kept to a minimum)

Bodmer, W.F. (1968). Demographic approaches to the measurement of differential selection in human populations. *Proc. Nat. Acad. Sci.*, *59*(3), 690-699.

Bodmer, W.F. (1972). Some reflections on mechanisms and prospects of genetic exchange. In Gerhard Raspé (Ed.), *Advances in the Biosciences*, 8. Workshop on mechanisms and prospects of genetic exchange. Pergamon Press. Vieweg. pp. 371-385.

Bodmer, W.F. and Cavalli-Sforza, L.L. (1972). Variation in fitness and molecular evolution. Proceedings of the Sixth Berkeley Symposium on Mathematical Statistics and Probability. University of California Press. pp.255-275.

Fisher, R.A. (1930). The Genetical Theory of Natural Selection. New York: Dover.
Haldane, J.B.S. (1957). The cost of natural selection. J. Genet., 55, 511-524.
Harris, H. (1966). Enzyme polymorphisms in man. Proc. Roy. Soc. B, 164, 298-310.
Kan, Y.W. and Dozy, A.M. (1978). Polymorphism of DNA sequence adjacent to human β-globin structural gene: Relationship to sickle mutation. Proc. Natl. Acad. Sci. U.S.A., 75(11), 5631-5635.
Kimura, M. (1968). Evolutionary rate at the molecular level. Nature, 217, 624-626.
Lewontin, R.C. and Hubby, J.L. (1966). A molecular approach to the study of genetic heterozygosity in natural populations. II. Amount of variation and degree of heterozygosity in natural populations of Drosophila pseudoobscura. Genetics, 54, 595-609.
Muller, H.J. (1958). Evolution by mutation. Bull. Amer. Math. Soc., 64, 137-160.
Nevers, P. and Saedler, H. (1977). Transposable genetic elements as agents of gene instability and chromosomal rearrangements. Nature (Lond.), 268, 109-115.
Raspé, G. (Ed.) (1972). Advances in the Biosciences, 8, Workshop on Mechanisms and Prospects of Genetic Exchange. Pergamon Press. Vieweg.

General References

Cavalli-Sforza, L.L. and Bodmer, W.F. (1971). The Genetics of Human Populations. W.H. Freeman and Company, San Francisco.
Lewontin, R.C. (1974). The Genetic Basis of Evolutionary Change. Columbia University Press.
Mayr, E. (1970). Population, Species and Evolution. Harvard University Press. pp. 453.
Nei, M. (1975). Molecular Population Genetics and Evolution. North-Holland, American Elsevier.

DISCUSSION

W. SZYBALSKI: Which is a more probable event in the course of evolution? To acquire a whole new function (by borrowing, stealing or taking away from another organism), or to develop it from scratch by accumulating point mutations? Common sense tells me that the latter mode could be a very slow process, whereas acquiring a complete set of genes from somewhere else should be a more frequent and a more efficient event. Thus, a process akin to the recombinant DNA technology should be quite important in natural evolution.

W.F. BODMER: I think I would disagree with you, at least for higher eukaryotes. My reason is that if you acquire a whole new function from another organism you put it into an already adapted system in a way which is likely to make it very difficult to integrate in the adapted system without further evolution. The question of acquiring new functions, which I did not discuss much, is a very interesting one. We all think that these mostly happen by gene duplication, which allows the original function to continue while a new function may develop. But I think there is increasing evidence that gene duplications may provide new functions, even after say a frame shift or some other change in the sequence which makes the amino acid sequence of what is produced from the new nucleotide sequence bear no particular evolutionary relationship to the previous sequence. I think one is posed a rather striking problem of how do you make use of new proteins which have no obvious homology with previous proteins. I think that does happen and I think it needs an explanation as to why it happens. It may be that a random, or almost random, amino acid sequence can fulfil some function, however imperfectly, and then that can very quickly be adapted to be more effective. But I think that picking up a whole function from somewhere else is an unlikely way to do anything useful without further change, because it won't fit into the adapted system. It poses the same problems as having a mutation in an organism - it won't fit into the adaptive structure of that organism. I am thinking here, on the whole, of highly adapted organisms. There may clearly be circumstances for simpler organisms such as bacteria where this is possible.

W. SZYBALSKI: The reason I asked this question is because in the early fifties, when I started to study the genetics of antibiotic resistance, most of the resistant mutants were caused by "point mutations". At present, however, in the real world of clinical practice, most of the drug resistance is transmitted as complete resistance modules, in the form of plasmids or transposons. Nature has developed new techniques for transferring whole genetic elements. By using the recombinant DNA technique, scientists are just copying the mechanisms used by microorganisms in their natural evolution.

W.F. BODMER: Evolution of drug resistance and that sort of phenomenon in bacteria is a case in point where you obviously can make use of other functions, and where you may not need the sort of adaptation I mentioned. But I think that isn't true for most of the complex evolutionary processes that we often think about.

M.H. RICHMOND: I was very surprised to hear you run down the importance of transposons and the ability to transfer blocks of pre-evolved DNA from one species to another as a major positive influence in evolutionary developments. It seems to me that we have enough information already available to suggest that this, indeed, can be very important.

W.F. BODMER: I suppose I am thinking of higher organisms and you are thinking of bacteria. I think there is absolutely no evidence for that in higher eukaryotes.

M.H. RICHMOND: I would agree there is little evidence, but at least we now do have a concept of how it may have occurred. The combination of a transposon on which it's clear from bacterial systems that you can accumulate genes of phenotyphic value with, for example, a virus or some vehicle which will allow that information to move over a relatively wide range of species, may in fact just provide the mechanism for making the quantum jumps in evolution, whose basis has mystified many people.

W.F. BODMER: I think that may be true for bacteria where such transfers are substitutes for diploidy and a proper sexual system. But it seems to me unlikely to be true for higher organisms, whose whole genetic organisation is quite different. In bacteria, for example, you have elements of control that place much more premium on the co-ordinate control of genes that are near to each other. In higher organisms, although you have, and I think they are extremely important, gene clusters, these tend to be of related genes used in different ways in different organs and different places during development. Now you need to co-ordinate the genetic activity associated with a particular function by controlling genes that are very widely dispersed. So I don't see how in the higher eukaryotes, at least, one can easily make much immediate use of transpositions of genetic material coming from a very different source. I am sure that in bacteria, on the other hand, that has been an important mechanism.

M.H. RICHMOND: But you have to organise, or the system has to be organised to get this co-ordinated control of functions in higher organisms. Presumably you can intervene at that point.

W.F. BODMER: I don't see how and I don't see any evidence that that has been the case. The understanding of the organisation of the control of gene activity in higher organisms is what a lot of the work that we should be talking about here is devoted to. Certainly a major transition must have taken place with the development of a nuclear membrane and different methods of processing message which led to a totally different mechanism for controlling genetic information, which is often more widely dispersed, than you have in bacteria.

R.H. PRITCHARD: Estimates of risk to the general population in the case of the hazards hypothesis require an assessment of the probability that a novel hybrid between two benign organisms would not only be a pathogen but have a positive adaptive value. It seems to me that there is a lot of evidence that could be used to assess the probability that a novel hybrid will have a positive adaptive value. We know, for example, that many bacterial species can become resistant to antibiotics either as a result of chromosomal mutations or the acquisition of plasmid-borne genes. There has been prolonged selection for such resistance against many antibiotics used in medicine and agriculture for up to 40 years. Despite that I know of no example of a naturally sensitive bacterial species in which resistance to any antibiotic has become the predominant phenotype in the absence of continued selection pressure. This resistance is expensive to maintain and in 20-40 years of selection no naturally sensitive bacterial species has succeeded in reducing its cost sufficiently to be able to compete successfully with its sensitive progenitors. This is a far more extreme test than that presented by the hazards hypothesis which supposes that a new hybrid which could survive and spread epidemically could be created at a stroke.

With respect to the assessment of the probability that a microbial hybrid between benign parents could be a pathogen, could one make use of data concerning the distribution of existing pathogens? I believe that syphilis (as an example) was introduced from the New World to the Old relatively recently. Clearly there was a niche waiting to be occupied in the Old World and an organism capable of occupying that niche is clearly designable. Could one use this kind of information to determine the probability that an unoccupied niche will be filled by the evolution of a new species? Such calculations would show the incredible difficulty of creating something new and workable, but could it be quantified?

ORGANIZATION AND EVOLUTION OF THE EUKARYOTIC GENOME

G. Bernardi

Laboratoire de Génétique Moléculaire, Institut de Recherche en Biologie Moléculaire, Université Paris VII, 2 Place Jussieu, 75005 Paris, France

The organization of the prokaryotic genome and the regulation of its expression are reasonably well understood at the present time. In contrast, these problems are still quite open in the case of the eukaryotic genome, in spite of the efforts of many laboratories in this area during the past few years. This situation is that much more regrettable since the issues under consideration are of capital importance for understanding evolution and differentiation.

In the present brief review I will attempt, first, to introduce the major questions concerning the organization of the eukaryotic genomes (without touching the problem of its regulation); second, to discuss what we have learned from the experimental approach which has been most widely used in recent years, namely the kinetics of DNA renaturation ; and third, to present a different approach to the problem.

Living organisms present a major discontinuity, separating prokaryotes and eukaryotes ; no intermediate forms are known. A comparison of prokaryotes and eukaryotes reveals that the major differences concern the size, the structure and the organization of the genome.

The genome size, namely, the amount of (nuclear) DNA per haploid cell, is constant in each eukaryotic species and covers an extremely wide range of values. Some fungi have genome sizes practically equal to those of some bacteria (the latter have a very limited range of genome sizes), whereas some animals and plants have genome sizes 10,000 times as large. A closer look at available data indicates that wide variations of genome size are often found within single orders, within single genera and even within interbreeding species. Since it is unlikely that these differences correspond to comparable differences in the amount of genetic information, the minimum genome sizes found in each order are usually considered, neglecting the interesting but less important problem of the variation of genome size within orders. Even so, a ratio of about 1000 is found between the smallest genome size of prokaryotes and the largest (minimum) genome size found in eukaryotes, that is, the genome size of mammals. A well-defined trend exists for the minimum genome size to increase with evolution. Very interestingly, if one plots such minimum genome size against the divergence time of different orders, it becomes evident that one can distinguish two phases in organ evolution: one in which genome size varied very little and one in which genome size strikingly increased. Very interestingly, the separation between these two phases corresponds to the appearance of multicellular organisms and of cellular differentiation.

The genome structure of eukaryotes is much more complex than that of prokaryotes. We will not discuss this point here. Suffice it to mention that at least three distinct structural levels have been recognized in eukaryotes, that of nucleosomes, that of chromomeres, and that of chromosomes. Another major difference between prokaryotic and eukaryotic cells is the segregation, in eukaryotes, of part of the genome into cytoplasmic organelles (mitochondria, chloroplasts).

Concerning the organization of the eukaryotic genome, the fundamental point here is that the genome size increase occurring in evolution has not been accompanied by a corresponding increase in the number of different polypeptide chains encoded. In general, it can be said that only a small percentage of the eukaryotic genome is expressed. For instance, in the early sea urchin embryo, only 4 % of the haploid genome appears to be expressed as polysomal mRNA ; in adult sea urchin tissues, this number drops to less than 1 % of the genome, a value lower than that of the DNA expressed in E.coli. These data stress what certainly is the most striking difference existing between the prokaryotic and the eukaryotic genomes. The former is made up simply of genes transcribed into mRNAs, rRNAs and tRNAs and of short regulatory sequences preceding each polycistronic transcription unit.The latter contains a large excess of DNA, compared to what is found in the final transcripts. Only a fraction of this excess DNA is accounted for. First of all,a certain percentage of eukaryotic DNA is present in simple highly repeated sequences, forming what are known as satellite DNAs ; these DNA segments are not transcribed and have a function which is still unknown. Second, some genes e.g. the rRNA, tRNA, and histone genes are present in multiple copies. Third, a number of eukaryotic genes contain non-coding sequences, the so-called intervening sequences, which may represent as much as 5 to 10 times the amount of DNA contained in the corresponding coding sequences.

The majority of the excess DNA is, however, not accounted for yet. The fact that most of it is in all likelihood non-coding has encouraged approaches in which the genome organization of eukaryotes is studied directly at the molecular level.

THE KINETICS OF RENATURATION OF EUKARYOTIC DNA

The main experimental approach used so far has been the study of the kinetics of renaturation exhibited by DNA fragments. The reannealing of separated complementary single strands of DNA ideally follows second order kinetics. For a given initial DNA concentration and a certain DNA fragment size, the half-time of reassociation should be proportional to the number of different types of fragments present and thus to the genome size. This expectation is exactly borne out in the case of viral and bacterial genomes, which are characterized by a unique DNA sequence. Eukaryotic DNAs, in contrast, show complex renaturation kinetics and can usually be resolved into fast, intermediate and slow-renaturing components. The latter represent in most cases 50-70 % of the genome, are formed by single copy sequences and comprise most eukaryotic genes and their intervening sequences ; the intermediate DNA is made up of repetitive sequences with degrees of reiteration comprising between 10 and 1000 copies, the fast DNA corresponds to satellite DNAs, the sequences of which are repeated over 100,000 times. In addition, some very fast renaturing material, following first order kinetics, has also been shown to exist in the eukaryotic genome ; these fragments can fold back on themselves since they contain palindromic nucleotide sequences ; they usually represent a few percent of eukaryotic DNA.

The relative arrangement of repetitive and non-repetitive (single-copy) sequences was investigated by reassociating to a low cot,(the product of the initial DNA concentration by renaturation time), labelled DNA, sheared to various fragment lengths, with excess short fragments of unlabelled DNA and by following the binding of labelled DNA to hydroxyapatite. Such analysis as applied to the Xenopus genome has shown that about 50 % of this DNA consists of closely interspersed repetitive and non-repetitive sequences (short-period interspersion). The average length of the repetitive sequence elements is 300 $\pm$ 100 nucleotides, while the non-repetitive sequences separating adjacent repetitive sequence elements average 800 $\pm$ 200 nucleotides. The remainder of DNA is mainly non-repetitive, though most of it contains rare interspersed repetitive elements spaced at a minimum of 4000 nucleotides apart (long-period interspersion).

Over 20 species, widely separated phylogenetically, have been shown to be endowed with the Xenopus interspersion pattern ; among insects, one dipteran(Musca domestica)and one lepidopteran (Antherea pernyi) show the Xenopus pattern, while another dipteran,(Drosophila melanogaster) and a hymenopteran (Apis mellifera) show a quite different pattern in which the repeated sequences are much more widely spaced from each other than in the short-period interspersion of Xenopus.

In summary, it can be said that the major contribution of this approach to our understanding of the organization of the eukaryotic genomes has been the demonstration that these genomes contain, in contrast to prokaryotic genomes,repeated sequences which are interspersed with the unique sequences. It has been speculated that the characteristic interspersed repeated sequences are correlated with the regulation of the expression of eukaryotic genes. Such speculation does not have, however, any experimental support and seems to be contradicted by the interspersion patterns of insect genomes. An alternative hypothesis, which seems more reasonable, is that the interspersed repeated sequences play a role in the unequal crossing-over phenomena, which were responsible, in all likelihood, for the process of evolutionary increase in genome size exhibited by eukaryotes. In any event, it seems that renaturation kinetics, at least as applied to unfractionated eukaryotic genomes, has provided all the information it can give and that new approaches are needed.

DENSITY GRADIENT FRACTIONATION OF EUKARYOTIC DNA

The approach which has been mainly used in our laboratory is based on the fractionation of eukaryotic DNA by density gradient centrifugation, in the presence of DNA ligands, mainly Ag^+ and an organic mercurial, bis-(acetato-methylmercuri) dioxane or BAMD. These techniques separate, in general, native DNA fragments containing short repeated nucleotide sequences according to their sequences and other DNA fragments according to base composition. Satellite DNAs and repeated genes, which have satellite-like sequences in their spacers,are easily separated, in general. For this reason, from now on, we will disregard them and consider the fractionation of the bulk of eukaryotic DNA, the so-called main-band DNA.When studying DNAs from eukaryotes widely distant from a phylogenetic point of view, we observed that symmetrical CsCl bands were exhibited by unicellular eukaryotes and invertebrate DNAs, as is the case for prokaryotic DNAs; the DNAs from fishes, amphibia and reptiles exhibited a very slight and increasing asymmetry on the heavy side of their CsCl bands ; the DNAs from warm-blooded vertebrates, birds and mammals, exhibited CsCl bands which were very asymmetrical on the heavy side. A fine analysis involving density gradient centrifugation in the presence of Ag^+ or BAMD led us to the recognition of 4 discrete DNA components in the main band of avian and mammalian DNAs. The existence of these discrete components has been confirmed by preparing them. The major DNA components exhibit, when run in CsCl, gaussian bands and a compositional heterogeneity very close to that of bacterial

DNAs. The relative amount and the buoyant densities of the major components of mammalian and avian genomes are very close to each other. In the case of the mouse genome, the four major components have buoyant densities equal to 1.699, 1.701, 1.704 and 1.708 g/cm^3 and represent about 26 %, 35 %, 18 % and 8 % of the genome, respectively. It should be noted that the major light DNA components of avian and mammalian DNAs are in the same buoyant density range as the DNAs of cold-blooded vertebrates, and that the major heavy components are responsible for the asymmetry of their CsCl bands.

Again in the case of the mouse genome, the renaturation kinetic properties show that most of the fold-back and interspersed repetitive sequences are present in the two light components, the two heavy ones being mainly formed by single copy sequences. In all cases investigated, the major components of warm-blooded vertebrates represent blocks of rather homogeneous base composition which are larger in molecular weight than 100 million daltons.

The existence of discrete major components in the genomes of warm-blooded vertebrates is of interest because it implies that different sections of these genomes are under different compositional constraints. While the reasons for such a situation are not yet clear, it may be relevant in this connection to mention very recent results of G. Cuny, M. Meunier and P. Soriano of our laboratory on the location of globin genes in the DNA components of rabbit, mouse and man (probes obtained from T. Maniatis, C. Weissmann, and B. Williamson were used). In all cases, the β-globin gene was found to be present in the 1.701 component ; preliminary results indicate that this is also the location of the human γ-globin gene, which is contiguous to the β-globin gene. In contrast, the α-globin gene has been localized in the 1.708 component of the mouse genome. These results are interesting for two main reasons : 1) α-globin and β-globin genes are the result of a gene duplication; a translocation of one of the two genes took place at a certain point in evolution, as witnessed by the different chromosomal and component location of the two genes. Now, the component location, but not the chromosomal location, indicates that it was the α-globin gene which was translocated from its original position; in fact, not only does the 1.708 component not exist in lower vertebrates, but also no DNA fragment having such a high density is detected. In contrast, the further duplications of the human β-globin gene remained in the component where the β-globin gene was and still is located, as witnessed by the location of the γ-globin gene. 2) The base composition of α- and β-globin mRNAs from rabbit, mouse and man are known, as well as that of human γ-globin genes. The α-globin mRNAs have a remarkably higher GC content (64 % for man and rabbit) than the β-globin mRNAs (51% for man, rabbit and mouse) and the γ-globin mRNA (51% for man). This is a surprising result if one considers that α-globin mRNA could have the same base composition as β-globin mRNA, and that its GC content has been increased at the price of having a large number of otherwise forbidden or avoided GC doublets. Considering that all intervening sequences studied so far have a lower GC content than the coding sequences (the β-globin gene of mouse has, for instance, 46 % GC _versus_ 51 % GC for its mRNA), it is likely that the GC contents of globin genes are close to the average base composition of the large DNA blocks in which they are embedded. If this conclusion is confirmed and extended to other genes, the compositional constraints seen to exist in the major components of the eukaryotic genome extend to the genes they contain.

REFERENCES

Concerning the kinetics of DNA renaturation , the reader is referred to the papers published by Britten, Davidson, and their colleagues. The analysis of globin cDNA has been reported by Konkel _et al_. (Cell, _15_. 1125-1132, 1978) and by Forget et al.at the ICN-UCLA Symposium on Eukaryotic Gene Expression (March 1979 ; paper in press). Previous papers from our own laboratory are given below.

Corneo G., Ginelli E., Soave C. and Bernardi G.
Biochemistry, 7. 4373 (1968)

Filispki J., Thiery J.P. and Bernardi G.
J. Mol. Biol., 80. 177 (1973)

Thiery J.P., Macaya G. and Bernardi G.
J. Mol. Biol., 108. 219 (1976)

Macaya G., Thiery J.P. and Bernardi G.
J. Mol. Biol.108. 237 (1976)

Cortadas J., Macaya G. and Bernardi G.
Eur. J. Biochem., 76. 13-19 (1977)

Macaya G., Cortadas J. and Bernardi G.
Eur. J. Biochem., 84. 179-188 (1978)

Cuny G., Macaya G., Meunier-Rotival M., Soriano P. and Bernardi G.
in Genetic Engineering (H.W. Boyer and S. Nicosia, eds.)
Elsevier, Amsterdam, 1978, pp. 109-115.

DISCUSSION

J.D. WATSON: On the point about intervening sequences, I don't think it's widely known that you can get great variations in DNA content ("C" values) not only over a long evolutionary period, but even within plant genera where you can get factors of ten variation. This was first emphasised by Stebbens and re-discovered by Joshua Lederberg, who was intrigued by the fact that plants that have a relatively small amount of DNA have the same number of chromosomes as those with tenfold more DNA. So, it's not a question of polyploidy lost in patches. Those species which have very short life cycles have the small DNA content, whereas those which have lots of DNA have long life cycles. If one combines these facts with the observation from Tonegawa's laboratory - that the intervening sequences occur between functional domains - the speculation arises as to whether the intervening sequences largely serve to promote recombination between functional domains. Perhaps the main reason for the vast increase in DNA is to promote recombination, as evolution occurs. To test this idea we shall need data as to whether the intervening sequences become much longer as the "C" value rises.

G. BERNARDI: I would like to stress two points. The first diagram I showed ends at man for a very simple reason, namely that I took the minimal value for each order, as had already been done by Britten. In fact, within certain orders like amphibia, you have a fantastic spread of genome sizes whereas you don't have them in other orders, like mammals or birds. The other point concerns the significance of interspersed repetitive sequences in eukaryotic genomes. This is the main discovery of renaturation kinetics. It's a pity that Britten and Davidson put so much emphasis on the regulatory role, for which there is not the least evidence, whereas the sequence may really play a role in that phenomenon of increase in genome size, which is so typical of eukaryotes and which doesn't exist in prokaryotes. In fact, all prokaryotes are within a factor of five at most in terms of genome sizes. Clearly there are two phases of evolution which can be distinguished, one in which evolution has taken place with increasing genome size and a longer one in which this has not occurred.

W.F. BODMER: Just a quick answer on the question of recombination and intervening sequences. I find that very implausible - one has to ask which came first, the chicken or the egg? I would think that the separate domains evolved and then were put together by the regions in between rather than the other way round. Recombination in higher eukaryotes is an extremely rare event at the DNA level. In higher organisms on average you've got about one crossover per chromosome per meosis which is an incredibly low frequency of recombination in terms of amounts of DNA. And, if recombination frequency were that important, you surely could easily adjust it by other ways. As in the case of mutation rates, recombination frequencies must have been carefully adjusted by recombination and there must be enormous scope for this without having to put increased amounts of DNA there to get increased recombination frequences.

G. BERNARDI: In agreement with what you say, one shouldn't forget that the intervening sequences are single copy sequences, which do not exist elsewhere in the genome. In this they are very different from the interspersed repetitive sequences which may have a role in recombination.

NATURAL MODES OF GENETIC EXCHANGE AND CHANGE

A. Campbell

Department of Biological Sciences, Stanford University, Stanford, California, 94305, U.S.A.

The ultimate source of natural genetic change is mutation, which may be narrowly defined as spontaneous alteration of the DNA base sequence. Mutation creates variation among individuals within a species. Starting with the primary variation created directly by mutation, genetic exchange among individuals of a species creates new combinations. In prokaryotes, genetic exchange has two aspects - (1) transfer of genes from a donor to a recipient and (2) establishment of the transferred genes, so that they are stably reproduced as part of the genome. In higher eukaryotes, both aspects are automatic consequences of the fusion of gametes that comprises an obligatory step in reproduction.

The emergence of new gene combinations does not in itself cause a significant change in the genetic composition of the species without the action of natural selection. Because wild populations are generally the products of extensive natural selection and therefore well adapted to their environments, most evolutionary change probably results from alterations in environmental conditions rather than from the appearance of genetically novel individuals.

The total number of gene combinations potentially constructible by the action of genetic exchange within known populations is enormous. Within the human population, for example, virtually every newly conceived child carries a unique constellation of genes never precisely duplicated during the previous history of the human race. Genetic novelty as such is a general rule of life, not a recent invention of biochemists.

I shall examine here some information, mainly derived from laboratory studies, on genetic exchange in prokaryotes, both intra- and inter-specific. My general conclusion will be that probably Nature can do everything that biochemists can; i.e., artificial gene splicers are not sampling from a pool of gene combinations that Nature never samples. My purpose will not be to sell that conclusion, but to provide some background on its factual and conceptual basis.

MODES OF GENE TRANSFER

Transfer of genes between bacteria can take place in the laboratory by at least four different mechanisms - transformation, transduction, conjugation, and cell fusion.

In transformation, free DNA, chemically extracted from cells of the genetic donor, is added to living cells of the recipient, which have generally been treated or preconditioned in some manner to enhance their ability to take up DNA from solution. Transformation has been demonstrated in many bacterial species as well as in the yeast *Saccharomyces cerevisiae*. Because DNA molecules are generally fragmented during extraction, an individual recipient generally receives only a small fraction of the donor genome.

In transduction, fragments of donor DNA are accidentally packaged within the protein coats of bacteriophages and later injected into the recipient cell using the same machinery that the phage uses in infection. Transduction can thus be considered an incidental side reaction to the infectious cycle of the phage. Besides packaging segments of pure host DNA, certain phages can also generate derivatives (specialized transducing phages) in which segments of host DNA are covalently joined to phage DNA. Such a modified phage can then reproduce indefinitely, thus providing a highly enriched source of the specific host DNA it contains.

Conjugation, like transduction, represents an incidental side reaction to the infectious transfer of a subcellular element - in this case that of conjugative plasmids, such as the *E. coli* fertility factor F. Transfer of conjugative plasmids requires direct contact of donor and recipient cells, and is mediated by specialized structures (pili) on the surface of the donor. Conjugative plasmids can effect the occasional transfer of chromosomal DNA by at least three mechanisms: (a) mobilization, in which the bacterial chromosome or a chromosomal fragment is accidentally transferred instead of the conjugative plasmid; (b) Hfr formation, in which the conjugative plasmid becomes spliced into the bacterial chromosome and then can serve as an origin for transfer of the whole chromosome; (c) F' formation, in which imprecise excision of plasmid DNA from an Hfr chromosome generates plasmids that resemble specialized transducing phages in carrying a small segment of host DNA covalently attached to the plasmid.

Specialized transducing phages and F' elements warrant special attention here because they represent the closest natural analogs of those products of artificial gene splicing routinely employed in DNA cloning.

For cell fusion, cells are treated with chemical or biological agents that modify their surfaces so as to promote fusion of their outer membranes, thus generating a single diploid or binucleate cell from two parental ones.

GENE TRANSFER IN NATURE

Because they constitute side reactions to the transfer of natural agents, we might expect that transduction and conjugation take place in Nature as well as in the laboratory. The expectation for transformation is less obvious, because two conditions are required - liberation of intact DNA from the donor and uptake by the recipient. In fact, transformation was discovered in experiments done under "natural" conditions (infection of laboratory animals) with no specialized treatment other than heating the donor bacteria to kill them (Griffith, 1928). Exchange under natural conditions by other mechanisms has also been observed (reviewed by Marmur, Falkow, and Mandel, 1963).

Cell fusion occurs naturally between the gametes of eukaryotes, and sometimes among other eukaryotic cells - e.g., in the fusion of hyphal tips to yield heterokaryotic mycelia that is observed when two genetically marked strains of the same fungal species are grown together. The production of such heterokaryons in similar experiments with filamentous bacteria (Actinomycetes, reviewed by Hopwood and Merrick, 1977) suggest that fusion may occur naturally in bacteria as well. The treatments used to promote fusion in the laboratory are sufficiently powerful to induce the fusion of cells from very diverse sources. The process thus appears to be quite artificial, but it might occur (even rather frequently) in Nature as well.

IMPORTANCE OF NATURAL GENE TRANSFER

Two methods of studying evolution are (a) to survey existing species or populations and to infer evolutionary pathways from their apparent relatedness and (b) to follow the succession of species (or variation within species) over time. The first method provides suggestive evidence that gene transfer has played some part in bacterial evolution (reviewed by Marmur, Falkow and Mandel, 1963 and Bodmer, 1970). Classically, the second approach has required examination of a fossil record, unavailable in suitable form for prokaryotes. However, this deficiency is partially compensated by the short generation time of prokaryotes, which makes it possible to monitor a significant amount of change on a time scale of years rather than eons.

There is one case in which the selective conditions of bacteria have been altered on a global scale, with subsequent monitoring of the way that the bacteria respond. This is the widespread dissemination of antibiotics by man. The mechanism of the response is clearcut: There is little if any natural selection for the type of antibiotic resistant mutants that arise in the laboratory by changes in the genes determining the target sites of antibiotic action. Instead, selection favors the spread, through conjugation, of plasmids that carry or acquire genes coding for enzymes that destroy the antibiotics. Some of these plasmids are themselves conjugative, whereas the transfer of others is mobilized by conjugative plasmids.

This result provides some insight into one reason that plasmids may exist in Nature: namely, to carry genes like the antibiotic resistance genes that are not needed by all individuals of a species all of the time, but which can be rapidly mobilized when needed. This expands the effective genome size of the species, while sparing the typical cell the burden of reproducing all of the genes potentially available to it. Clowes (1972), Campbell (1972), and Reanney (1976) have discussed various ramifications of this idea.

Thus many of the genes that are typically plasmid-borne confer resistance to deleterious agents that are encountered only occasionally in Nature. An instructive example of another type is provided by bacteria of the genus *Pseudomonas*. Collectively, these bacteria can oxidize a diverse array of organic compounds, any one of which is available only at certain times and places. Many of the genes for specific oxidative pathways are carried by different plasmids, which can be acquired or shed as needed.

Natural selection thus may favor partitioning the genes of a bacterial species between chromosome and plasmids according to whether their functions are required continually (or frequently upon short notice) rather than only occasionally. This selection is not based on any systematic qualitative difference between the two sets of genes, other than how their functions relate to the natural ecology of the species. However, it is likely that, once a plasmid location is chosen, the selective pressures governing subsequent evolution will differ from those acting on chromosomal genes. The critical difference results from the frequent intercellular transfer that plasmids must undergo, if the capacity for transfer is their main reason for existence. Plasmids typically can pass by conjugation among various strains of a bacterial species which have undergone some evolutionary divergence from each other. To be effective, the products of plasmid genes must be superimposable upon the physiological patterns of all these strains. Therefore, plasmid genes will generally evolve to determine products showing fewer highly specific interactions with other cellular components than a typical cellular protein may. The residence time of plasmid genes within a strain is too short; they will soon move to another strain whose components are a bit different. Thus, hybrids combining the chromosomal genes from different wild strains are frequently poorly adapted because their components do not match well. The operator of one strain may not recognize the repressor from another, for example. Yet plasmids pass freely between the two strains and function effectively in both.

The natural importance of plasmid transfer is simpler to justify than is the transfer of chromosomal genes. But remember that, wherever plasmid transfer takes place, some opportunity exists for transfer of chromosomal genes as well.

INTERSPECIFIC GENE TRANSFER

If plasmid-borne genes function effectively in diverse natural strains, we might expect that some of them would also do so when transferred between bacteria too distantly related to fall within the arbitrary boundaries that delineate a bacterial species. In fact, conjugative plasmids frequently can cross species boundaries. The F factor of *Escherichia coli*, for example, can be transferred to bacteria of other genera within the family *Enterobacteriaceae*, such as *Salmonella*, *Shigella*, or *Serratia*. A conjugative plasmid from *Pseudomonas aeruginosa* (R1822) has an extremely broad host range, being transmissible to many different families of Gram-negative bacteria (Datta and others, 1971; Olsen and Shipley, 1973).

As far as we know, whenever a plasmid or virus can transfer its own DNA from one host strain to another, there is some chance for accidental transfer of host DNA as well. The expectation therefore is that whenever two species are interconnected by phage or plasmid transfer, every gene of the first species eventually will find its way into some cell of the second. The individual species retains its identity in the face of these opportunities for miscegenation because the component genes of a species have co-evolved to function in harmony with each other, so that foreign genes are usually a liability rather than an asset.

If much or all of the bacterial world shares access to a common gene pool, what about gene transfer between cells of even more distant taxonomic groups, such as between animals and plants, or between prokaryotic and eukaryotic cells? Existing data provide no assurance of significant opportunities for exchange (but do not negate that possibility either). There are viruses capable of infecting cells of both higher plants and insect vectors; but these are all RNA viruses, whose capacity to transfer host genes is undocumented. DNA from a plasmid harbored by bacteria of the species *Agrobacterium tumefaciens* is efficiently transferred into plant cells, which are thereby rendered neoplastic (Chilton and others, 1977). And certain fungi are reported to harbor latent viruses capable of growth on bacteria (Tikchonenko, 1978). These limited facts permit no useful assessment of the frequency with which heterologous gene transfer might take place.

ESTABLISHMENT OF TRANSFERRED DNA

After donor DNA has entered the recipient, further steps may be necessary to establish it as a permanent part of the recipient genome. Establishment may involve either a net addition of new DNA to the recipient, or replacement of some recipient DNA by donor DNA. Addition may either entail acquisition of a plasmid that replicates autonomously, or insertion of exogenous DNA into the chromosome (as in lysogenization by bacteriophage λ). Replacement of a segment of recipient DNA by the corresponding segment of the donor is the most common mechanism of establishment in interspecific crosses, because the DNA transferred in conjugation, transduction or transformation is usually a chromosomal fragment lacking the means to replicate or to insert. In cell fusion, the partners cannot be classified as "donor" or "recipient." Establishment requires either the ability (usually transient in bacteria) for the diploid cell to reproduce as such or the formation, by homologous exchange, of haploid recombinants carrying some genes from each parent.

Homologous exchange, required for establishment by replacement, depends on general recombination systems that require substrate molecules that closely resemble each other in base sequence - like a DNA sequence from wild type *E. coli* and a derivative differing from it by a few point mutations. A DNA fragment with no counterpart in the recipient, and which lacks a replication origin and an insertion mechanism, will generally not become established. This is the usual (but not invariable) fate of DNA transferred between unrelated hosts.

BARRIERS TO TRANSFER AND ESTABLISHMENT

Whereas genes can sometimes be transferred between distantly related organisms, exchange among close relatives is more common. The factors limiting exchange in the former case can be considered as barriers to free exchange. These include properties of the cell surface (which must exhibit rather specific chemical configurations to allow virus penetration and, to a lesser extent, recipient ability in conjugation); restriction enzymes, which degrade foreign DNA; incompatibility between entering and resident plasmids; and lack of homology in base sequence. The first two can be considered barriers to transfer and the latter two as barriers to establishment.

One may ask whether such barriers exist for a reason or whether their ability to restrict gene flow is accidental. The question is meaningful only when carefully defined. Failure to do so has degraded the quality of the discussions of the barrier(s) to eukaryote-prokaryote exchange to the level of apparent conceptual nonsense (Wade, 1976). As this discussion has been widely disseminated, it is worth taking a moment to analyze it.

Charles Darwin convinced most biologists that the concept of purpose in biology cannot be taken literally. Evolution has no foresight, and cannot prepare organisms for future contingencies, except as selection can act on past experience. Modern biologists sometimes save time by speaking of "purpose" or "function" as convenient shorthand for specification of the nature of the selective forces that have caused organisms to have a particular property (teleonomy rather than teleology); a practice that facilitates communication only when all participants understand the conventions employed. Thus a barrier to exchange that exists for a purpose is one that is maintained because exchange is selected against. The barrier must occasionally be crossed (otherwise there could be no selection for its maintenance), and the results of such crossing must be generally disadvantageous to the resulting progeny. Such barriers are seen in the sexual isolating mechanisms of higher organisms. Horses prefer to mate with other horses rather than with donkeys; the mutant maverick that ignores this rule begets sterile progeny, and his genes are lost to posterity.

It is clear that many factors limit the rate of genetic exchange between eukaryotes and prokaryotes. The notions that a barrier exists "for good reason," that its purpose is "to protect the genetic material of higher cells from prokaryotic takeover" (Wade, 1976) or to avoid any other catastrophes that have not yet occurred seem to rest on conceptual bases at least 100 years out of date.

TESTS FOR RARE GENE EXCHANGER

If we wanted to test experimentally whether exchange between very distant species could ever occur, we would anticipate that the frequency, being the product of a very low rate of transfer and a very low probability of establishment, would be immeasurably low. One way to estimate the frequency is to measure transfer and establishment separately.

Genetic studies with prokaryotes strongly indicate that, though the rate of establishment of DNA may be very low, it is never zero. This is because various mechanisms allow DNA rearrangement, even when genetic homology is absent. Some of these will be discussed by Dr. Starlinger at this meeting. Also, Chang and Cohen (1977) showed that the *Eco*RI restriction system can catalyze the integration of added DNA precut by the *Eco*RI enzyme. The rate of transfer is also unlikely to be zero, because large amounts of host DNA do come in contact with bacteria of the gut (Hoskins, 1978). Even though nuclease degradation is rapid, some of this DNA probably finds its way into bacterial cells. Transfer could be estimated more directly by artificially implanting a prokaryotic sequence in a eukaryotic organism, then measuring its rate of return to the prokaryote of origin (Campbell, 1978). This experiment potentially tells us whether intact DNA from eukaryotes can get into prokaryotes, because we choose a DNA whose rate of establishment will be high once it gets in.

The reverse question, "Does intact prokaryotic DNA get into eukaryotic cells?" can be answered by experiments of the type performed by Chan and others (1979), in which an animal virus, SV40, is cloned in *Escherichia coli*, then added back to an animal host. Whereas the frequency of infection was very low under the conditions employed, the experiments encourage the belief that the DNA of enteric bacteria must occasionally penetrate the cells of their eukaryotic hosts. This conclusion derives from the fact that intact phage particles containing SV40 dimers inserted into phage DNA are infectious, at a low but detectable rate. As lysogenic bacteria liberate phage particles within the animal intestine, and as phage particles collectively can package every part of the bacterial genome, there is unlikely to be any gene of *E. coli* which has not, at some time in history, found its way into cells of mouse and man. Experiments designed to estimate the rate more precisely should be feasible.

CONCLUSION

One of the ideas behind the Asilomar Conference - that biochemists might synthesize gene combinations that were unlikely to arise by natural means - was and is questionable. Gene transfer has been demonstrated among many bacterial species that appear unrelated by most criteria. The assumption that transfer implies the possibility of establishment, while not directly demonstrated in every case, is strongly indicated by everything we know about genetic mechanisms. Construction of "exchanger lists" that require direct demonstration on a case by case basis (as prescribed by the 1978 NIH Guidelines) is in my judgment a waste of time. Some transfer between even more distant groups, such as from Gram-positive to Gram-negative bacteria, or from eukaryotes to prokaryotes, appears probable, but is based on indirect arguments. More information on transfer frequencies will be needed if we are to have definite knowledge about the natural opportunities for heterologous transfer.

REFERENCES

Bodmer, W. F. (1970). The evolutionary significance of recombination in procaryotes. *Symposia Soc. Gen. Microbiol.*, 20, 279-294.

Campbell, A. M. (1972). Episomes in evolution. In H. H. Smith (Ed.), *Evolution of Genetic Systems*. Gordon and Breach, New York. pp. 534-561.

Campbell, A. (1978). Tests for gene flow between eucaryotes and procaryotes. *J. Infect. Diseases*, 137, 681-685.

Chan, H. W., M. A. Israel, C. F. Garon, W. P. Rowe, and M. A. Martin (1979). Molecular cloning of polyoma virus DNA in *Escherichia coli:* lambda phage vector system. *Science*, 203, 887-892.

Chang, S. and S. N. Cohen (1977). *In vivo* site-specific genetic recombination promoted by the *Eco*RI restriction endonuclease. *Proc. Natl. Acad. Sci. USA*, 74, 4811-4815.

Chilton, M.-D., M. H. Drummond, D. J. Merlo, D. Sciaky, A. L. Montoya, M. P. Gordon and E. W. Nester (1977). Stable incorporation of plasmid DNA into higher plant cells: the molecular basis of crown gall tumorigenesis. *Cell*, 11, 263-271.

Clowes, R. C. (1972). Molecular structure of bacterial plasmids. *Bacteriol. Rev.*, 36, 361-405.

Datta, N., R. W. Hedges, E. J. Shaw, R. B. Sykes, and M. H. Richmond (1971). Properties of an R factor from *Pseudomonas aeruginosa*. *J. Bacteriol.*, 108,1244-1249.

Griffith, F. (1928). Significance of pneumococcal types. *J. Hygiene*, 27, 112-159.

Hopwood, D. A. and M. J. Merrick (1977). Genetics of antibiotic production. *Bacteriol. Rev.*, 41, 595-635.

Hoskins, L. C. (1978). Host and microbial DNA in the gut lumen. *J. Infect. Diseases*, 137, 694-698.

Marmur, J., S. Falkow, and M. Mandel (1963). New approaches to bacterial taxonomy. *Adv. Genet.*, 1, 329-372.

Olsen, R. H. and P. Shipley (1973). Host range and properties of the *Pseudomonas aeruginosa* R factor R1822. *J. Bacteriol.*, 113, 772-780.

Tikchonenko, T. I. (1978). Viruses of fungi capable of replication in bacteria (PB viruses). In H. Fraenkel-Conrat and R. R. Wagner (Eds.), *Comprehensive Virology*, Vol. 12. Plenum Press, New York. pp. 235-271.

Wade, N. (1976). Recombinant DNA: a critic questions the right to free inquiry. *Science*, 194, 306-313.

DISCUSSION

S.D. EHRLICH: I think it would be fair to say that transformation is a process that occurs in Nature. You did state it but very mildly, although it was one of the processes of genetic exchange that was first observed and it was also fairly well demonstrated with *Bacillus* in soil.

A. CAMPBELL: Yes, I agree. One can think of potential problems in terms of destruction of the DNA by donor nucleases and things like that. However, one should not lose sight of the fact that the initial demonstration of transformation by Griffith, in fact, did not involve deliberate extraction of the DNA of the donor and it occurred under more or less natural conditions. The only thing that was done was to kill the donor by heat.

W. SZYBALSKI: You have quoted an interesting example from the NIH Guidelines. On the one hand you presented your views - that whenever plasmids transfer genes, chromosomal DNA could go with it - and I agree. On the other hand, there is a view expressed in the NIH Guidelines - and you are one of the authors of the NIH Guidelines as the member of RAC - which is just the opposite. How is this possible? Are the NIH Guidelines a scientific document or, in part, a misguided political manifesto?

A. CAMPBELL: The question refers to the decision by the Director of NIH on the relation of plasmid transfer to chromosomal gene transfer in the preamble of the 1978 NIH Guidelines, Appendix A. This is an administrative decision by the Director. I don't envy him the responsibility of making that decision, and I would not use this meeting as a forum for appealing it.

The problem is that an administrative decision is properly responsive to many inputs - political and legal, as well as scientific - but the document itself is written in the form of a scientific discussion which draws a conclusion based on published evidence. In this sense, it might be said to constitute an irregular addition to the scientific literature. Whereas administrative decisions must be made, scientific issues cannot be settled by administrative decree. In the past few years I have sometimes worried that the worst damage the NIH Guidelines may have inflicted on science and society is not only to impede research and threaten freedom of inquiry but to degrade the level of discourse among scientists in areas where statements of scientists relate to such administrative decisions.

The extent of promiscuous gene transfer in Nature is an important and exciting scientific question. Having devoted a number of years to studying the natural incorporation of chromosomal genes into bacteriophages as well as the reverse process, I have given much thought to the logical implications of those processes. I would conclude that wherever phage or plasmid DNA can pass from one strain or species to another, host DNA can pass as well. But just as scientific issues are not decided by administrative decree, the pontifications of experts and specialists need not be infallible either. I would encourage you to read the references and make your own scientific judgment.[1]

---

[1]These comments were expanded for clarification by A. Campbell in the published version.

W. SZYBALSKI: I read the literature and did many experiments on DNA transfer within and between species. I agree with your conclusions and could add that the DNA transfer is quite ubiquitous between species, even when there is no known transfer by phage or plasmids. Just the efficiency is low, but there is still plenty of exchange in Nature on the global scale.[2]

P. CLARKE: I wonder if in concentrating on the antibiotic resistance plasmids you might not be underestimating the role of plasmids during evolution. One of the things that we tend to forget about prokaryotes is that, although they started evolving a long time ago they are still evolving. They have had to cope with a world in which the plants and animals started evolving with them and needed to evolve their own metabolic capabilities. It was very important indeed for them to evolve, as the plants started throwing out all sorts of complicated organic compounds. Perhaps having plasmids, on which point mutations could allow new enzyme activities to develop, and a transfer agent to transfer them more rapidly through the population, was the primary way in which these metabolically very active bacteria coped with their evolving environment, so that the transfer of genetic material was very important.

A. CAMPBELL: Yes, I couldn't agree more with that. The only reason for concentrating on the antibiotic resistance is, it's the one case where one really has hard evidence for a massive change in the population in response to a change in selective conditions.

J. SCHELL: The basis for my comment is the following. At the summary of your talk you said that more research was needed to determine whether or not natural transfer of DNA between prokaryotes and eukaryotes occurs. If that statement reflects the fact that transfers of this kind were unexpected and therefore not much studied, I would agree with you. If your statement implies that such transfers are unlikely, then I would disagree. For example, in the one well-established case where there is now good evidence that a bacterial plasmid has a highly involved mechanism for transferring prokaryotic DNA into eukaryotes, this is also a conjugative plasmid. Taken together with your description of how the whole bacterial world is in fact genetically interlinked by transferable plasmids, then if you conceptually involve plasmids like the Ti-plasmid, any prokaryotic gene can end up in eukaryotes as well.

A. CAMPBELL: Yes, that example is in my written summary. My point was really the former one, I think for most eukaryotic/prokaryotic exchange other than very special cases I, at least, would need more data to get some feeling for whether the frequency is really infinitesimal. Those are my only doubts.

---

[2]These comments were inserted by W. Szybalski into the published version for clarification.

TRANSPOSONS AND INSERTION SEQUENCES

P. Starlinger

Institut für Genetik, Universität zu Köln, Weyertal 121, 5000 Köln 41, F.R.G.

These days the "recombinant DNA" technology has aroused much public interest and even anxiety. For any biology high school teacher, who received his training some thirty years ago, this must be an astonishing phenomenon. After all, genetic recombination is one of the basic facts of genetics that he has been reading about in any university and high school textbook ever since he entered university. What then is the great excitement all about?

As a starting point, I should like to make clear that genetic recombination is quite different from the new technique which more appropriately should be called the technique of formation of new combinations of DNA rather than the recombination of pre-existing arrangements. While the basic arrangement of DNA is left unchanged in recombination, and only minor alterations, so-called alleles, are being exchanged, the new technique is capable of combining DNA from any source. This creation of new combinations has led to a fear expressed by some scientists that new dangers may be at hand against which we have not had the time to evolve any natural defences.

Why do some people believe that new combinations are not created naturally, and that it needs the sophistication of a biochemist to achieve something that Nature has been unable to do by itself? Let me quote two reasons for this belief.

First, DNA from different organisms is not mixed easily, due to natural species barriers. Second, even if present within the same cell, DNA molecules are thought to assume an attitude of benign neglect towards each other unless they are largely identical and differ only in minor ways. If it could be shown that these assumptions are incorrect and that new combinations of DNA are constantly formed in Nature,the main reason to fear this research would be removed and we could concentrate on the remaining details of public health concern, e.g. in those cases when DNA molecules known to encode genes for toxins are used in a particular experiment.

While others will undoubtedly concentrate on the question of how tight the species barriers really are, I hope to show that the second of the above assumptions, the lack of interaction of unrelated DNA molecules is incorrect. I will choose my examples mainly from bacteria, where the study of the processes by which unrelated DNA molecules interact naturally, has proceeded furthest. I will in the end, however, mention that similar mechanisms most probably also occur in higher organisms.

Let me start with the description of what is called insertion sequences (abbreviated to IS elements) in *E.coli*. IS elements are short DNA sequences, approximately the size of one gene, which are normal residents of the *E.coli* chromosome. Unlike other DNA segments, however, they are not fixed to a certain position on the *E.coli* chromosome. The cells possess mechanisms, most probably of an enzymatic nature, which can transpose the IS elements to new positions on the *E.coli* chromosome. If the cells harbour additional chromosomes, e.g. the so-called plasmids or the DNA of viruses, the IS elements can be transposed to these also (Starlinger and Saedler, 1972, 1976; Bukhari and co-editors, 1977).

At their new locations, the IS elements exert influences on the DNA adjacent to them. If the IS element is inserted into a gene, this gene can no longer function properly, and the result is a null mutation of that gene. However, effects on adjacent genes are also observed. More often, these genes are inactivated but in some cases a previously silent gene is activated.

While these effects may well be considered quite local and not of much interest to the general question of the creation of novel DNA combinations, more excitement must be aroused by the next, slightly more complicated class of transposable DNA elements, the so-called transposons. In most properties, transposons resemble closely the IS elements, so much so that in some instances transposons use IS elements as their building blocks. There is one important difference, however. In addition to the DNA sequences involved in the transposition process, the transposon carries one or more genes that can be expressed and lead to the formation of an enzyme(s). In many instances these enzymes inactivate a certain antibiotic, and thus render the cell possessing the transposon resistant towards the antibiotic (Cohen, 1976; Kleckner, 1977; Starlinger, 1978).

The formation of new resistance plasmids often conferring multiple resistance towards a variety of antibiotics is, in most instances, due to the location of the resistance genes on transposons. Therefore, several transposons can be picked up by a single plasmid, even if this plasmid did not before share any sequence homology with this resistance gene and, therefore, could not possibly have acquired it by general genetic recombination. The multiple drug resistances are of considerable medical concern, so much so in fact that the early discussions on the new recombinant DNA technology were heavily burdened with advice on how not to create artificially novel combinations of antibiotic resistance which might be incorporated into bacteria, that might escape from the laboratory and create problems for the public health services. Nobody any longer is specifically concerned with this question. It is well known that the transposition of these genes is a perfectly natural event, which has to be combatted at the level of selection by, for example, using antibiotic therapy judiciously and maintaining hygenic conditions in hospital wards and by not feeding antibiotics to livestock in order to squeeze an additional percent of yield and profit.

Thus, if a gene is introduced into a transposon, it becomes as movable as if it were in the hands of a recombinant DNA technologist. But what about genes other than those conferring antibiotic resistance? First, it must be stressed that the genes known to be carried on transposons are very varied. Genes for resistance against heavy metals, genes for sugar fermentation, (Cornelis and co-workers, 1978) and even genes for an *E.coli* enterotoxin (So and co-workers, 1979) have been found on a transposon. By a little extrapolation we might expect that any gene, if placed between two nearby IS elements, will behave as if it were on a transposon, provided it is not introduced into another bacterial species lacking the genes encoding the transposition apparatus.

Secondly, and possibly even more important, however, is the discovery that much more pronounced rearrangements of unrelated DNA can be brought about by IS elements. Possibly, the most interesting is the discovery of replicon fusion that was reported by Gill and co-workers (1978). These authors showed that instead of transposing a certain transposon from one plasmid to another, a fused structure can be formed under certain circumstances which consists of one copy each of the two plasmids, and of two copies of the transposon. It looks as if the transposon has been duplicated during the process of transposition and the molecules have been trapped in an intermediate stage of the transposition process.

This is an exciting possibility. If it were true that transposition does not occur via an excision and re-insertion mechanism, but rather by copying the transposon and leaving one copy in the old position while inserting the other copy at a new place: if it were also true that this process occurs while the donor and the recipient molecule are components of a fused structure, then the transposition of an IS element or transposon may be a very general mechanism to join DNA segments from any source, if they are present in the same bacterial cell. It is even conceivable that this joining mechanism is the more ancient one, and the disentangling of the fused structure that is necessary, if we are to observe a simple transposition, has to be brought about by an additional recombination mechanism, that may also be absent. The observation by Gill and co-workers that the replicon fusion is still possible in deletion mutants that cannot carry out ordinary transposition, and that separation of the fused structure can be achieved by genetic recombination, points in this direction. Is it conceivable that in the early days of evolution the formation of new combinations of DNA was a more common event than today, and that genetic recombination evolved later, when valuable DNA combinations had been created, which were worthwhile preserving? This is only speculation, but, the experimental evidence at least does not contradict it.

What is known about the formation of novel DNA combinations in higher organisms? The high school teacher, whom I referred to at the beginning, could easily remind us that chromosomal aberrations, translocations, the formation of dicentric and acentric chromosome fragments and of ring chromosomes have been common knowledge among cytogeneticists for decades. He might have a harder time to explain by which mechanism these aberrations are created. Using the information that the cytogeneticist of his day told him, he will talk about mechanical breaks in chromosomes and about a stickiness of freshly broken ends and their consequent coming together in new combinations. This does not appeal to us latter-day molecular biologists. Can we instead understand some of those events in terms of the transpositions and abortive transpositions that I described for bacteria? This may well be possible. We must first ask, whether transposable DNA sequences are present in eukaryotes. This is not easily answered because in many instances our knowledge of the appropriate biochemistry is still limited. If, however, we ask about transposable genetic elements, that are defined by genetic experiments rather than by biochemistry, the answer is a clear yes.

The most detailed experiments come from the work of Barbara McClintock, who has described such elements in maize (McClintock, 1950; Fincham and Sastry, 1974). It is astounding, in how many respects, even in minute detail, these elements behave similarly to the IS elements discovered much later (Nevers and Saedler, 1977). There is not space for me to discuss these questions in detail. It should be mentioned, however, that in addition to the transposition of these elements themselves, they have been shown to cause chromosome breaks. The question can at least be asked, whether other gross chromosomal rearrangements in maize may, in some instances, be related to a mechanism linking unrelated DNA molecules, and making use in this process of relatively small DNA sequences that are capable of interacting with any unrelated DNA molecule, because they are recognized, cleaved and ligated again by appropriate enzymes.

It will be some while before we know the answers, but now that the questions are being asked, and bearing in mind the astonishingly fast progress in this field of research in the last few years, this time may not be too far away. In any case, we can already be sure that the joining of seemingly unrelated DNA sequences is a common event in Nature. Indeed, we are forced to assume a more humble attitude, when we see that Nature has evolved enzymes, such as nucleases, DNA ligases etc., not only for the use of clever biochemists, but is using them for her own purpose. At the same time we no longer need be afraid of our own role in this realm of human undertaking, rather we might reserve our fears for other topics that are really man-made.

REFERENCES

Bukhari, A., J.A. Shapiro and S. Adhya (1977). *DNA Insertion Elements, Plasmids and Episomes*, Cold Spring Harbor Laboratory.

Cohen, S.N. (1976). Transposable genetic elements and plasmid evolution. *Nature (London)*, 263, 731-738.

Cornelis, G., D. Ghosal and H. Saedler (1978). Tn*951:* A new transposon carrying a lactose operon. *Molec. Gen. Genet.*, 160, 215-224.

Fincham, J.R.S., and G.R.K. Sastry (1974). Controlling elements in maize. *Ann. Rev. Genetics*, 8, 15-50.

Gill, R., F. Heffron, G. Dougan and S. Falkow (1978). Analysis of sequences transposed by complementation of two classes of transposition-deficient mutants of Tn*3*. *J. Bacteriol.*, 136, 742-756.

McClintock, B. (1950). The origin and behaviour of mutable loci in maize. *Proc. Natl. Acad. Sci. USA*, 36, 344-355.

Nevers, P. and H. Saedler (1977). Transposable genetic elements as agents of gene instability and chromosomal rearrangements. *Nature (London)*, 268, 109-115.

So, M., F. Heffron and B.J. McCarthy (1970). The *E. coli* gene encoding heat stable toxin is a bacterial transposon flanked by inverted repeats of IS*1*. *Nature (London)*, 277, 452-456.

Starlinger, P. (1978). *In vivo* formation of recombinant DNA molecules by IS elements and transposons. In A.M. Chakrabarti (Ed.) *Genetic Engineering*. CRC Press Inc., West Palm Beach, p. 123.

Starlinger, P. and H. Saedler (1972). Insertion mutations in microorganisms. *Biochimie*, 54, 177.

Starlinger, P., and H. Saedler (1976). IS-elements in microorganisms. *Current Topics Microbiol. Immunol.*, 75, 111-152.

DISCUSSION

A. CAMPBELL: One of the features of the structure of transposons like Tn*9* is that you have two IS elements with the gene inbetween. There has been a good deal of talk about the idea that if you have any gene between two IS elements (as any gene on the chromosome is, because there are several IS elements in the chromosome) that this whole segment should be transposable like a transposon. This, I think, is very relevant to the present discussion because it provides a mechanism by which that whole chunk could move into a recipient chromosome. Would you comment on the present factual status of that idea?

P. STARLINGER: There is not very much experimental evidence where this has been done deliberately. What is known now is that there are two transposons where two different genes are put between two IS*1* copies. One of those transposons is the one you have just mentioned which confers chloramphenicol resistance, the other has been published by So and co-workers and is a transposon encoding an enterotoxin of *E. coli*. Now, by extrapolation, we could assume if another gene would be put in-between two IS*1* copies, it would also be transposed. This will soon be tested. However, I think I was cheating when I said the only difference between IS elements and transposons is the presence of the gene for something unrelated. I think it will come out that there's another difference and that is that a transposon probably brings along the machinery for its transposition because these transposons are known to be on resistance plasmids, which are often very capable of moving between species. Now, an IS element which is only known within its own species may have this function somewhere else on the chromosome and so if you would take one of these and make a transposon out of it, you will also have to bring in these transposition functions, at least if you want to leave the species and I think this is the direction in which work is now being done.

M.C. VAN MONTAGU: Just to extend the communication that you gave of Heffron and Falkow's work, this is surely not an isolated case. For example, all the fusion plasmids that we isolated between the broad host range plasmid RPg and other plasmids like the Ti plasmid are of this type. There it's the kanamycin locus that acts as a kind of defective transposon; it no longer transposes, but a segment still is able to duplicate and to insert DNA inbetween this duplication. So all the fusion plasmids obtained have such a duplication of the kanamycin region and then one finds repeated segments exactly as in the structure you show. So it's probably a very general event.

W. SZYBALSKI: To answer your question, I should refer to unpublished data of Mike Malamy, who was able to either excise or transpose some genes located between two IS*3* sequences.

And now a second comment. This discussion, starting with Bodmer, reminds us how very important it is for evolution to have new genetic material. Nature would have been very negligent not to use already pre-formed genes for that purpose. Therefore, Nature has developed all kinds of specific mechanisms. We heard from Allan Campbell about transducing phages and about the IS and Tn elements from Peter Starlinger. I am sure that organisms use also the restriction enzymes and ligase, the *in vivo* recombinant DNA technique, for natural evolution, i.e. to borrow or steal the already developed genetic modules. And while examining the higher organisms, and their complex genomes, I wonder how they came into existence? Why is there so much extra DNA? Could it be a reflection that whenever evolution needed one new gene, a whole big chunk of the transposing DNA was brought in with it? I mean a whole lambda phage genome or another episome was used to bring in one useful gene. May be the big eukaryotic genomes are just put together from many transposition elements that carry only small pieces of DNA that were required at given stages of evolution. Thus, most of our own genome will be a chain of lambda - Mu - episome - IS - Tn - like elements each carrying some required genetic module. I wonder!

P. STARLINGER: Umm....

H. KORNBERG: That was a masterly concise comment! Any other questions?

M.H. RICHMOND: I think it's important, and I am sure you would agree, Peter, to stress that just because you do get DNA between a pair of insertion sequences it doesn't necessarily mean to say you will get transposition. One example that we have looked at is a duplication of an IS sequence within a tetracycline gene as part of a tetracycline transposon. So you now have, as it were, three ISs, two of which must be in direct repeat and that's a relatively stable situation. The other point I would make is that you don't get the incidence of transposition that one might expect. The chromosome is very big and the transposition frequences that you get between plasmids, if applied to the chromosome, would lead you to expect that transposons would be highly mutagenic. In practice they don't seem to be so. So, there must be protective devices as well.

P. STARLINGER: I should like to comment on that as well. I agree and I think it even has to be extended. If we look at two chromosomes like those of *E. coli* and *Salmonella typhimurium* then these have separated quite a while ago and this can easily been seen in their base sequence which has evolved to such an extent that I think about one sixth of the bases are now different between the two species. At the same time we know of mechanisms that reshuffle the chromosome and we do know about the frequencies with which this occurs and judging from these laboratory frequences we would assume that the general genetic arrangement of genes along the chromosome should be very different not only between *E. coli* and *Salmonella* but also after a short while within *E. coli* strains. This is not found, this arrangement is preserved much better than the base sequence and so there must be some selective force which somehow asks for the present arrangement to be retained. I think there are some possibilities but one can summarize them by saying we are really ignorant of what the selective force is. So there is something to be discovered.

LABORATORY GENETIC MANIPULATIONS

G. Bertani

Microbial Genetics Laboratory, Karolinska Institutet, S-10401 Stockholm, Sweden

As part of the general theme of this session, I am expected to review the classical, that is, *in vivo* genetic manipulations of organisms that have developed over the years. I assume a genetic manipulation to be something more advanced or more complicated than an ordinary cross or hybridization. Since I am a microbial geneticist, and *Escherichia coli* with its viruses and plasmids is the organism known in the finest detail from the genetic point of view, I will limit my considerations to bacteria.

I will first recall some features of *E. coli* genetics that seem to apply to numerous other bacterial types.

All known types of recombination in bacteria are directional, in the sense that one can always recognize a donor and a recipient organism, and that only a part, often a very small part, of the total genome is transferred. Exceptions are some experimentally contrived cell fusions in *Bacillus* (Schaeffer and co-workers, 1976; Fodor and Alfoldi, 1976).

The transfer may be simply the uptake by a bacterium of a piece of DNA set free through the disintegration of another bacterium, followed by integration within the recipient's chromosome. This is called transformation. It occurs naturally in certain types of bacteria, which have even developed special and to some extent specific mechanisms for the uptake (Raina, Metzer and Ravin, 1978). In many other types of bacteria it can be produced artificially by treating the recipient with high concentrations of calcium, or by stripping the bacterium of its cell wall.

The transfer may occur via a virus, a bacteriophage, of the not fully virulent kind, particularly those that are able to establish lysogeny, the so-called temperate bacteriophages. Their DNA can implant itself in the infected bacterium without killing the latter and replicate with it.

The DNA of the virus can be found in the bacterium as an autonomously replicating DNA ring. This is the case for the phages of the P1 class. At some later stage the phage DNA may be incorporated into newly made virus particles. As a byproduct of this process, it happens that equally large fragments of DNA of the host bacterium also become packaged in mock virus particles. These particles will infect other bacteria, and will deliver to the new recipient the DNA fragment of the bacterium in which they were formed. Integration of the fragment into the chromosome of the recipient will produce a permanent genetic change in the latter.

This is called transduction, or more precisely, generalized transduction, since it does not distinguish between different segments of the bacterial chromosome, which all have the same chance of being incorporated into a virus-like particle.

With other temperate phages, the DNA of the virus inserts itself into the bacterial chromosome with relatively high frequency. This can occur at very specific sites, which have been mapped precisely. This is the case for the phages lambda and P2. The phages possess enzymes that very specifically cut the DNA of the host bacterium and that of the virus at fixed sites, and recombine the two DNA molecules as in a reciprocal genetic exchange. This can also work in reverse, and the DNA of the virus will then be set free from the bacterial chromosome and start replicating autonomously. New virus particles will be formed eventually, and they will infect new bacteria. At times, the excision process might not work properly and cuts the DNA at the wrong place. As a consequence, one can obtain bacteria that have lost a small bit of chromosome, next to the site of specific attachment for that particular virus. This process has been called eduction. If the virus multiplies in such cases, it carries with itself, as a part of its own DNA, the DNA fragment removed from the bacterial chromosome. Upon infection of a new bacterium, this fragment may also be integrated in the chromosome of the recipient, which will thus be altered genetically. We call this phenomenon specialized transduction, since it concerns only those regions of the bacterial chromosome neighbouring a site of attachment of the virus DNA. Purified preparations of such phages supply homogeneous DNA, enriched for a given bacterial gene or set of genes.

Still other temperate viruses, those of the Mu or mutator type, are unspecific in their choice of attachment site on the host chromosome. When within a gene, they will inactivate its function. They have additionally the property that when they replicate autonomously their DNA replicates while embedded in the host cell DNA. The mechanism is not yet understood, but it is quite clear that hybrid structures containing the virus DNA and relatively long stretches of host bacterium DNA are formed routinely in the course of the virus replication. It seems that the cutting out of these hybrid structures and unselective integration of the virus DNA at a variety of places on the bacterial chromosome may occur simultaneously, so that a mobilization of whole segments of the host DNA occurs while the virus multiplies. These displaced segments can be recovered if the infected cell can be used at the same time as a donor in a cross (Faelen and Toussaint, 1976).

Other semi-autonomous genetic elements found in many bacteria are the plasmids (Meynell, 1972). They can be isolated from the cells as small DNA rings. Some of these plasmids are able to take over the control of the synthesis of parts of the cell wall of the host bacterium. Structures can be formed on the outside of the bacterium which allow it to establish contact or conjugate with some other bacterium. At this point a copy of the plasmid DNA can transfer itself from the original host to the recipient bacterium. These plasmids are called self transmissible because of this property. The prototype of this class is the F factor. In the same class, the very common R factors are found carrying genes that affect the resistance of the host cell to a variety of inhibitors and antibiotics (Falkow, 1975). Self transmissible factors can also insert themselves in the host chromosome at a variety of places. They can still make the bacterium conjugate. In such cases, however, when they would be transferring copies of themselves to a recipient cell, they mobilize also the chromosome of the host bacterium, dragging it along, beginning with the region nearest their site of attachment. The bacteria capable of doing so are called HFr for high frequency recombination and have been very useful in charting the structure of the *E. coli* chromosome. Still another interaction with the bacterial chromosome is the formation of a new plasmid carrying, attached to the F factor, a small segment of the host DNA. These elements are called F', and can of course transfer the attached bacterial genes to a recipient cell with high frequency.

There are also smaller plasmids which are not able to transfer themselves from cell to cell, although they are able to replicate autonomously. They still can be transferred with the help of another, self transmissible plasmid. They can also be transferred as DNA, in a variation of the transformation process. These are the plasmids most commonly used for gene cloning. Even a superficial recounting of these processes reveals the fact that we are not playing with one strain of one organism, but with a variety of different and more or less autonomous genetic elements. These elements are of common occurrence. If a set of natural isolates of enteric bacteria is examined by fairly routine methods, it is not unusual to find more than half of them carrying a phage or a plasmid. Occasionally more sophisticated experiments done for other purposes have revealed in the chromosome of a bacterial strain the presence of fragments of genetic material that are obviously homologous to a known phage (Cohen, 1959; Gottesman and co-workers, 1974). What we are handling then, in doing bacterial genetics, is not one biological object but a constellation of objects, all having in common the property that at some juncture they can interact genetically with each other.

There is another point which is essential for understanding the progress made in bacterial genetics. It is obvious that, in working with bacteria and viruses, we can easily handle large numbers of individuals. For some characters, resistance to drugs and to viruses, for example, we can thus select directly for rare types, and isolate them even if they are present in the population in very small numbers, as low as one in 10^{10} or even less. In addition, over the years, a variety of indirect selective methods have been worked out, based on the increasing knowledge of the physiology and genetics of the organism. The classical example is the penicillin technique for the isolation of nutritionally requiring mutants. A parallel development has been the progress made in the area of mutagenesis. During the last thirty years a variety of powerful mutagenic chemicals have come into use in addition to radiation. In genetic work sometimes one needs to introduce into a strain a mutation by recombination with another strain that carries it. With *E. coli* it is often less laborious to introduce anew the desired genetic marker by mutagenesis. Mutagenisation of a strain for this purpose is a routine operation in bacterial genetics.

(Example. We want to move mutation A^- from strain I into strain II, Character A, however, cannot be selected for directly. If a gene B, with easy selection properties, is known to be sufficiently near to A, we can mutagenise strain II, isolate a B^- mutant, then transduce with phage Pl from I into II B^-, selecting for B^+. A^- will come along often enough because of its linkage to B, and the B^+ state of II will be re-established.)

Mutagenic treatments usually have no genetic specificity, in the sense that they attack equally strongly different regions of the exposed chromosomes. One consequence of this is that with powerful mutagenic treatments one runs the risk of introducing a number of different mutations in the same chromosome. This can be avoided by applying another selective technique, localised mutagenesis (Hong and Ames, 1971).

(Example. We want to obtain a set of new mutations in gene A, closely linked to B. We first isolate a mutant B^-, second mutagenise very heavily a preparation of phage Pl grown on the original strain, and third infect with it the B^- strain, selecting for B^+. Because of the close linkage recipient bacteria that have been transduced to B will have also received often enough a damaged, i.e. mutated, gene A. The rest of the recipient chromosome will be undamaged.)

There is a class of mutational events which are extremely useful in manipulating the genome of *E. coli* and its viruses. These are deletions and duplications. We do not know how they occur. Some types of deletions can be selected for directly on the basis of the known effects they produce on the bacterium. The simultaneous loss of the functions of two unrelated, adjacent genes immediately suggests a deletion.

(Example. Defect in or loss of gene *tonB* makes *E. coli* resistant to phage T1. Adjacent to *tonB* are the genes *trpA, trpB*, etc. needed for the synthesis of tryptophan. Quite often a mutant selected for resistance to T1 is found to be simultaneously dependent on externally supplied tryptophan.)

There is one situation in which deletions can be selected for directly, independently of their genetic effects, provided they are viable. This is based on the interesting relationship between amount of DNA in a virus particle and the particle's stability to heat. Within certain limits, the shorter the segment of DNA packaged in the particle, the more resistant is the latter to heating. Phage deletion mutants can be isolated quite easily by looking at the survivors in a population heated at an appropriate temperature. The range of temperatures covered is so large that it is possible to use this technique in a quantitative way, and measure the amount of DNA deleted in such a mutant from the survival of the mutant at different temperatures.

Gene duplications are often isolated as a result of selection for increased functionality of a gene or set of genes. For example (see Normark and co-workers, 1977), *E. coli* produces small amounts β-lactamase, not sufficient to protect it against penicillin. Among penicillin resistant mutants, sets are obtained exhibiting stepwise increased lactamase production. They can be shown to result from tandem duplication of the lactamase gene.

A result of deletions and duplications may be gene fusion (Franklin, 1978). Gene fusions are useful in the study of the regulation of gene expression (a gene A might be put under the control of the regulation system of gene B) and of structural and functional features of proteins (fusion of structural genes A and B may result in synthesis of a protein having the N-end of protein B and the C-end of protein A).

For a deletion to be viable, the deleted genetic material must be non-essential. This may well be the case over short distances. If widely separated genes are to be fused, they have first to be translocated so as to reduce the amount of intervening DNA. Another purpose in producing gene transpositions may be to bring certain genes near a phage attachment site, so that a specialised transducing phage can be obtained.

(Example. Very close to *tonB*, on the side opposite to the *trp* genes, is *att80*, the specific attachment site for the lambdoid phage *ϕ80*. After *tonB* is deleted, the *trp* genes may be sufficiently close to *att80* to be eventually incorporated into *ϕ80* phage.)

(Example. If we want to incorporate genes XY into phage *ϕ80*, we must first look for integration of an F factor in the vicinity of XY. The situation is improved if the F is defective, unable to replicate autonomously at high temperatures (Gottesman and Beckwith, 1969). From this strain we isolate an F' carrying genes XY. If the latter is transferred to a recipient bacterium, selecting simultaneously for XY and resistance to T1, at high temperature, transpositions of the F'XY plasmid into the *tonB* gene will be obtained. The XY genes may now be sufficiently close to *att80* for *ϕ80* phage to incorporate them, perhaps following deletion of some of the DNA that still separates them.)

(Example. An all purpose technique for placing an "indicator" gene whose product is easy to detect, under the controls belonging to another set of genes, is based on the property of phage Mu to establish itself anywhere in the *E. coli* chromosome (Casabadan, 1976). One uses here a phage which carries, as the result of complicated manipulations, a DNA fragment including the "indicator" gene, and a fragment of phage Mu DNA, in addition to the phage's own material. Through mutation this phage is also defective, being unable to utilize its normal attachment site on the host chromosome. If one wants the "indicator" gene placed under the controls of gene X, one first looks for a lysogenic bacterium carrying phage Mu DNA within gene X. The strain is infected with the complex phage described above. Normal, homologous recombination will occur between the Mu fragment present in the complex phage and the Mu DNA inserted within gene X. As a result, all the DNA of the complex phage, including the "indicator" gene, will be established as an insertion within gene X. Deletions of the intervening material will bring about the desired fusion between the gene elements controlling X and the "indicator" gene.)

A general point can now be made. When a virus DNA inserts itself into the DNA of the host bacterium, or when imperfect excision of the former removes a bit of chromosome from the bacterium, or when deletions and duplications are formed, novel DNA joints are created. These are in principle equivalent to the novel joints produced by *in vitro* DNA recombination. One difference is that we do not control in any direct way where such novel joints will take place, rather, we let them occur and then, exploiting our knowledge of the genetics and the physiology of the bacterium, we select for them and isolate them. These novel structures, while occurring spontaneously, are thus selectively amplified by our laboratory manipulations.

The obvious limitation of the *in vivo* manipulations is that the genes to be fused must be present in the same cell. If genes belonging to different strains have to be operated upon, they must be first brought together in the same cell. This can be done relatively easily between fairly different bacterial strains, often classified under different generic names. If one follows the more objective definition of the species as the natural grouping within which genetic exchange takes place efficiently, then it must be quite clear that the traditional species of bacteria established by taxonomists represent in reality categories that are much smaller than the "natural" species. In general, geneticists have tended to work with isogenic material, i.e. to avoid hybridization between less closely related strains, in order to secure firmer interpretations from their experiments. It is likely, however, that as interest in genetic interactions between distant organisms increases, rational methods will be found for overcoming in the laboratory some of the isolating barriers found in Nature, even without using *in vitro* recombination techniques.

REFERENCES

Casabadan, M.J. (1976). *J. Mol. Biol.*, 104, 541-555.
Cohen, D. (1959). *Virology*, 7, 112-126.
Faelen, M. and A. Toussaint (1976). *J. Mol. Biol.*, 104,525-540.
Falkow, S. (1975). *Infectious Multiple Drug Resistance*. Pion Limited, London.
Fodor, K. and L. Alfoldi (1976). *Proc. Natl. Acad. Sci. USA*, 73, 2147-2150.
Franklin, N.C. (1978). *Ann. Rev. Genet.*, 12, 193-221.
Gottesman, M.M., M.E. Gottesman, S. Gottesman and M. Gellert (1974). *J. Mol. Biol.*, 88, 471-487.

Gottesman, S. and J.R. Beckwith (1969). *J. Mol. Biol.*, 44, 117-127.
Hong, J. S. and B.N. Ames (1971). *Proc. Natl. Acad. Sci. USA*, 68, 3158-3162.
Meynell, G.G. (1972). *Bacterial Plasmids*. Macmillan, London.
Normark, S., T. Edlund, T. Grundström, S. Bergström and H. Wolf-Watz (1977). *J. Bacteriol.*, 132, 912-922.
Raina, J.L., E. Metzer and A.W. Ravin (1978). *J. Bacteriol.*, 133, 1224-1231.
Schaeffer, P., B. Cami and R.D. Hotchkiss (1976). *Proc. Natl. Acad. Sci. USA*, 73, 2151-2155.

DISCUSSION

P. STARLINGER: I have a question for you and Allan Campbell regarding the excision of class II F' plasmids, in this instance, the bacterial DNA is cleaved outside of the F particle on either side, so what is really fused are two pieces of bacterial DNA, neither of which is a special sequence like F or an insertion or a lambda sequence. I think this is the only published case where any random DNA sequences of *E. coli* are being joined together apart from the production of deletions. Do you know anything about the mechanism by which this is being done or is there any speculation on it?

A. CAMPBELL: The nice thing about the IS and transposon work that you presented is that it gives us one specific mechanism that can generate some events of this kind, deletions and so forth. I think that there must be other mechanisms in the cell as well which have been known for a long time and whose properties lead to somewhat the same conclusion in terms of the possibilities for re-arrangements, but it's much nicer to have a specific example that one can talk about.

W. SZYBALSKI: I have a general question to all speakers, as it relates to the subject of this symposium. We learned here about many natural modes of transfer of genetic materials between species or within species. I fail to comprehend why the recombinant DNA technique should be treated differently from many other laboratory techniques used for the transfer of genetic material. I would conclude on the basis of information presented in this session that the recombinant DNA technique, being more precise, should be safer than any other random-mating technique. Is there any *a priori* reason to single out and to regulate the recombinant DNA technique? I see no reasons.

G. BERTANI: I see no reason to single out this technique in particular. The key difference is in the size of the taxonomic gaps that can be bridged by *in vitro* DNA recombination. One knows too little about the barriers, and people worry about the worst that could happen. I mentioned mutagenesis, a little on purpose, in the sense that it does introduce new "things" in the bacterium, but we certainly don't worry about creating, say, a new toxin when we do that in *E. coli*.

W. SZYBALSKI: Why didn't we have guidelines for chemical mutagenesis thirty years ago? I guess because nobody worried about the dangers of genetics just after the war.

G. BERTANI: Most new mutations may be considered to be damaging. On the other hand, mutagenesis is also "creative". We know that from theory, and also from experience in the laboratory. It is possible to develop new enzymatic activities selecting for mutants in an appropriate way. There is also a report of changes in the base composition of a bacterium in the presence of high mutation pressure.

W. SZYBALSKI: There are no reports of any dangers created by making bacterial mutants using chemical mutagenesis. On the other hand, bacterial mutants were of great benefit to scientific research and to society in general.

G. BERTANI: No, I don't know of any case of dangerous microorganisms produced by mutagenesis in the laboratory.

H. KORNBERG: According to the programme before me, I am supposed to provide a Chairman's summary. I think that, to use the phraseology at least of the science of which we have been talking, if I were to attempt to summarise the excellent presentations that we have heard this morning, I would be guilty of terminal redundancy - so I will not attempt to do that. However, I feel it would be improper for me to close without excercising the prerogative of the Chairman and saying something which has occurred to me through listening carefully to this morning's speeches.

It is clear that, from a biological perspective, the kind of fears and anxieties that originally motivated people to alert society in general to possible hazards inherent in recombinant DNA technology may very well have been over-estimated. We know that most of the manipulations that we regard as novel, in fact, occur already *in vivo*. We also know that the chances of the survival of a deleterious mutation may be small. Indeed, as I said in my introduction, our capacity to upset the global gene pools may be so limited that we may be like the fly on the axle-tree of the chariot wheel. But, as responsible scientists, we must also remember a number of other factors and here I am now speaking as impartial Chairman, in the sense that an Irish judge once defined impartiality as treading the fine line between justice and injustice. In that sense, if I may voice concerns of those who are not at this meeting, and remind us of political realities, it is to express my opinion that the possible attractions of saying "Well, there is really nothing to worry about, let us get rid of all guidelines" would be as naive as were the over-reactions that we manifested, or some of us manifested, in the early days of this technology. In the first place, one cannot put the genie back into the bottle. It is *we* who first alerted the public to possible risks, and it is due to us that there is public concern: the public now has the right to a critical assessment of the validity or otherwise of the dangers. Absence of evidence is not evidence of absence: in this matter the onus of proof may largely rest with us.

Let me just remind you that it was only last year that we had a second outbreak of smallpox in this country. Admittedly, this has nothing whatever to do with the kinds of technique which we are discussing at this meeting, but the incident, nevertheless, alarmed society to a possibly exaggerated extent to the dangers that are inherent in working with biological materials. I have not seen the paper today but certainly yesterday, as I came down in the train, the newspaper was full of stories of a nuclear accident at a power station in Pennsylvania where the public had been assured there were only negligible risks of accident. It is these kinds of considerations that we, as members of society and users of public funds, have to bear in mind when we ask, as I hope we will, for relaxation of the stringest rules that at present handicap our work.

We must also remember that the discussions we shall have tomorrow morning, on the practical benefits of recombinent DNA research, imply that very soon we will see the kind of technology that we are developing applied commercially, on a scale which is far beyond that which would normally apply in a laboratory, perhaps in countries where the safety regulations applying to Europe and the USA are not so stringent, and where the chances of accident or hazard may be very much greater. I think it is therefore prudent for us for the moment to reserve judgement, to wait for the discussions, that I personally am looking forward to very much, of the options that are open to us and the perception of risks, in order to enable us to achieve a more balanced perspective and to formulate balanced conclusions from this meeting.

So, my view, therefore, as the Chairman of this session, is that I am immensely grateful to have had this morning such an interesting, well-informed and concise illustration of the biological background, but as one who has been involved with the matters under discussion even before I joined Lord Ashby's working party, I must state my personal position that I am not yet ready to advocate that we should recommend the *total* abolition of all guidelines.

ACHIEVEMENTS OF RECOMBINANT DNA RESEARCH

EXPERIMENTAL TECHNIQUES AND STRATEGIES FOR DNA CLONING

S. N. Cohen

Departments of Genetics and Medicine, Stanford University School of Medicine, Stanford, California 94305, U.S.A.

The task I have been assigned is to review briefly the techniques that collectively are known as "recombinant DNA methodology", and to describe some of the experimental strategies used in this research. Although recombinant DNA is commonly referred to as "gene splicing", "gene cloning" would seem a more appropriate term. There are four general requirements: 1) a replicon (cloning vehicle or vector) able to propagate itself in a recipient organism, 2) a method of joining another DNA segment to the vector, 3) a procedure for introducing the composite molecule into a biologically functional recipient cell, and 4) a method of selecting or identifying those cells that have acquired the hybrid DNA species.

The earliest DNA cloning experiments utilized bacterial plasmids, which are circular extrachromosomal elements of DNA that are capable of autonomous replication. Most simply, a plasmid can be viewed as a replication system to which another segment of DNA has been linked by natural recombination mechanisms. For plasmids isolated from natural sources, the added segment commonly provides the host carrying the plasmid with a selective advantage, so that the plasmid is propagated in the bacterial population. Plasmids are widespread among bacteria and encode a variety of traits, including resistance to antibiotics and heavy metals; production of enterotoxins, virulence factors or antibiotics; fertility functions; production of restriction and modification enzymes; resistance to UV irradiation; a capacity to metabolize polycyclic hydrocarbons, such as carbon and octane; and tumorigenicity in plants.

The use of plasmids as vectors in DNA cloning experiments depends on the ability to introduce plasmid DNA molecules into bacterial cells by transformation. The development of a procedure for transformation of E. coli K-12 with plasmid DNA has made it possible to establish clones of bacterial cells that carry the progeny of a single plasmid DNA molecule and this in turn has enabled studies with plasmids in ways that previously had been practical only with bacteriophages. Because transformation enables the cloning of individual plasmid DNA molecules, my colleagues and I reasoned some years ago that if a foreign DNA segment could be inserted within a plasmid DNA molecule while it was outside of the cell, the replication apparatus of the plasmid might be used to replicate the foreign DNA segment in the same way that the replication systems of naturally formed plasmids can replicate various segments that have been linked to them by natural biological processes. To do this, the exogenous DNA segment would have to be inserted at a site that does not interfere with the ability of the plasmid to replicate or with the expression of genes required for selection of transformant cells that had acquired the hybrid plasmid DNA molecule. Of course, it was not possible to know before the initial DNA cloning experiments were done whether linkage of a bacterial plasmid replicon to a foreign DNA segment would result in a viable combination. Could an E. coli plasmid replication system in fact propagate and amplify a piece of DNA that had been derived from a very dissimilar biological species?

The task of breaking plasmids open at specific sites was much simplified when the properties of Type II restriction enzymes were discovered. Such enzymes recognize specific DNA sequences that are identical in the 5' to 3' direction on each of the two DNA strands. One enzyme that was among the earliest to be studied (the EcoRI endonuclease) cleaves its recognition sequence asymmetrically so that projecting single-strand termini are formed. These "cohesive ended" termini thus formed can be rejoined by hydrogen bonding of the overlapping nucleotide bases and ligated to yield covalently linked DNA segments. Since the six base pair recognition sequence of the EcoRI enzyme is always cleaved identically by the endonuclease, identical cohesive ends are formed on DNA segments derived from different biological sources, and these segments can readily be linked. A number of other endonucleases with similar properties but different recognition sequences have been identified.

Alternatively a nucleotide "tail" consisting of a polymeric run of identical nucleotides (e.g., dA) can be added to one DNA species and a tail of complementary nucleotides (e.g., dT) can be added to another; the two DNA species can then be joined by hydrogen bonding of the complementary nucleotide bases and subsequent ligation. It has also become apparent that DNA termini with projecting ends are not required for joining; even blunt-ended DNA molecules can be linked using the ligase encoded by bacteriophage T4.

In the work that Chang, Boyer, Helling and I reported several years ago, EcoRI-generated segments of several different bacterial plasmids were inserted into the pSC101 vector, and the composite molecules were introduced into E. coli K-12 by transformation. Since the site of insertion (i.e., the EcoRI endonuclease cleavage site of pSC101) did not interfere with the capacity of the plasmid to replicate or express its tetracycline resistance gene, the vector could be used to clone the introduced DNA fragments. In subsequent experiments, Chang and I found that DNA derived from a plasmid indigenous to a quite unrelated bacterial species (Staphylococcus aureus) could be cloned in E. coli K-12 using similar procedures, and moreover that the bacterial genes derived from the Gram-positive coccus could be expressed in the Gram-negative species that served as the new host. It has since become apparent that heterospecific gene expression can also be accomplished between various other bacterial species combinations, but that phenotypic expression in some heterospecific gene transplants is a more complex matter.

Since I expect that Ken Murray will be discussing the use of bacteriophages as vectors in DNA cloning experiments, I need not cover this subject except to note that following the initial demonstration that a bacteriophage λ derivative (which had been mutated to remove most of the cleavage sites for the endonuclease being used) could be employed as a vector, a large number of recombinant DNA experiments have been carried out using phage vectors, and more recently plasmid-phage combinations.

It is worthwhile pointing out that the essential tools for recombinant DNA work (plasmid or phage replicons plus enzymes that cleave DNA at specific sites) are natural biological products. About two years ago, it was demonstrated that the EcoRI endonuclease and DNA ligase can accomplish in vivo the cleaving and joining processes mediated by these enzymes during in vitro recombinant DNA experiments. While such in vivo events occur at low frequency and require special experimental techniques for detection, they nevertheless show the qualitative similarity between the *in vitro* and *in vivo* processes. Moreover, from the considerations discussed earlier in this meeting by Bodmer and others, one would expect that low frequency events that provide a bacterial host with a natural biological advantage will be selected during the course of natural evolution, whereas combinations that occur at higher frequency but provide no selective advantage will be lost during evolution.

The transposable genetic elements discussed earlier by Peter Starlinger can also be used for carrying out "recombinant DNA" experiments *in vivo*. In experiments designed to study the process of transposition, my co-workers and I recently introduced a plasmid-derived replication system into an endonuclease cleavage site located within one of the transposable elements, Tn3. The "replicating transposon" thus formed was shown to be capable of undergoing translocation to other DNA species present concurrently in the same cell; such recipient genomes can be cleaved in a variety of ways, and then introduced by transformation into a host cell, where the *in vivo* action of the DNA ligase accomplishes recircularization of the fragment. Thus, constructs that are functionally analogous to "recombinant DNA" molecules formed *in vitro* can be made *in vivo* using a replicating transposon as a cloning vector.

During the past several years, a number of new methods have been devised to identify cells that have acquired particular DNA species. In instances where direct selection for phenotypic properties expressed by the cloned DNA is not feasible, *in situ* hybridization procedures can detect DNA introduced into either plasmid or bacteriophage vectors. More recently developed *in situ* immunological procedures enable identification of cells that express protein products encoded by a cloned DNA segment.

The strategies for obtaining expression of eukaryotic genes in bacterial cells will be discussed in greater detail by other speakers. However, some general comments are appropriate at this point. Shotgun cloning of mechanically sheared chromosomal DNA has been carried out using either "tailing" procedures or synthetic "linkers" that include specific endonuclease recognition sequences, and libraries of clones that contain fragments of complex mammalian chromosomes have been established and used in a variety of studies. However, so far as expression is concerned, the presence of intervening sequences, or "introns", between structural genes interrupts the continuity of structural gene sequences and limits the use of such methods for obtaining functional expression in bacteria of chromosomal DNA from higher eukaryotes.

A second procedure that has been used extensively is the isolation of messenger RNA (mRNA) from certain organs or tissues that produce relatively large quantities of a particular mRNA species. A double-stranded DNA copy of this mRNA is then made using the RNA-dependent DNA polymerase (reverse transcriptase). The resulting double-stranded DNA molecule can then be linked to a vector and cloned. A third approach involves the use of knowledge of the amino acid sequence of a specific protein to synthesize chemically a double-stranded DNA species that encodes the protein. These methods will be discussed in further detail by Bill Rutter (see this volume).

Expression in bacteria of cloned genes that encode products of higher eukaryotes can be accomplished in several ways. One of these methods utilizes the transcriptional and translational start signals of a prokaryotic gene to initiate transcription and translation, which then proceeds into a eukaryotic DNA segment that has been inserted within the prokaryotic gene. While inclusion of the eukaryotic peptide as part of a hybrid protein may help protect it from degradation by bacterial proteases, the hybrid peptide may not display the biological activity characteristic of its eukaryotic component, and general applicability of this method may require a procedure for specifically cleaving the hybrid. A recently reported approach circumvents the latter problem by using a polycistronic message initiated at a prokaryotic transcriptional start signal coupled with an appropriately located ribosomal binding site and translational start signal, preceding the eukaryotic gene. In this way, formation of a separate peptide of eukaryotic origin may be accomplished.

Using these various strategies, it has been possible to obtain immunological expression of a number of mammalian proteins in bacterial cells, and in at least one instance, to accomplish biologically functional phenotypic expression of a mammalian enzyme (the mouse dihydrofolate reductase) in bacteria. It seems likely that additional strategies that accomplish such expression more efficiently will be forthcoming.

APPLICATION OF RECOMBINANT DNA TECHNOLOGY TO THE MOLECULAR GENETICS OF PROKARYOTES

Noreen E. Murray and K. Murray

Department of Molecular Biology, University of Edinburgh, Edinburgh EH9 3JR, Scotland, U.K.

The recombinant DNA technology adds to an already powerful range of methods for manipulating and analysing chromosomes in *E. coli*. Segments of DNA from a variety of prokaryotic sources can now be transferred via plasmid or phage vectors to *E. coli* where advantage may be taken of genetic techniques previously applicable only to *E. coli*. For some Gram-negative bacteria *in vivo* techniques have recently become available for the transfer of DNA from one species to another (see, for example, Faelen and co-workers, 1977).

Some of the new insights into prokaryotic genetics offered by recombinant DNA technology are best illustrated with examples. The few chosen demonstrate a variety of uses. Labelled probes identify homologous DNA sequences and elucidate their organisation within the chromosome. Purified DNA sequences are ready substrates for *in vitro* mutagenesis while sequences within a λ vector become amenable to efficient analysis by genetic recombination. Novel genetic fusions are made and used in the study of gene expression and in the amplification of gene products.

DETECTION OF HOMOLOGOUS SEQUENCES

The power of the *in vitro* recombination methods has been greatly enhanced by parallel developments in methods for fractionating and analysing DNA fragments and, of course, for rapid determination of nucleotide sequences, methods which in turn have depended upon the use of restriction enzymes. A particularly powerful method of analysis, developed by Southern (1975), involves fractionation of DNA fragments by electrophoresis in agarose gels, followed by denaturation of the fragments in the gel and transfer to cellulose nitrate membranes for hybridization reactions with radio-actively labelled RNA or DNA probes. This very sensitive technique for detecting homologous sequences has been fundamental to progress in the isolation and characterization of fragments from eukaryotic chromosomes but it also adds an extra dimension to microbial genetics. DNA sequences may now be identified by their hybridization with a specifically chosen probe rather than by their function.

The following example uses hybridization techniques in the identification, physical characterization and analysis of defective phage genomes resident in the chromosome of *E. coli*. This approach was suggested by an experiment in an EMBO course held at Basel in 1976 in which transducing derivatives of phage λ were constructed from restriction enzyme digests of *E. coli* and vector DNAs. DNA was prepared from phages that included the *trp* genes of *E. coli* and used to make probes. Digests of DNA from wild-type and *trp*-deletion strains of *E. coli* were fractionated by gel electrophoresis and transferred to cellulose nitrate membranes for hybridization with the probes derived from the DNAs of λ*trp* transducing phages. The hybridization experiments indicated regions of homology in addition to those provided by the bacterial DNA within the λ transducing phage.

This approach has been used in further studies and identifies two, if not three, areas of the *E. coli* chromosome that share homology with the DNA of phage λ (Kaiser and Murray, 1979; K. Kaiser, personal communication). Commonly used laboratory strains of *E. coli* all showed some homologous sequences when digests of their DNAs were fractionated on an agarose gel and hybridized with a radio-active preparation of phage λ DNA (Fig. 1). Much of the difference between strain AB1157 (Fig. 1, track c) and the other two K-12 strains (Fig. 1, tracks a and b) reflects the absence of a defective prophage whose presence in most K-12 strains was suggested on the basis of genetic evidence (Low, 1973). Detailed elucidation of the organization of this prophage genome within the *E. coli* chromosome has relied on hybridization studies using appropriate labelled probes (Diaz and co-workers, 1979; Kaiser and Murray, 1979). The remaining homology identifies residual DNA sequences from a second and probably even a third prophage.

The same technique is being applied to a comparative analysis of the genes that determine the restriction specificities of *E. coli* strains. The host specificities are determined by chromosomal genes (*hsd)*; *E. coli* K-12 and B, for example, are characterized by their different host specificity systems which result from a complex protein comprising three different polypeptides. The information relevant to the DNA sequence specificity resides in the subunit determined by the *hsdS* gene and the restriction and modification subunits of the K and B specificity systems are interchangeable. It seems probable that the *hsd* genes of the two strains will have much sequence homology and divergence in the *S* gene should reflect the different sequence recognition specificity. A probe containing the specificity gene of the K system hybridizes with the region including the B specificity gene. Comparative sequence studies will identify the extent of the homology and may reveal the parts of the polypeptide involved in the recognition of a particular DNA sequence.

CONSTRUCTION OF MUTANTS

Restriction enzymes have been used in a variety of ways to make deletion mutants. The region of the *hsd* complex that codes for the specificity protein has been identified by deletion analysis. The DNA fragment carrying the host specificity determinants includes two targets for R.*Hind*III separated by 1.8 kilo bases of DNA. This intervening fragment was removed by digesting the DNA of a λ*hsd* transducing phage with R.*Hind*III and then joining the flanking pieces of DNA. Loss of this DNA is accompanied by a loss of host specificity (B. Sain and N.E. Murray, unpublished observation).

A similar experiment has been used to make a particularly important deletion derivative of the commonly used plasmid vector pBR322 (A. Twigg and D. Sherratt, personal communication). These workers made partial digests of pBR322 DNA with R.*Hae*II and recovered derivative plasmids by transformation. The loss of one particular *Hae*II fragment gave a plasmid that was no longer sensitive to the mobilisation products of a transmissible *col* plasmid present in the same cell. The non-mobilisable derivative of pBR322 has lost a site at which gene products essential for plasmid mobilisation interact and therefore this vector provides extra biological containment.

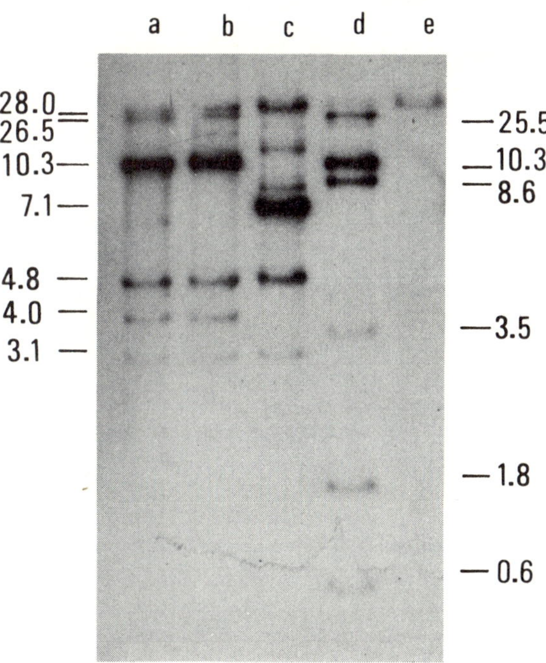

Fig. 1. Autoradiograph showing those R.*Hind*III fragments in several *E. coli* DNAs which share homology with λ DNA. Electrophoresis was carried out in a 1% agarose gel. Sizes of identified bands are given in kilobase pairs (kb). Tracks (a) and (b): CR63 and C600 DNA respectively. This pattern is typical of *E. coli* K-12 strains in general. Track (c): AB1157 DNA. Three bands (28.5 kb, 4.8 kb, 3.1 kb) are of the same size and relative intensity as bands in C600 DNA digests. Two C600 DNA fragments (26.5 kb and 4.0 kb) are not found in AB1157 DNA digests. A 10.3 kb fragment in C600 DNA digests appears to have been altered to give a 7.1 kb fragment in AB1157 DNA. The two unidentified bands in AB1157 DNA are due to incomplete digestion of this DNA with the enzyme. Tracks (d) and (e): *E. coli* B and *E. coli* C DNA respectively.

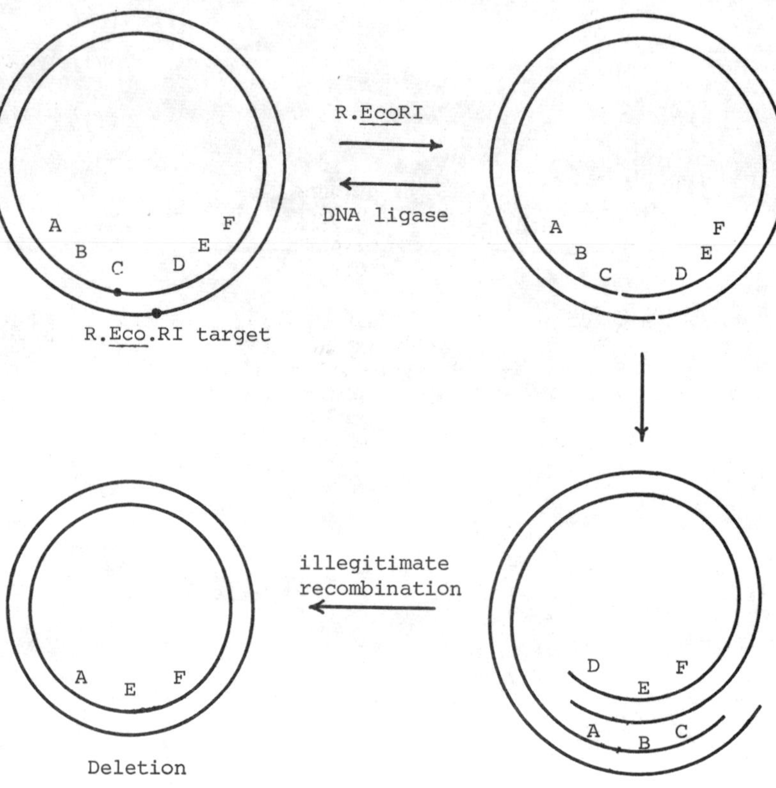

Fig. 2. Scheme for the generation of deletions by illegitimate recombination. As illustrated an intramolecular recombination event leads to joining associated with deletion of DNA.

An alternative use of restriction enzymes in the isolation of deletion mutants relies on the ability of the cell to join DNA molecules. Essentially, linear DNA molecules may be circularised within the cell either by joining cohesive ends, should they be available, or by an alternative pathway in which joining is accompanied by loss of some DNA (see Fig. 2). Such illegitimate recombination events introduce deletions at particular sites, but the deletions have variable end-points. This mechanism was demonstrated for *E. coli* (Murray and Murray, 1974) but was used most effectively as a source of deletion derivatives of SV40 (Lai and Nathans, 1974; Carbon and co-workers, 1975). The efficiency with which deletion mutants are isolated is, of course, increased by enzymatic removal of cohesive ends, an approach that has been used recently to isolate deletions affecting the transposition of Tn*3* (Heffron and co-workers, 1977).

Heffron and colleagues (1978) have described an elegant extension of this method whereby mutations are made by the insertion of an R.*Eco*RI recognition sequence. A plasmid in which the only target for R.*Eco*RI was within Tn*3* was the substrate. The plasmid DNA was treated with DNAase 1 in the presence of Mn^{++} to yield linear molecules with short 5' single strand extensions. This digest was then treated with *Eco*RI methylase to protect the target for the R.*Eco*RI before filling in the cohesive ends by extension of the 3' ends to complete the double-stranded structures (Fig. 3). Synthetic "linker fragments", octanucleotides that contain the hexanucleotide sequence recognized by R.*Eco*RI were then joined to both ends using T4 DNA ligase in the so-called "blunt-ended" ligation reactions. Treatment of the products with R.*Eco*RI generated cohesive ends from the attached "linker fragments" and these ends were joined to give a circular molecule containing a new *Eco*RI target. In this way mutations were introduced wherever the plasmid was broken by the initial DNAase treatment and these sites may be mapped with precision by digestion of the DNA with R.*Eco*RI. This technique was used to produce a fine-structure map of transposon Tn*3*; some mutations, mapping within a narrow region of TN*3*, result in an increased frequency of Tn*3* transpositions, but others abolish transposition entirely.

These new methods for the re-arrangment of large segments of DNA sequences and for introduction of mutations in specific regions have obvious application in studies of the effect of nucleotide sequence on structure and biological function. In a recent publication, Grosschedl and Hobom (1979) gave nucleotide sequences in regions of the λ chromosome involved in replication which show that parts of this sequence, which it is suggested are required for the so-called inception and initiation of replication (and hence binding of DNA polymerase), can be deformed into extensive hairpin or cruciform structures of the type proposed by Gierer (1966). The techniques for mutation and re-arrangement of specific regions of the chromosome by appropriate restriction experiments provide opportunities to alter the sequence such that this type of structure would be changed, and then to study the effect of this change upon biological function.

Finally, an important addition to the range of methods for introducing mutations at specific sites is provided by the substitution of a base analogue for a normal nucleotide in polymerase reactions *in vitro* as developed by Weissmann and his colleagues (Domingo and co-workers, 1976; see also the article by van Ooyen and co-workers, this volume). Since it may be directed at a specific nucleotide, this approach is the most precise of all.

ANALYSIS BY GENETIC RECOMBINATION

The most sensitive quantitative analyses by genetic recombination have been achieved for coliphages, particularly phage T4, (Benzer, 1961; Crick and co-workers, 1961). When segments of DNA are transferred to a λ vector they, like the phage

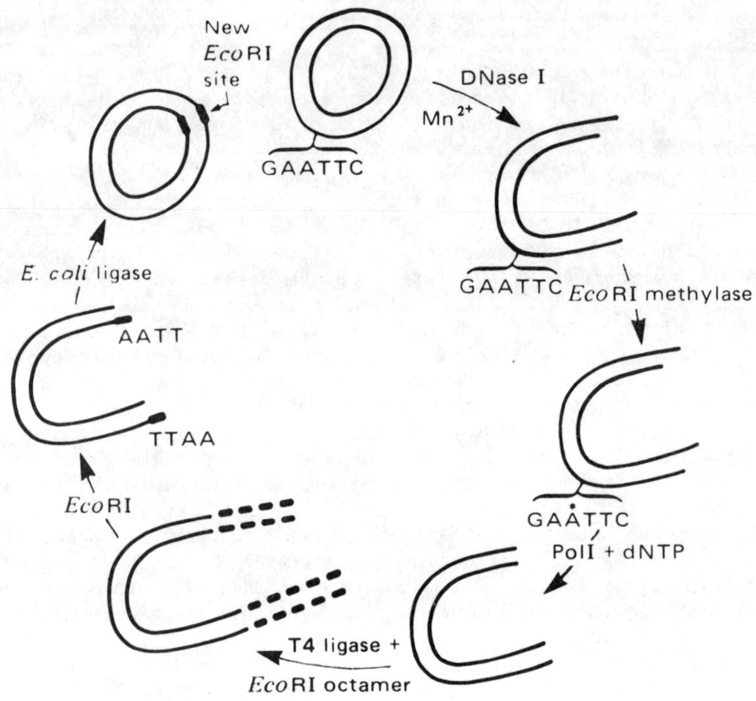

Fig. 3. The construction of mutant plasmids by the insertion of an *Eco*RI restriction target. (Reproduced with permission from F. Heffron and co-workers (1978), *Proc. Natl. Acad. Sci. USA,* 75, 6012-6016.)

chromosome itself, may be subjected to rigorous genetic analysis. The included DNA is a substrate for the bacterial and phage recombination systems and flanking markers may be used to assay, or select, recombinants. In the absence of the general recombination systems of both host and phage, site-specific recombination is detected in phage λ and this system is the paradigm for current progress in elucidating a mechanism by which cells can insert one DNA molecule into another (see, for example, DNA Insertion Elements, Plasmids and Episomes, ed. by Bukhari and and co-workers, 1977).

Recently a hot spot for genetic recombination has been identified within the genome of coliphage P1 following the transfer of fragments of the DNA of this phage to genetically marked λ vectors (Sternberg, personal communication). This hot spot for recombination persists in the absence of the λ general (*red*) and site-specific (*int*) recombination systems and is independent of the major bacterial recombination pathway. Phage P1 is anomalous in that, despite terminally redundant chromosomes, the genetic map remains linear. Analyses of recombination between λ-P1 hybrids, constructed by joining fragments of DNA generated by digestion with R.*Eco*RI, show that a specific fragment of the P1 chromosome is the substrate for a relatively efficient recombination system and that this same fragment includes genes from both ends of the genetic map of P1. The nature of this site-specific recombination system, which may be encoded by the included fragment, is now open to molecular and genetic investigation.

GENE FUSIONS AND GENE EXPRESSION

Gene fusions have been used for many years and are particularly useful in studying the control of expression of genes whose products are difficult to assay. To overcome this limitation the regulatory elements of the gene under investigation are fused to another gene whose product is easy to assay. Extensive use has been made in this way of the *lacZ* gene of *E. coli* since there is a simple and sensitive assay for its product, β-galactosidase. Although such fusions have usually been effected *in vivo*, they can obviously be produced in many cases by the *in vitro* recombination methods.

The structures of the two major operator and promoter regions of λ(O_LP_L and O_RP_R) have been elucidated by Ptashne and Pirrotta and their colleagues (Ptashne and co-workers, 1976; Walz and co-workers, 1976) and each comprises three segments that bind the λ repressor (the *cI* gene product), but with varying affinities. These observations led to the proposal of a model for the autoregulation of transcription of the repressor gene (*cI*) from the P_{RM} promoter (Fig. 4). The operator for rightward transcription is represented by three repressor binding sites O_R1, O_R2 and O_R3. This operator region includes P_R, the promoter for rightwards transcription, and P_{RM} from which leftwards transcription of the *cI* gene is initiated. At low concentration of repressor, O_R1 is preferentially bound because it has the highest affinity for repressor and this prevents RNA polymerase from binding at P_R and initiating transcription of the *cro* gene. Binding of repressor to O_R1, however, does not block transcription from P_{RM} and may indeed enhance it. When high concentrations of repressor are present the other two binding sites are filled and transcription of the repressor gene from P_{RM} is prevented.

Ptashne and co-workers (1976) used a fusion of the *cI* and *lacZ* genes in order to assay expression from P_{RM}. The vehicle for this fusion was a λ prophage in which the normal λ control system has been replaced by that of a different lambdoid phage (*imm*^{21}). In the absence of the promoter for the *lacZ* gene, all transcription of this gene was initiated from P_{RM} (Fig. 4). The level of λ repressor was varied and β-galactosidase was assayed. The results (Fig. 4) show the number of molecules of β-galactosidase produced from strains carrying either this prophage or a

a) THE STRUCTURE OF THE RIGHT OPERATOR REGION OF λ

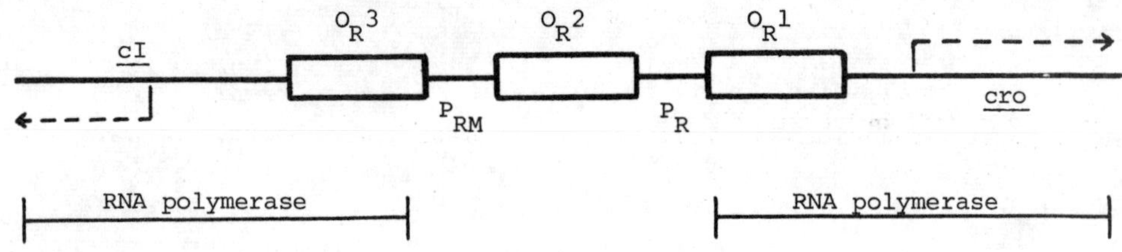

b) SYNTHESIS OF β-GALACTOSIDASE DIRECTED BY P_{RM}^{λ}

Lysogen	Approximate No. of Molecules per Cell β-galactosidase	λ repressor
(lac cI^+)	2,650	200
(lac cI^-)	100	0
(lac cI^+)/pKB252	210	10,000

c) TRANSCRIPTION FROM P_{RM}^{λ} IN THE λimm^{21} PROPHAGE

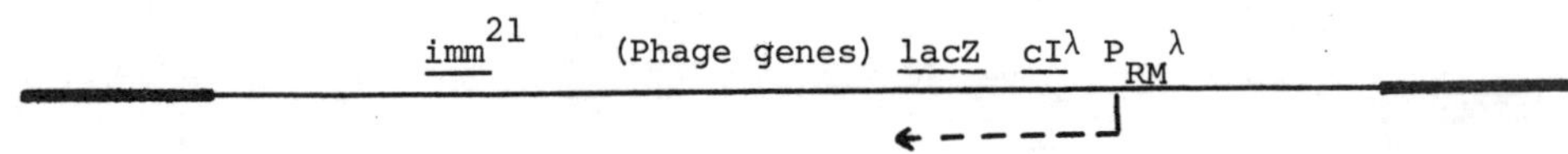

Fig. 4. a) The organization of the right operator region of phage λ showing the three binding sites for the λ repressor, the regions protected when RNA polymerase is bound, the P_{RM} and P_R promoters and the transcripts initiated from these promoters.
b) The cellular levels of β-galactosidase produced from the *lacZ*-P_{RM}^{λ} fusion in response to changes in the concentration of λ repressor.
c) The *lacZ*-P_{RM}^{λ} fusion within a λimm^{21} prophage.

derivative that is unable to make λ repressor (i.e. *cI*$^-$). β-Galactosidase production was stimulated at low concentrations of repressor, but in contrast, high levels of repressor reduced it. High levels of λ repressor were achieved by transforming the lysogenic strain with plasmid pKB252, a plasmid that over-produces λ repressor (Backman and co-workers, 1976). Collectively, these results support the proposed model of autoregulation and illustrate ways in which gene fusions and *in vitro* recombination methods can be combined to study the control of gene expression.

GENE EXPRESSION AND ITS AMPLIFICATION

When foreign genes are inserted into *E. coli* by either a plasmid or phage vector the new gene products may sometimes be detected by their enzymic properties or by immunological methods. Sensitive analytical techniques are also available for the identification of new polypeptides (Adler and co-workers, 1976; Jaskunas and co-workers, 1975).

Frequently it is advantageous to use further genetic manipulations to amplify the level of gene product, sometimes by several orders of magnitude. This may be achieved by increasing the number of gene copies, by using fusions that increase the efficiency of transcription, or even by combination of these two approaches. In addition it may be necessary to increase the efficiency of translation which can sometimes be achieved by appropriate fusions (Roberts and co-workers, 1979). The two promoter systems most commonly used to increase transcription efficiency are the promoters of phage λ or that of the *lacZ* gene of *E. coli*. Both can provide efficient, controlled transcription.

The example chosen to illustrate this aspect of the recombinant DNA technology is particularly relevant since it concerns T4 DNA ligase, an enzyme that is central to the technique.

The gene coding for T4 DNA ligase was transferred to a λ vector together with a relatively inefficient promoter sequence (Wilson and Murray, 1979). The λ vector was used to increase the number of copies of the ligase gene and to take advantage of a λ promoter to enhance further the levels of T4 DNA ligase produced.

A map of the λ genome is shown in Fig. 5. Genes for head and tail proteins are on the left while the major control system, *cI* and the operator regions are to the right. Genes *N* and *Q* are important genes whose products provide positive regulation. The gene *N* protein is essential for the efficient expression of all genes, while gene *Q* protein is required to effect expression of the late genes, including gene *S*, whose product is essential for host cell lysis, and all the head and tail gene products. Transcription of these genes from a single promoter is possible because the phage genome circularises following infection of *E. coli*.

The ligase gene of phage T4 was inserted into the central region of the phage λ vector in place of non-essential genes. The vector retains the site at which recombination takes place with the *E. coli* chromosome so that this phage may be integrated and propagated as part of the *E. coli* chromosome. On induction of the lysogen, the prophage is excised and replicates to give 100 or more DNA copies per cell. However, in the normal course of events, the cell soon lyses liberating phage and proteins. Some years ago it was shown by Müller-Hill and co-workers (1968) that lysis can be prevented by making gene *S* defective. An amber mutation in gene *S* was therefore used to prevent cell lysis. Transcription from the promoter of the T4 ligase gene was augmented with transcription from the late promoter of λ. This transcription is therefore dependent on the product of gene

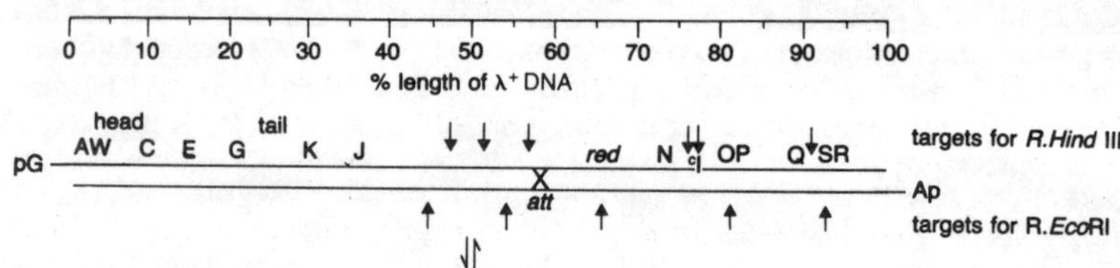

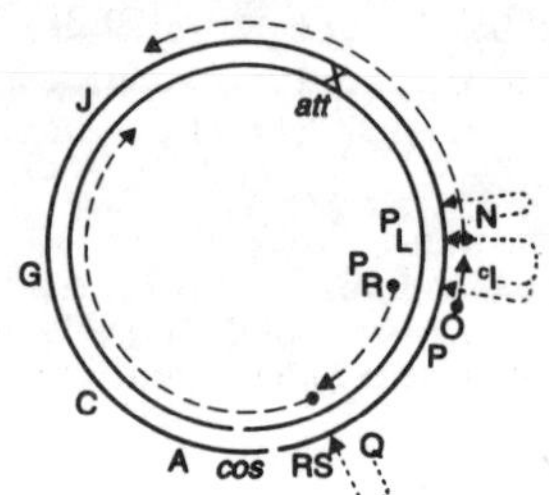

Fig. 5. A simplified version of the λ chromosome showing some of its more important control functions. The mature phage contains a single, linear duplex DNA molecule with a molecular weight of 3.1×10^7. At the 5' ends of the DNA are single-stranded projections of 12 bases with complementary sequences. Upon injection into the host cell, the DNA circularizes by base pairing of these cohesive ends. Chromosomes of lambdoid phages are normally drawn in the linear form; here genes are located at the positions corresponding to the percentage of the length of the wild-type phage DNA. Genes on the left of the linear map code for head and tail proteins of the phage. Much of the central region is inessential and can be deleted without seriously impairing phage growth. *Red* represents the phage recombination system. *O* and *P* are concerned with replication of the phage DNA, and *S* and *R* code for proteins that lyse the host cell when the phage products have been assembled into infectious particles. The *cI* gene codes for a repressor protein which interacts at the sites shown by the dotted arrows in the lower part of the figure to prevent expression of the phage genes. Removal of the repressor permits expression in both directions from P_L and P_R as shown by the broken arrows. *N* and *Q* are positive regulatory genes, the products of which interact at the positions shown by dotted arrows. *Q* is necessary for the expression of genes *S* and *R* and genes to the left of these as indicated by the long broken arrow inside the circle. Thus, after circularization of the chromosome, gene *Q* activates the expression of genes *A, C, E,* etc. (i.e., those on the left of the linear map) as well as genes *R* and *S*. Also shown in the figure is the attachment site by which the phage chromosome may be inserted into its host chromosome (where it may be stably replicated along with the host), and the positions of targets for the restriction enzymes R.*Hind*III and R.*Eco*RI in the wild-type chromosome. Some of these targets must be removed by deletion or mutation in order to make phage derivatives that can be used as receptors. (Reproduced with permission from K. Murray (1976), *Endeavour*, 35, 129-133.)

Q. To prevent the packaging of phage DNA into heads and the consequent production of phage particles, additional amber mutations were used, in particular one in gene *E* whose product is the major capsid protein.

This system gave high yields of T4 DNA ligase, possibly as much as 2% of the soluble cell protein (Murray and co-workers, 1979). However, isolation of the T4 ligase gene from other T4 genes may be as important as the amplification of expression. Phage T4 codes for many nucleases all of which are detrimental to experiments designed to join DNA molecules. Nuclease-free T4 DNA ligase is readily isolated following induction of the *λlig* prophage and the enzyme is capable of effecting the ligation of "blunt-ended" DNA fragments.

Other systems for the amplification of useful enzymes, including DNA polymerases, are already available and those providing amplification of restriction enzymes are likely to follow.

ACKNOWLEDGEMENTS

Previously unpublished work from this laboratory was supported in part by grants from the Medical Research Council and the Science Research Council.

REFERENCES

Adler, H.I., W.D. Fisher, A. Cohen and A.A. Mardigree (1967). *Proc. Natl. Acad. Sci. USA*, 57, 321-326.
Backman, K., M. Ptashne and W. Gilbert (1976). *Proc. Natl. Acad. Sci. USA*, 73, 4174-4175.
Benzer, S. (1961). *Proc. Natl. Acad. Sci. USA*, 47, 403-415.
Bukhari, A.I., J.A. Shapiro and S.L. Adhya (Eds.) (1977). *DNA Insertion Elements, Plasmids and Episomes.* Cold Spring Harbor Laboratory, New York.
Carbon, J., T. Shenk and P. Berg (1975). *Proc. Natl. Acad. Sci. USA*, 72, 1392-1397.
Crick, F.H.C., L. Barnett, S. Brenner and R.J. Watts-Tobin (1961). *Nature (London)* 192, 1227-1232.
Diaz, R., P. Barnsley and R.H. Pritchard (1979). *Molec. gen. Genet.*, in press.
Domingo, E., R.A. Flavell and C. Weissmann (1976). *Gene*, 1, 3-25.
Faelen, M., A. Toussaint, M. Van Montagu, S. van den Elsacker, G. Engler and J. Schell (1977). In *DNA Insertion Elements, Plasmids and Episomes*, Cold Spring Harbor Laboratory, New York. pp. 521-530.
Gierer, A. (1966). *Nature (London)*, 212, 1480-1481.
Grosschedl, R. and G. Hobom (1979). *Nature (London)*, 277, 621-627.
Heffron, F., P. Bedinger, J.J. Champoux and S. Falkow (1977). *Proc. Natl. Acad. Sci. USA*, 74, 702-706.
Heffron, F., M. So and B.J. McCarthy (1978). *Proc. Natl. Acad. Sci. USA*, 75, 6012-6016.
Jaskunas, S.R., L. Lindahl, M. Nomura and R.R. Burgess (1975). *Nature (London)*, 257, 458-462.
Kaiser, K. and N.E. Murray (1979). *Molec. gen. Genet.*, in press.
Lai, C.J. and D. Nathans (1974). *J. Mol. Biol.*, 98, 179-193.
Low, K.B. (1973). *Molec. gen. Genet.*, 122, 119-130.
Murray, K. (1976). *Endeavour*, 35, 129-133.
Murray, N.E., S.A. Bruce and K. Murray (1979). *J. Mol. Biol.*, in press.
Murray, N.E. and K. Murray (1974). *Nature (London)*, 251, 476-481
Müller-Hill, B., L. Crapo and W. Gilbert (1968). *Proc. Natl. Acad. Sci. USA*, 59, 1259-1264.

Ptashne, M., K. Backman, M.Z. Humayun, A. Jeffrey, R. Maurer, B. Meyer and R.T. Sauer (1976). *Science*, 194, 156-161.

Roberts, T.M., R. Kacich and M. Ptashne (1979). *Proc. Natl. Acad. Sci. USA*, 76, 760-764.

Southern, E.M. (1975). *J. Mol. Biol.*, 98, 503-518.

Walz, A., V. Pirrotta and K. Ineichen (1976). *Nature (London)*, 262, 665-668.

Wilson, G.G. and N.E. Murray (1979). *J. Mol. Biol.*, in press.

NEW INSIGHTS INTO THE MOLECULAR GENETICS OF ANIMALS

R. Breathnach

Laboratoire de Génétique Moléculaire des Eucaryotes du CNRS Unité 184 de Biologie Moléculaire et de Génie Génétique de l'INSERM, Faculté de Médecine 11, rue Humann 67085, Strasbourg, France

Application of the techniques known as genetic engineering has brought new insight into the structure and mode of expression of eukaryotic genes. Several eukaryotic genes have been shown to contain intervening sequences, or introns, which interrupt the DNA sequences, termed exons, which code for the gene's final RNA product. The first gene shown to be split was the 28S rRNA gene of *Drosophila melanogaster*. Subsequently several adenovirus early and late genes were shown to contain introns (for discussion of these cases and references see Chambon,1978), and these observations were rapidly followed by the demonstration of introns in the mouse (Tilghman and co-workers, 1978) and rabbit (Jeffreys and Flavell, 1977) β-globin genes, the chicken ovalbumin gene (Breathnach, Mandel and Chambon, 1977) a mouse immunoglobulin light chain gene (Tonegawa and co-workers, 1978) and some yeast tRNA genes (Goodman, Olson and Hall, 1977); Valenzuela and co-workers, 1977). Further examples of genes with introns are discussed in a review by Wahli and Dawid (1979b). This article will deal only with the organisation and expression of some mRNA-coding cellular genes with introns.

GLOBIN GENES

Studies of the globin gene family have allowed some interesting conclusions as to the evolutionary age of intervening sequences to be drawn, and provided some information about the evolution of duplicated genes. The mouse α- (Leder and co-workers, 1978), β-major (Tilghman and co-workers, 1978) and β-minor (Tiemeier and co-workers, 1978) globin genes, the rabbit β-globin gene (Jeffreys and Flavell, 1977) and the human β- (Lawn and co-workers, 1978) and γ-globin genes (Smithies and co-workers, 1979) have all been shown to have two intervening sequences, interrupting the protein coding regions in exactly analogous positions. If the multiple globin genes of a species have arisen as believed by a series of gene duplications, these results suggest that the intervening sequences were present before the first such duplication occurred, estimated to have been more than 500 million years ago (Ohno, 1968).

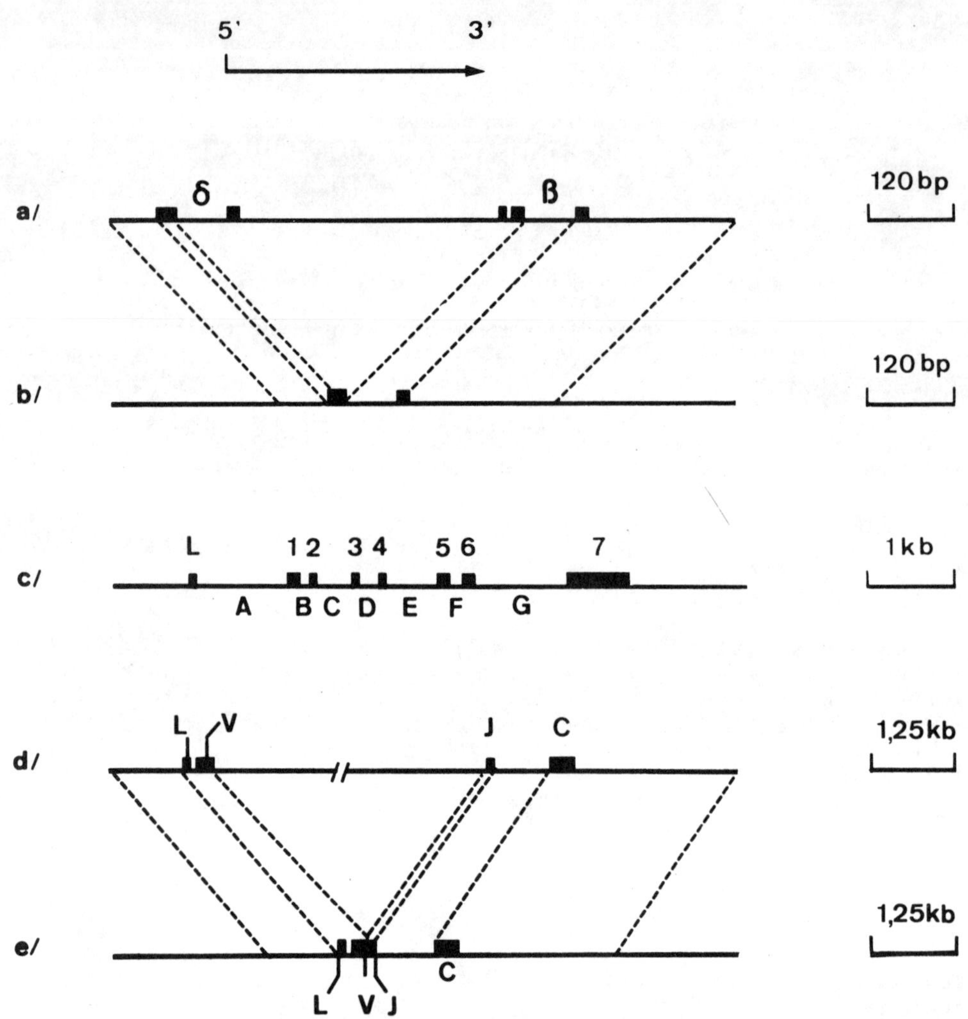

Fig. 1. The organisation of some split genes

(a) Close linkage of the human δ-and β-globin genes in the human genome. The δ-gene may contain an additional intervening sequence.

(b) Arrangement of δ-and β-globin sequences in the Lepore gene. Dotted lines linking (a) and (b) indicate the origin of particular regions of the Lepore gene.

(c) Structure of the chicken ovalbumin gene.

(d) (e) Structure of a mouse immunoglobulin λI light chain gene in embryonic (d) or producer (e) cells. Dotted lines linking (d) and (e) indicate the rearrangement discussed in the text.

Sources of data are given in the text.

After a gene duplication,the flanking sequences surrounding a duplicated gene may diverge extensively, as heteroduplex studies of the mouse β-major and β-minor globin genes (Tiemeier and co-workers, 1978) have shown that sequence homology between the genes, apart from the coding sequences, is limited to a few hundred nucleotides flanking them and including only those portions of the intervening sequences immediately adjacent to the coding sequences. These workers used their data to argue that such flanking sequence divergence may be necessary to reduce the recombination target area and to fix the duplicated genes. The small residual homology of the β-major and β-minor gene flanking sequences may then represent a minimum necessary for their correct expression.

Comparison of the mouse α- and β-major globin genes has revealed a 200 base pair region of homology, lying 1.5 kb away from the 3'-terminus of these genes (Leder and co-workers, 1978). This homology region may extend to the β-minor gene; its significance, if any, is not clear.

The molecular arrangement of the human δ- and β-globin genes has been elucidated [Fig. 1 (a)] by a combination of restriction enzyme mapping of the genomic DNA (Mears and co-workers, 1978; Flavell and co-workers, 1978) and studies on cloned globin gene fragments isolated from a human genomic library (Lawn and co-workers, 1978). These studies have confirmed the close physical linkage of the δ- and β-genes predicted from genetic evidence, and also that the Hb Lepore condition (leading to a polypeptide containing the N-terminal animo-acids of δ-globin and the C-terminal acids of β-globin) is due to a gene fusion (Fig. 1b).

CHICKEN OVIDUCT GENES

The genes for ovalbumin, conalbumin, ovomucoid, and lysozyme are under hormonal control in the chicken oviduct. The best characterized gene is the ovalbumin gene (Breathnach and co-workers, 1978; Catterall and co-workers, 1978; Gannon and co-workers, 1979), which consists of eight exons, separated by seven intervening sequences (Fig. 1c). The first exon (L) codes for the first 47 nucleotides of the ovalbumin mRNA, all of which are in the 5'-non coding region of the messenger: these 47 nucleotides thus comprise a leader sequence similar to those of adenovirus and SV40 mRNAs. Molecular cloning has recently resulted in the isolation of large fragments overlapping the chicken ovalbumin gene (Royal and co-workers, 1979). The region thus characterised contains the complete ovalbumin gene and two other genes X and Y, of as yet unknown function, all of which are oriented in the same direction for transcription, and whose expression in the chicken oviduct is under hormonal control. Genes X and Y each have 8 exons; the first 7 exons of X and Y are similar in size to the corresponding exons of the ovalbumin gene. It seems likely that genes X and Y are related to the ovalbumin gene by a gene duplication process. Preliminary results suggest that there is little sequence similarity between the intervening sequences of the 3 genes, but that the exons are partially homologous (Royal and co-workers, 1979; R. Heilig and J.L. Mandel, personal communication). This arrangement of genes with similar exons but non-homologous intervening sequences is very reminiscent of the arrangement of the globin genes discussed above. Interestingly, the precursor of the major yolk proteins of the frog, vittellogenin, is also encoded by a small family of genes (Wahli and co-workers, 1979a).

Electron microscopy of hybrids between cloned DNAs and their corresponding messengers has shown that the genes for conalbumin (Perrin and co-workers, 1979) and ovomucoid (Chambon and co-workers, 1979) both contain at least seven intervening sequences. The lysozyme gene has been less well characterized, but contains several intervening sequences, one of them in the 3'-untranslated region of the messenger (Nguyen-Huu and co-workers, 1979).

GENE REARRANGEMENTS AND IMMUNOGLOBULIN GENES

From the limited available evidence, gene rearrangement does not seem a common method for the control of transcription in eukaryotes. Thus the environments of the rabbit β-globin gene (Jeffreys and Flavell, 1977) the chicken globin gene (Engel and Dodgson, 1979), the chicken ovalbumin (Breathnach, Mandel and Chambon, 1977) and lysozyme genes (Nguyen-Huu and co-workers, 1979) and the silk fibroin gene (Manning and Gage, 1979) appear identical irrespective of the tissue investigated. On the other hand, gene rearrangement does occur during the course of differentiation of immunoglobulin producing cells. For example, in mouse embryo DNA, a λ_I light chain polypeptide is encoded by four separate exons (L, V, J and C of Fig. 1d) which code for most of the leader peptide (L), the rest of the leader peptide and the variable region (V), a region of 13 residues at the V-C junction (J) and the constant region (C). In the λ_I chain producer genome, however, the exons V and J of the embryonic DNA have been brought into contiguity by an exact recombination event (Bernard and co-workers, 1978) (Fig. 1e).

The structure and rearrangements of individual mouse κ-light chain genes seem very similar to those described for the λ_I chain gene (Seidmann and Leder, 1978). However, whereas the multiple λ_I variable regions observed in myelomas seem to have been generated by somatic mutation from a single germ-line variable gene (Brack and co-workers, 1978), multiple κ variable genes are encoded by the mouse germ-line genome (Seidmann and co-workers, 1978). These exist as many small closely related subgroups sharing close homology only within a subgroup. The existence of multiple κ variable region genes encoded by the germ-line genome, any one of which could in principle be linked to any of the as yet 13 identified J segments, could go a long way towards accounting for the observed repertoire of κ-light chain variable regions (Seidmann and co-workers, 1978; Weigert and co-workers, 1978).

In marked contrast to the mouse globin genes discussed above, comparison of two cloned κ variable genes from the same subgroup shows sequence homology extending thousands of bases into the sequences flanking the coding regions (Seidmann and co-workers, 1978). The resulting large recombination target size could account for the observed genetic flexibility of κ variable genes, by favouring expansion or reduction of specific variable region subgroups by unequal crossing over between homologous but non-allelic variable region sequences. Unlike the situation for the globin genes, this process may be useful to the organism as it generates additional antibody diversity, particularly when linked to mismatch repair (see Seidmann and co-workers, 1978 for discussion).

Recently Sakarno and co-workers(1978) have isolated a clone containing sequences coding for the constant region of the mouse immunoglobulin $\gamma1$, heavy chain. They found that the three constant region protein domains and the hinge region are encoded in separate DNA segments, and suggest that this arrangement could have arisen by triplication or quatriplication of one ancestral exon.

SPLICING

Studies of the major late adenovirus transcriptional unit have led to the conclusion that genes with introns are transcribed into colinear precursors (see e.g. Goldberg, Nevins and Darnell, 1978) from which the intron transcripts are excised, followed by ligation of the exon transcripts in a process termed "splicing".

The most direct evidence for the splicing model for cellular species comes from the study of yeast mutants (Hopper, Banks and Evangelidis, 1978) which accumulate some tRNA precursors. These precursors are encoded by tRNA genes containing a short intron (Goodman, Olson and Hall, 1977; Valenzuela and co-workers, 1977), and they contain a full transcript of the intron. When treated with a yeast extract, they may be converted to mature tRNA by excision of this intron transcript (Knapp and co-workers, 1978; O'Farrell and co-workers, 1978).

Evidence for splicing during the maturation of cellular mRNAs is less direct and relies on the demonstration of RNA molecules colinear with the cellular gene, and/or which contain linked intron and exon transcripts. This has been done for globin genes (Tilghman and co-workers, 1978; Smith and Lingrel, 1978) ovalbumin genes, (Roop and co-workers, 1978; Chambon and co-workers, 1979), immunoglobulin light chain (Gilmore-Hebert and co-workers, 1978; Schibler, Marcu and Perry, 1978) and heavy chain genes (Schibler, Marcu and Perry, 1978).

In a colinear transcript of a gene with several introns, the order of splicing out of the intron transcripts could be random or could, for example, proceed sequentially in a 5'→3' direction. It has been suggested that the 5'-proximal splice occurs first during the maturation of late adenovirus-2 mRNAs (Berget and Sharp, 1978). On the other hand, characterisation of a number of discrete processing intermediates of the chicken ovalbumin primary transcript has led to the conclusion that excision of intron transcripts B-F (Fig. 1c) does not follow a rigid order as, for example, molecules have been observed which still have the intron D transcript, but not the intron E transcript, and vice-versa. It has not been ruled out that the 5'-terminal splice (intron A, Fig. 1c) occurs first, however (Chambon and co-workers, 1979).

INTRON-EXON BOUNDARIES

For correct splicing to occur, the splicing enzyme must presumably recognise some feature at the boundary of the exon and intron transcripts. A large number of exon-intron junctions have now been sequenced on cloned DNAs coding for mRNAs. (Fig. 2), and the results allow the following conclusions to be drawn (Breathnach and co-workers, 978; Catterall and co-workers, 1978) :

1) there is no evident way of forming base-paired structures bringing adjacent exon transcripts into close proximity and looping out intron transcripts for excision.
2) directly repeated nucleotides at the beginning and end of a given

intron (boxed in Fig. 2) prevent unique definition of intron-exon boundaries.

3) the 5'-extremities of introns are all related to the sequence 5' -TCAGGTA- 3' and the 3'-ends to 5' -TXCAGG- 3'.
4) in all cases the intron may start with the GT of the sequence 5' -TCAGGTA- 3' and end with the AG of the sequence 5' -TXCAGG- 3' the GT and AG being invariantly present at intron-exon junctions. In this case a unique splice point could be defined for all the genes shown in Fig. 2, despite the repeated nucleotides mentioned above. To date there is only one exception to this "GT-AG" rule, a heavy chain immunoglobulin intron (Sakano and co-workers, 1979).
5) tracts rich in pyrimidines are always found immediately at the 5'-side of an exon.

The model sequences of 3) probably play a part in determining the specificity of splicing, but as they occur elsewhere within the introns and exons of the ovalbumin gene, they cannot alone be sufficient as splicing signals. The most strongly conserved segments of corresponding introns of the rabbit and mouse globin genes are located within 10 nucleotides of the intron-exon junctions (van den Berg and co-workers, 1978). This is also the case for a corresponding intron of the ovalbumin gene and the ovalbumin-like gene X (R. Heilig, and J.L. Mandel, unpublished) and seems to emphasise the importance of the sequences at the junctions for correct splicing. The lack of homology between corresponding introns of genes linked by a duplication process might also suggest that secondary structure of the intron transcript is not important for splicing. Secondary structures involving exon transcripts may however be important for the splicing specificity.

The intron-exon boundaries of the yeast tRNA genes do not fit the general patterns discussed above for mRNA coding genes (Valenzuela and co-workers, 1977; Goodman, Olson and Hall, 1977). The tRNA precursors may however be handled by a different type of splicing enzyme to those involved in the processing of mRNA precursors. This enzyme must also be present in higher eukaryotes, as a yeast tRNA gene with an intron has been shown to be correctly processed in *Xenopus* oocytes (de Robertis and Hall, 1979).

POSSIBLE TRANSCRIPTIONAL CONTROL REGIONS

Comparison of the nucleotide sequences of a number of prokaryotic promoters and promoter-mutants has shown that the E.coli RNA polymerase recognises two features of a DNA duplex: the recognition site, centred about 32 bp before the beginning of the mRNA, with a sequence related to 5' -TGTTGACATTT- 3' and a second sequence of the form 5' -TATAATG-3' (The "Pribnow box") preceding the transcription-initiation side by 5-7 bp (see Scherrer, Walkingham and Arnott, 1978 for discussion). The availability of cloned eukaryotic genes makes it possible to search for sequence homologies around transcription initiation sites which might represent eukaryotic RNA polymerase recognition sequences.

Some sequences preceding the presumed starts of transcription of several genes transcribed by RNA polymerase II (which transcribes genes coding for mRNAs) are shown in Fig. 3. A common feature is an AT rich region flanked by GC rich regions, roughly 30 bp from the start of transcription. This feature, first observed for the histone genes

Transcription
5' ⟶

		5' sequence		3' sequence		Type
OVALBUMIN	Leader	... T C A A A A G G T C A C T C ...	Intron A	... G C T C T A G A C A A C T ...	Exon 1	Type 1
	Exon 1	... A A A T A A G G T G A G C C ...	Intron B	... A T T A C A G G T T G T T ...	Exon 2	Type 3
	Exon 2	... A G C T C A G G T A C A G A ...	Intron C	... T A T T C A G T G T G G C ...	Exon 3	Type 1
	Exon 3	... C C T G C C A G T A A G T T ...	Intron D	... T T T A C A G G A A T A C ...	Exon 4	Type 2
	Exon 4	... A G A A A T G G T A A G G T ...	Intron E	... C T T A A A G G A A T T A ...	Exon 5	Type 2
	Exon 5	... G A C T G A G G T A T A T G ...	Intron F	... G C T C C A G C A A G A A ...	Exon 6	Type 1
	Exon 6	... T G A G C A G G T A T G G C ...	Intron G	... C T T G C A G C T T G A G ...	Exon 7	Type 1
SV40 Late 19S (213-476)		... T C A G A A G G T A C C T A ...	Intron	... T T T C C A G G T C C A T ...		Type 3
SV40 Late 19S (291-476)		... C G T T A A G G T T C G T A ...	Intron	... T T T C C A G G T C C A T ...		
SV40 Late 19S (444-476)		... T T A A C T G G T A A G T T ...	Intron	... T T T C C A G G T C C A T ...		Type 2
SV40 Late 16S (444-1381)		... T T A A C T G G T A A G T T ...	Intron	... C T T C T A G G C C T G T ...		
SV40 Early T (4837-4490)		... A A C T G A G G T A T T T G ...	Intron	... A T T T T A G A T T C C A ...		Type 1
SV40 Early t (4556-4490)		... C T A T A A G G T A A A T G ...	Intron	... A T T T T A G A T T C C A ...		
Mouse β Globin Small Intron		... T G G G C A G G T T G G T A ...	Intron	... T T T T T A G G C T G C T ...		
Rabbit β Globin Small Intron		... T G G G C A G G T T G G T A ...	Intron	... T T C T C A G G C T G C T ...		
Mouse β Globin Large Intron		... C T T C A G G G T G A G T C ...	Intron	... C C C A C A G C T C C T G ...		
Rabbit β Globin Large Intron		... C T T C A G G G T G A G T T ...	Intron	... C C T A C A G T C T C C T ...		
Ig 99 λI Small Intron / Ig 303 λI Small Intron		... A G C T C A G G T C A G C A ...	Intron	... T T T G C A G G G G C C A ...		
Ig 13 λII Small Intron		... T G C T C A G G T C A G C A ...	Intron	... T T T G C A G G A G C C A ...		
Ig 303 λI Large intron		... G T C C T A G G T G A G T C ...	Intron	... T C C T G A G G C C A G C ...		
Possible model sequence		T C A G G T A (8 8 14 18 19 19 9 / 19)		T X C A G G (14 11 17 17 10 / 17)		

Fig. 2. Comparison of DNA sequences at exon-intron boundaries of various genes. Boxed nucleotides represent direct repeats (see text). The vertical broken line shows how the excision-ligation events could occur in all cases at unique positions with respect to the invariant di-nucleotides GT and AG (see text). The frequency with which a given nucleotide appears in each position of the model sequence is shown. The ovalbumin sequences are from Breathnach and co-workers (1978) and Gannon and co-workers (1979). The SV40 sequences are from Ghosh and co-workers (1978); the globin sequences from Konkel and co-workers (1978) and Weissmann and co-workers (1978); the immunoglobulin sequences are taken from Sakarno and co-workers (1979).

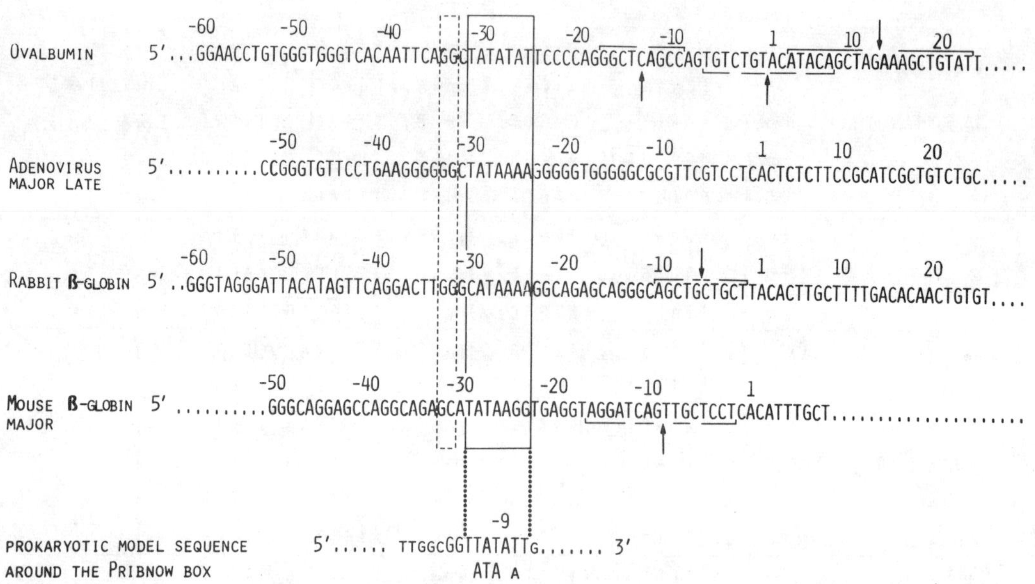

Fig. 3. DNA sequences upstream of transcription initiation sites.

Sources of data : the ovalbumin gene (Gannon and co-workers, 1979), the mouse β-globin major gene (Konkel and co-workers, 1978), the rabbit β-globin gene (Weissmann and co-workers, 1979) and adenovirus-2 major late genes (Ziff and Evans, 1978). Sequences are aligned to emphasise similarity. The "Hogness box" and flanking GC rich regions are shown boxed with continuous and dotted lines respectively. The "Pribnow box" is given for comparison in the form discussed by Scherrer and co-workers (1978). Lines above or below sequences indicate regions of symmetry. The number 1 indicates the start of transcription as defined by the first nucleotide of the corresponding mRNA.

(see Kedes, 1979 for review) has been termed the "Hogness box", and may play a role similar to that of the "Pribnow box" in prokaryotes.

Apart from the genes shown in Fig. 3 and the *Drosophila* and sea urchin histone genes, the "Hogness box" occurs for the chicken conalbumin gene (R. Hen and L. Maroteaux, personal communication) and the silk fibroin gene (Tsujimoto and Suzuki, 1979).

ACKNOWLEDGEMENT

The studies on the chicken oviduct genes which were carried out in Strasbourg by C. Benoist, R. Breathnach, M. Cochet, F. Gannon, P. Gerlinger, R. Heilig, R. Hen, A. Krust, L. Maroteaux, M. LeMeur, J.P. LePennec, J.L. Mandel, K. O'Hare and F. Perrin were supported by grants to P. Chambon from the CNRS (ATP 2117), the INSERM (C.R.A.T. 76.5.462 and 76.5.468) and the Fondation pour la Recherche Medicale Francaise. Those which were carried out in Paris by B. Cami, A. Garapin, W. Roskam and A. Royal were supported by grants to P. Kourilsky from the CNRS (ERA 201, and ATP No. 3558).

REFERENCES

van den Berg, J., A. van Ooyen, N. Mantei, A. Schambock, G. Grosveld, R. Flavell, and C. Weissmann (1978). Comparison of cloned rabbit and mouse β-globin genes. *Nature*, 276, 37-44.

Berget, S.M., and P. Sharp (1979). Structure of late adenovirus-2 nuclear RNA. *Cell*, in press.

Bernard, O., N. Hozumi and S. Tonegawa (1978). Sequences of mouse immunoglobulin light chain genes before and after somatic changes. *Cell*, 15, 1133-1144.

Breathnach, R., J.L. Mandel, and P. Chambon (1977). Ovalbumin gene is split in chicken DNA. *Nature*, 270, 314-319.

Breathnach, R., C. Benoist, K. O'Hare, F. Gannon and P. Chambon (1978). Ovalbumin gene : Evidence for a leader sequence in mRNA and DNA sequences at exon-intron boundaries. *Proc. Natl. Acad. Sci.* USA, 75, 4853-4857.

Caterall, J.F., B.W. O'Malley, M. Robertson, R. Staden, Y. Tanaka, and G. Brownlee (1978). Nucleotide sequence homology at 12 intron-exon junctions in the chick ovalbumin gene. *Nature*, 275, 510-513.

Chambon, P. (1977). Molecular biology of eukaryotic genome is coming of age. In *Cold Spring Harbor Symp. Quant. Biol.* 42, 1209-1244.

Chambon, P., C. Benoist, R. Breathnach, M. Cochet, F. Gannon, P. Gerlinger, A. Krust, M. LeMeur, J.P. LePennec, J.L. Mandel, K.O' Hare, and F. Perrin (1979). Structural organization and expression of ovalbumin and related chicken genes.*Proceedings of the 11th Miami Winter Symposium*, Academic-Press, (in press).

Flavell, R.A., J.M. Kooter, E. de Boer, P.F.R. Little, and R.Williamson (1978). Analysis of the β-δ-globin gene loci in normal and Hb Lepore DNA. *Cell*, 15, 25-41.

Gannon, F., K.O'Hare, F. Perrin, J.P. LePennec, C. Benoist, M. Cochet, R. Breathnach, A. Royal, A. Garapin, B. Cami, and P. Chambon (1979). A cloned complete ovalbumin gene. *Nature*, in press.

Ghosh, P., V.B. Reddy, J. S scoe, P. Lebowitz,and S.M. Weissmann (1978). Heterogeneity and 5'-terminal structures of late SV40 RNAs. *J. Mol. Biol.*, 126, 813-846.

Gilmore- Hebert, M., and R. Wall (1978). Immunoglobulin light chain mRNA is processed from large nuclear RNA. *Proc. Natl. Acad. Sci. USA*, 75, 342-345.

Goldberg, S., J. Nevins , and J. Darnell,(1978). Evidence from UV transcription mapping that late adenovirus type 2 mRNA is derived from a large precursor molecule. *J. Virology,* 25, 806-810.

Goodman, H.M., M. Olson and B.D. Hall (1977). Nucleotide sequence of a mutant eukaryotic gene. *Proc. Natl. Acad. Sci. USA,* 74, 5453-5457.

Hopper, A.K., F. Banks, and V. Evangelidis (1978). A yeast mutant which accumulates precursor tRNAs. *Cell,* 14, 211-219.

Jeffreys, A.J., and R. Flavell (1977). The rabbit β-globin gene contains a large insert in the coding sequence. *Cell,* 12, 1097-1108.

Kedes, J. (1979). *Annual Reviews of Biochemistry,* in press.

Knapp, G., J. Beckmann, P.F. Johnson, S.A. Fuhrmann and J. Abelson (1978). Transcription and processing of intervening sequences in yeast tRNA genes. *Cell,* 14, 221-236.

Konkel, D.A., S.M. Tilghman and P. Leder (1978). The sequence of the chromosomal mouse β-globin major gene. *Cell,* 15, 1125-1132.

Lawn, R.M., E.F. Fritsch, R.C. Parker, G. Blake, and T. Maniatis (1978). The isolation and characterisation of linked δ-and β-globin genes from a cloned library of human DNA. *Cell,* 15, 1157-1174.

Leder, A., H.I. Miller, D.H. Hamer, J.G.Seidmann, B. Norman, M. Sullivan and P. Leder (1978). Comparison of cloned mouse α-and β-globin genes. *Proc. Natl. Acad. Sci USA,* 75, 6187-6191.

Manning, R., and L. Gage (1978). Physical map of the *Bombyx mori* DNA containing the gene for silk fibroin. *J. Biol. Chem.,* 253, 2044-2052.

Mears, J.G., F. Ramirez, D. Leibowitz and A. Bank (1978). Organisation of human δ-and β-globin genes in cellular DNA and the presence of intragenic inserts. *Cell,* 15, 15-23.

Nguyen-Huu, M.C., M. Stratmann, B. Groner, T. Wurtz, H. Land, K. Giesecke, A. Sipppel and G. Schutz (1979). Chicken lysozyme gene contains several intervening sequences. *Proc. Natl. Acad. Sci USA,* 76, 76-80.

O'Farrell, P.Z., B. Cordell, P. Valenzuela, W.J. Rutter and H.M. Goodman (1978). Structure and processing of yeast precursor tRNAs. *Nature,* 274, 438-445.

Perrin, F., M. Cochet, P. Gerlinger, B. Cami, J.P. LePennec, and P. Chambon. The chicken conalbumin gene. *Nucleic Acids Research,* submitted.

Roop, D., J. Nordstrom, S. Tsai, M. Tsai, and B.W. O'Malley (1978). Transcription of structural and intervening sequences in the ovalbumin gene. *Cell,* 15, 651-685.

Royal, A., A. Garapin, B. Cami, F. Perrin, J.L. Mandel, M. LeMeur, F. Bregegere, F. Gannon, J.P. LePennec, P. Chambon, and P. Kourilsky (1979). The ovalbumin gene region. *Nature,* in press.

Sakarno,H., J. Rogers, K. Huppi, C. Brack, A. Traunecker, R. Maki, R. Wall, and S. Tonegawa (1979). Domains and the hinge region of an immunoglobulin heavy chain are encoded by separate DNA segments. *Nature,* 297, 627-633.

Scherer, G., M. Walkinshaw, and S. Arnott (1978). Computer aided oligonucleotide analysis provid a model sequence for RNA polymerase-promoter recognition in *E. coli* *Nucleic Acids Res.*, 5, 3759-3773.

Schibler, U., K. Marcu, and R.P. Perry (1978). Synthesis and processing of messenger RNAs specifying heavy and light chain immunoglobulins in MPC-11 cells. *Cell*, 15, 1495-1509.

Seidman, J., and P. Leder (1978). The arrangement and rearrangement of antibody genes. *Nature*, 276, 790-795.

Seidman, J.G., A. Leder, M. Nau, B. Norman and P. Leder (1978). Antibody diversity. *Science*, 202, 11-17.

Smith, K., and J. Lingrel (1978). Sequence organisation of the β-globin mRNA precursor. *Nucleic Acids Research*, 5, 3295-3301.

Smithies O., A.E. Blechl, K. Denniston-Thomas, N. Newell, J. Richards, J. Slightom, P. Tucker, and F. Blattner (1978). Cloning human foetal γ-globin and mouse α-type globin DNA. *Science*, 202, 1284-1289.

Tiemeier, D.C., S.M. Tilghman, F. Polsky, J. Seidmann, A. Leder, M. Edgell, and P. Leder (1978). A comparison of two cloned mouse β-globin genes, *Cell*, 14, 237-245.

Tilghman, S.M., D. Tiemeier, J. Seidman, B.M. Peterlin, M. Sullivan, J. Maizel and P. Leder (1978). Intervening sequence of DNA identified in the structural portion of a mouse β-globin gene. *Proc. Natl. Acad. Sci., USA*, 75, 725-729.

Tonegawa, S., A. Maxam, R. Tizard, O. Bernard, and W.Gilbert (1978). Sequence of a mouse germ line gene for a variable region of an immunoglobulin light chain. *Proc. Natl. Acad. Sci, USA*, 75, 1485-1489.

Tsujimoto, Y., and Y. Suzuki (1979). Structural analysis of the fibroin gene. *Cell*, 16, 425-436.

Valenzuela, P., A. Venegas, F. Weinberg, R. Bishop, and W.J. Rutter (1978). Structure of yeast phenylalanine-tRNA genes. *Proc. Natl. Acad. Sci., USA*, 75, 190-194.

Wahli, W., and I. Dawid (1979a). Vitellogenin in *Xenopus laevis* is encoded in a small family of genes. *Cell*, in press.

Wahli, W., and I. Dawid (1979b). Application of recombinant DNA technology to questions of developmental biology in a review. *Developmental Biology*, in press.

Weigert, M., L. Gatmaitan, E. Loh, J. Schilling and L. Hood (1978). Rearrangement of genetic information may produce immunoglobulin diversity. *Nature*, 276, 785-790.

Weissmann, C., N. Mantei, W. Boll, R. Weaver, N. Wilkie, B. Clements, T. Taniguchi, A. van Ooyen, J. van den Berg, M. Fried and K. Murray (1979). Expression of cloned viral and chromosomal plasmid-linked DNA in cognate host cells. *Proceedings of the 11th Miami Winter Symposium*, in press (Academic Press, N.Y.).

Ziff, E. and R. Evans (1978). The promoter and capped 5'-terminus of RNA from the adenovirus-2 major late transcription unit. *Cell*, 15, 1463-1475.

DISCUSSION

W.F. BODMER: Obviously the whole question of the evolution of the flanking and intervening sequences is going to be of great interest. You mentioned the notion that the change in the sequence of the flanking region may be to prevent recombinant products being produced. I think one has to distinguish very carefully the frequency with which a recombinant product of that sort may be produced and the frequency with which it gets incorporated into the population. The latter only happens if the initial product increases in frequency. The frequency of the production of the event, which is equivalent to a mutation frequency, is likely to be of very little consequence relative to the chance of it getting incorporated into the population, which is going to depend, probably, on natural selection. At least, if you are thinking that the recombination frequency is the determining mechanism, one should consider the evolutionary balance between the rate at which these products arise and the rate which they are lost. I think there is often a tendency to confuse the production of some new product with its eventual incorporation into the population.

R. BREATHNACH: All I would like to say is I think it must, at least, be not disadvantageous when you have a duplication of this kind for the flanking sequences to diverge.

J.K. SETLOW: What is the information about the order in which the introns are cut out? On one of the slides it looked as though it was almost random.

R. BREATHNACH: Yes, as far as we can tell there is no fixed order. However, this may not be true for the intron transcript A, which is the first intron transcript to be removed, and this may possibly be of interest because the first exon is labelled L as it encodes sequences of the messenger which are in fact solely in the 5'-untranslated region of the messenger and so is perhaps formally analogous to the leader sequences that have been described for SV40 and adenovirus; I think there is perhaps some evidence from the lab of Phil Sharp that in the adenovirus case what generally happens is that the first intron transcript in the major late adenovirus precursor is removed first, and it is possible that this is the same for the ovalbumin gene, but we are not really very sure about it yet. The rest of the intron transcripts seem certainly to be removed in a more-or-less random fashion.

P. STARLINGER: In the beginning you mentioned that, probably, the duplications of the ovalbumin gene happened first and that then the introns evolved quicker than the exons so that now they have sequence divergence. Is there any basis to that statement as opposed to the opposite possibility that what has been transposed are the separate exons and that they are integrated into a matrix of DNA, which then becomes introns? Certain of these arrangements may be selected for and perhaps this is a way in which new and useful combinations of DNA can be made. In this case maybe something has to be preserved and therefore the sequence of the exons of the ovalbumin gene is conserved, but in other cases new combinations could be formed.

R. BREATHNACH: There is no formal evidence against that, I suppose, but it seems to me a little unlikely that you would then make three genes which have exactly the same exons arranged like this, because if you are putting these exons in, why put them in always in a series right next to each other? Why not put some in one place and some elsewhere? Why have this triplication exactly in a row?

P. STARLINGER: This may be selection?

R. BREATHNACH: Yes, possibly.

M.G.P. STOKER: Could I bring out a general point because it relates to the meeting as a whole. You threw away the word cloning here and there. Can I ask you to make it clear that this exciting revolution in molecular genetics, although it did not depend on cloning for its initial discovery, would not have been possible to develop in detail in the way in which you have described so nicely, if it hadn't been possible to clone? Am I correct in this?

R. BREATHNACH: Yes, certainly. Essentially all the work I have described has been carried out using clones. Obviously, to get sequence data out you need cloned DNA. You can only look for very interesting things like *in vivo* expression of genes if you have them cloned. You can only do the electron microscopy if you have cloned genes. If there had been no cloning, we would have come to the conclusion that genes are split, using the technique of Southern, and beyond that we could have got restriction maps of individual genes - but anything further - that would have been totally impossible without the rapid progress in the genetic engineering techniques.

M.G.P. STOKER: Certainly.

W. SZYBALSKI: Does cloning of the ovalbumin gene require, in your country, any special containment? And, if it does, why? What are the risks?

R. BREATHNACH: The answer for France is yes, it requires containment.

W. SZYBALSKI: What are the risks?

R. BREATHNACH: I don't know. That's what this meeting is meant to find out, I suppose. Perhaps in the final session we can have a more detailed discussion. I can see no reason for this containment.

W. SZYBALSKI: But I couldn't see any risks of propagating *E. coli* K-12 which carries a complex ovalbumin or its fragment.

J. SUBAK-SHARPE: Assuming there exist special enzymes which excise introns, there should be a set or grouping of nucleotide sequences, members of which are found at the joins of exon and intron, which must be forbidden from the rest of the exon genetic material which actually constitutes the messenger. Otherwise one might get unrequired cuts. The minimum size of these sequences, compatible with your data, and their size range, must have been computed by you. Could you tell us what they are?

R. BREATHNACH: No, we have not looked at this possibility. If we assume that the same sequences of say 7 or 8 nucleotides are forbidden in the exons of mouse, chicken, SV40, etc., one could look for these in the globin, ovalbumin and SV40 messengers, whose sequences are available. Alternatively, the chicken conalbumin lysosyme and ovomucoid genes have been cloned in the laboratories of Chambon, Kourilsky and O'Malley and no doubt a very large number of additional chicken boundary sequences will shortly be available to test your suggestion.

RECOMBINANT DNA AND ANIMAL VIRUSES: PROGRESS AND PROSPECTS

J. Sambrook

Cold Spring Harbor Laboratory, P.O. Box 100, Cold Spring Harbor, New York, 11724, U.S.A.

Perceptions of recombinant DNA research using the genomes of animal viruses have changed greatly during the last 8 years. In some respects these changes have been prophetic for they often have foreshadowed those which were to occur in other areas of recombinant DNA research.

The most convenient starting place to begin a historical review is 1972, when the Berg group declared publicly their intention to construct and propagate in E. coli, a recombinant consisting of prokaryotic DNA sequences coupled to those of SV40. SV40 is a virus, which is found in the kidneys of rhesus monkeys to whom it causes no apparent harm. However, it has the capacity to cause tumors in newborn laboratory animals and to cause similar changes in cells grown in culture. It was the idea of a bacterium which is a common inhabitant of the human gut carrying the genes of a tumor virus which led to a furore at a Cold Spring Harbor meeting in 1973, to a letter of disquiet signed by the majority of those who attended the 1973 Gordon Conference on nucleic acids, and by March 1974 to the establishment of a highly effective moratorium.

Between then and Asilomar, one year later, there occurred at NIH two or three meetings of a small group of animal virologists to discuss the problem. These meetings were entirely inconclusive, because we could see little evidence of danger, certainly no more danger than that which came from handling the viruses themselves. We had been working with these viruses for many years with no apparent problems. But, if we were to say that recombinant DNA involving their genomes were dangerous then we would invite regulation, not only of recombinant DNA, but also of animal virology. Unwilling to accept a radical position which found little basis in experience, we went to Asilomar indecisive and sceptical and with a rather dilute working paper.

However there was one precedent for regulation of research on animal viruses that we had overlooked and which formed a major part of the discussion of animal viruses at Asilomar. This precedent involved a set of rather esoteric viruses called adeno-SV40 hybrid viruses. The history and properties of these viruses have been discussed in detail elsewhere (see Tooze, 1979) and will not be reworked here. Suffice it to say that they are laboratory artifacts created by a rare recombination event which links together the genomes of two entirely unrelated viruses - SV40 and human adenoviruses. When the adeno-SV40 hybrid viruses were discovered during the late 1960's, some people felt that they may possess some novel quality which would cause them to be more pathogenic than either of their parents. Subsequent investigation has failed to support this idea. Nevertheless as a consequence of the initial fears, fairly rigorous physical containment is required for work with these viruses, and a document must be signed which stipulates that the workers assume full moral and legal responsibility for the agents in their laboratory. This was the first MUA and it preceded Asilomar: Its existence established that research could be governed by guidelines and that such guidelines could deal with perceived risks as well as proven dangers. In the absence of strong counter-opinion at Asilomar these precedents formed the basis of the restrictive guidelines which emerged afterwards.

Within a very few months it became obvious that the guidelines were prohibitive: No work could go on, at least in the US. Pressure gradually grew to revise the guidelines, which found an outlet at the NIH open hearings in December 1977, Washington, when a second working group was set up to reconsider the problem of recombinant DNA involving animal virus genomes. This second group which consisted of 27 European and US scientists met at Ascot in January last year and did what should have been done three years earlier at Asilomar. It worked through the various scenarios, both probable and improbable, and with remarkably little dissension concluded that the probability that *E. coli* carrying viral DNA inserts could represent a significant hazard to the community was so small as to be of no practical consequence. We could find no case in which recombinant DNA posed a threat greater than the viruses themselves, in fact there were very many dangerous viruses which we felt could be studied in great safety by recombinant DNA techniques.

The recommendations of the Ascot meeting were accepted and after a while translated in the United States into the set of guidelines which came into force three or four months ago and which have reduced markedly the containment required to do experiments. Instead of P4 for most experiments we can now use P2. Instead of EK3 we can use EK2 or even EK1. Now we have not been operating under these new rules long enough to generate any results which will stick to your ribs. But, there is a vast amount of work going on with great potential, and in addition there is a very small number of experiments which were begun and completed under the old guidelines. I will describe very briefly both sorts of experiments so that you can get a feel of what people are trying to do.

The experiments which have started since the guidelines went into effect fall into several categories. First recombinant DNA is being used to propagate viral genomes which up to now have been propagated poorly or not at all in the lab - a good example being the hepatitis viruses. This group consists of three or more unculturable viruses which are known to be important causes of viral hepatitis and probably account for a vast majority of hepatitis. Recently the non A, non B virus, which causes most cases of post-transfusion hepatitis and a large proportion of sporadic hepatitis in the population, was transmitted to chimpanzees, but neither it nor Type A nor Type B viruses have been grown in cell culture. However, several groups of research workers financed both publicly and privately have cloned various forms of hepatitis virus in prokaryotic vectors and it now seems that the development of practical means of controlling hepatitis through immunization, many from cloned recombinant DNA copies rather than from more traditional methods of virus culture.

If hepatitis is the best example of recombinant DNA being used in this way it is only marginally so, for there are many others. For example, EBV, Epstein-Barr virus, is associated with Burkitts lymphoma in Africa, with nasopharyngeal cancer in Asia and is the cause of infectious mononucleosis in the United States. It is extremely difficult to grow. Consequently we know virtually nothing about its molecular biology. It has a very large genome and is being cloned, again in several laboratories. Inevitably therefore our knowledge of this important virus will increase, and very fast.

Secondly there are viruses which grow well in vitro but which are so dangerous to handle that one does not want actually to propagate them in the laboratory: in that group, of course, is Lassa fever, Marburg virus and smallpox. The World Health Organization at the moment, for instance, is faced with the problem of trying to preserve the smallpox virus genome in some form while not having the virus propagated; one of their options is to preserve the virus as a series of recombinant DNA clones. Splitting the genome up with restriction endonucleases would be a way of totally disarming the virus and there is no doubt in my mind that if he had been handling recombinant clones instead of infectious virus, Henry Bedson - and Janet Parker - would be alive today.

Thirdly recombinant DNA has great potential in viral diagnosis. Usually virus diseases are diagnosed by symptom or on the basis of serological tests. While these approaches are satisfactory for many acute virus diseases, they are relatively poor for viruses with extremely long incubation times or which establish long-term latent infections. For these, people are trying to devise new methods of diagnosis involving nucleic acid hybridization with recombinant DNA clones. By this means they should be able to detect small amounts of viral DNA which otherwise would go undetected.

Fourth, animal virology is certain to benefit now that methods are becoming available to obtain expression of recombinant DNAs in their prokaryotic hosts. It seems obvious that once the correct genes have been cloned we will be able easily to manufacture cleaned-up vaccines, as well as to obtain substances like interferon which are of potentially great medical importance.

Fifth, animal viruses will be used as vectors to reintroduce into eukaryotic cells, foreign DNA sequences which can be expressed. That such an idea was feasible is demonstrated recently by two groups who showed that it is possible partially to replace the late coding region of SV40 with a copy of the coding sequences for globin. The inserted globin sequences are correctly expressed in infected cells, albeit under the control of the normal viral signals. This single experiment, involving lytic infection of cells in tissue culture is a long way from gene therapy in animals. But it points to a road which may lead us to develop that capacity, should we so wish.

As I said previously, very few recombinant DNA experiments with animal virus genomes were carried out under the old guidelines in the United States. However the restrictiveness acted as a great spur to international cooperation and I want to end my talk by telling you about a piece of work which was begun last year in London as a joint project between my laboratory and the Imperial Cancer Research Fund. It concerns the arrangement of SV40 DNA sequences integrated in the line of transformed rat cells known as 14B. These cells, like all others transformed by SV40 contain integrated viral DNA: it is the expression of the viral sequences which causes the cells to display a malignant phenotype. Some years ago, we showed (Botchan, Topp and Sambrook, 1976) that 14B cells contained a single insertion of SV40 DNA. Using the crude methods available at that time we were able to provide low resolution maps of the integrated viral genome. The advent of recombinant DNA technology now enables to analyze the molecular details of the junctions between viral DNA and chromosomal DNA in transformed cells. We used the lambda bacteriophage Charon 4A to clone an EcoRI restriction fragment of 14B DNA that contains the entire viral insert along with flanking chromosomal sequences on both the 5' and 3' sides. Figure 1 shows a heteroduplex molecule constructed by annealing the DNA strands of Charon 4A with a recombinant clone of viral DNA and linear molecules of SV40 DNA. The SV40 sequences are placed very asymmetrically with respect to EcoRI sites in the chromosomal DNA fragment. We have begun sequence analysis of the SV40 junction from the side of the insert that contains very little cell DNA. The results show that there are 72 nucleotides of cell DNA extending from the EcoRI site and that these sequences simply merge into SV40 without any interspersion or mutation of viral sequence at the joint. No homology is evident between the inserted viral sequences and the adjoining cellular DNA.

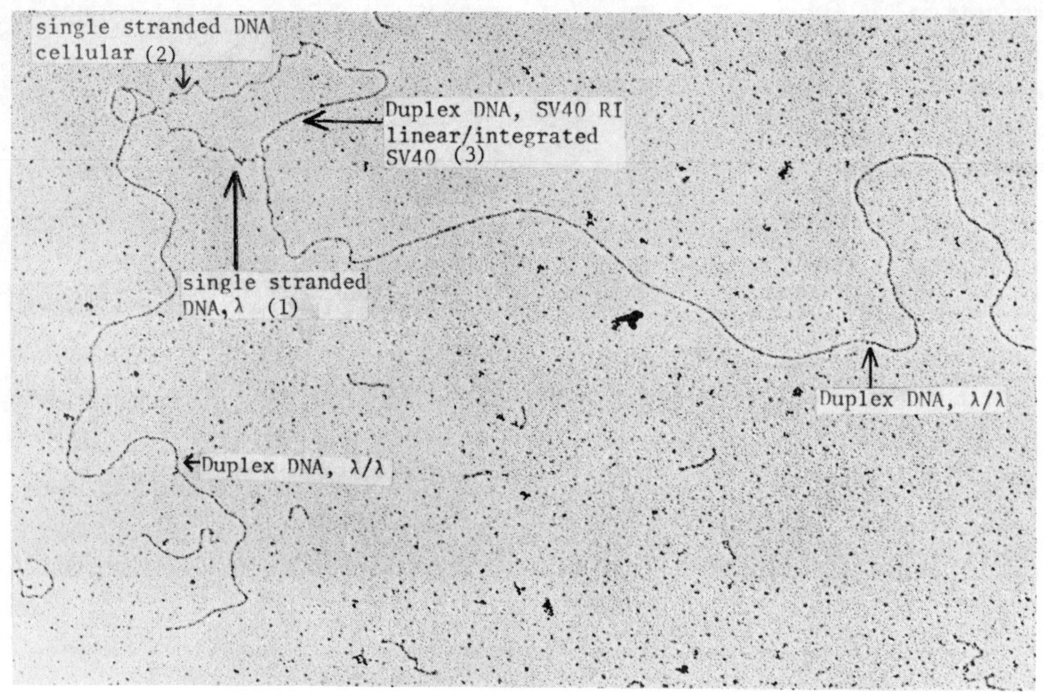

Fig. 1. Heterotriplex DNA consisting of (1) a λ Charon 4A DNA strand, (2) a strand of λ Charon 4A DNA into which the cellular DNA EcoRI restriction fragment that contains integrated SV40 DNA has been inserted, and (3) a strand of SV40 DNA linearized by cleavage with EcoRI.

From these data and from those published previously (Botchan & co-workers, 1976) it now seems that integration of SV40 DNA into the genome of mammalian cells is different from that of bacteriophages like λ into their prokaryotic hosts. Instead of a precisely designed mechanism designed to insert the viral genome at a specific chromosomal location, the possibility now becomes strong that mammalian cells may have the ability to integrate viral DNA by a non-homologous recombination at any one of a large number of sites.

While these results promise to lead to the solution of a nagging problem, they are not highly significant when taken in isolation. They are important to this meeting only because they exemplify what I believe will be the chief benefit of recombinant DNA to animal virology. In addition to the practical benefits, some of which I discussed earlier, recombinant DNA promises to be the tool which will lead most quickly to a greatly increased understanding of animal viruses. It is this expansion of our knowledge that will enable us to deal more rationally with the viruses themselves, and to penetrate the workings of the cells that they infect.

REFERENCES

Botchan, N.M. (1976), W. Topp and J. Sambrook. *Cell*, 9, 269-287.
Tooze, J. (Ed.)(1979). *The Molecular Biology of DNA Tumour Viruses*. Cold Spring Harbor.

DISCUSSION

O.W. PROZESKY: There is a human polyoma virus available, BK virus, which occurs in 90% of people because they have antibody and, as in the monkey, there don't seem to be any ill effects. Why do you continue with SV40 which is a foreign virus to human cells, if eventually you want to put genes into a human cell?

J. SAMBROOK: There is an underlying assumption in your question that there is a danger involved in using papovaviruses as vectors. I would reiterate that such dangers are conjectural. There is, therefore, no reason to believe that one virus would be more or less safe than another. The major factor in favour of SV40 is the wealth of knowledge about its molecular biology that has accumulated over a period of years and has penetrated down to the level of nucleotide sequence. It is a matter of convenience rather than principle to my mind.

O.W. PROZESKY: The rotaviruses are double stranded multi-segmented RNA viruses. How do you get the copy to put into the DNA?

J. SAMBROOK: There is no apparent reason why one should not make a copy by melting the RNA and copying one strand after the other.

F.S. ROLLESTON: The Ascot workshop statement was to the effect that the hazards of cloning eukaryotic viruses in an *E. coli* K-12 system are no greater than the hazards of working with the viruses themselves. One of the many inconsistencies in present guidelines, Canadian ones included, is that that statement has been juxtaposed with other statements, such as, the hazards of working with human DNA are considerably greater than the hazards of working with humans themselves. Why did not the NIH, in accepting the Ascot workshop recommendation, make it general such that the word "viral" is replaced by "source of DNA"?

J. SAMBROOK: I can't pretend to speak for the NIH or whoever it was who made this decision. But it does seem to me to be that the recommendation to drop the virus guidelines which came from Ascot was, in fact, influential in causing the guidelines for other DNAs to drop: once the virus problem was tucked safely away then the other guidelines dropped also. But I agree that many inconsistencies remain in the NIH Guidelines - their only virtue is that they now allow us to work; previously they did not.

F.S. ROLLESTON: Certainly, I would accept that. But I wonder whether, since the Ascot workshop was targeted toward *E. coli* as the host-vector system, one would go further to the generalised statement that, with respect to eukaryotic host-vector systems, the hazards of performing recombinant DNA experiments are seen effectively as zero, apart from the hazards of the DNA and the host-vector system which, of course, is the implication of the Ascot workshop recommendation.

J. SAMBROOK: I can't see any danger in eukaryotic host-vector systems.

A. CAMPBELL: Could I just comment on that? After the Ascot workshop many of the members of the Recombinant Advisory Committee felt that it would be appropriate and consistent to re-do the whole guidelines on the basis of the kind of thinking about hazards that the Ascot workshop had brought about. We also felt that to do that at that stage would have delayed substantially any further revisions and therefore that wasn't pushed - I think that's the historical answer to the question.

W. SZYBALSKI: I believe that the possibility of making vaccines, very pure and safe vaccines, by the recombinant DNA technique should be mentioned. Such vaccines will be safer, since they will consist of pure antigens (or antibodies) and be free of viruses. This will be especially important for viruses suspected to be carcinogenic. We have no vaccine against Herpes II, which causes a VD-like disease, extremely painful, especially to women. The development of such vaccines is virtually prohibited or strongly discouraged and delayed by the regulations.

J. SAMBROOK: It is of course a hopeful possibility to use as vaccines virus-coded products synthesized in *E. coli;* but it does seem to me that we need to get much better expression of foreign genes in *E. coli* in order to realise it. Furthermore, there would remain significant problems of purifying the viral product away from undesirable bacterial substances e.g. endotoxin. Finally, we could use the system only for viruses which code for discrete products that by stimulating the immune system confer protection.

W. SZYBALSKI: But, it should lead to much safer vaccines than those we have now. Correct?

J. SAMBROOK: Yes.

W. SZYBALSKI: Let me try to answer Dr. Prozesky's question as to why we couldn't use a safer human SV40-like virus. One of the answers is, I believe, that because of the regulations based on the guidelines, we cannot readily try new, often potentially safer, approaches.

O.W. PROZESKY: Could I just answer that last point. That was why I asked the question - why are the human ones not in the guidelines as SV40 is? SV40 has been specifically excluded from all human vaccines with great vigour over many years. The people who make polio vaccine and all other vaccines are very afraid of SV40. I know it's been put into a lot of people and they have been followed up and nothing serious had happened, but if we exclude SV40 from all human vaccines, I don't think we should put it into humans in other forms if there are human viruses as an alternative available.

M.A. MARTIN: I think there are several confusing issues that are being raised. In terms of cloning in *E. coli,* which is the question Dr. Szybalski asked, under the United States guidelines one can now clone any class 1 or class 2 animal virus DNA in *E. coli* K-12 host-vector systems. In some instances class 3 animal viruses such as VSV can be cloned in *E. coli* systems. There are also a few viruses currently classified as class 3 agents, which, under certain circumstances, are treated like class 1 or 2 agents depending on the geographical location of the laboratory or whether the virus is used in animal studies. At present *all* class 1 and 2 viruses can be cloned in *E. coli* K-12 systems.

In terms of the other kind of experiment which Dr. Prozesky mentioned, that is, the cloning in eukaryotic cells, the current NIH Guidelines list six specific categories of animal viruses or segments of animal viruses which can be used as vectors in animal cells. There is a catch-all category called "segments of viral genomes" and under that particular category one can use up to a quarter of any animal viral genome to clone a foreign DNA segment. At present one can use polyoma, SV40, mouse adenovirus, human adenovirus, segments of herpes viruses or 25% of any eukaryotic virus as a potential vector.

E. LUND: I would like to ask about the use of human viruses as vectors. You seem to be quite sure yourself that it would be all right. My problem in this connection arises from the fact that we, as far as I know, do not know what it is in the genome that makes the difference between a virulent and a non-virulent strain of a virus. How can we be sure that during the work nothing happens that will change a relatively moderate virus into a real troublemaker?

J. SAMBROOK: The conclusion we reached at Ascot was that in most cases virulence was a multi-genic character requiring co-ordinate and full expression of the viral genome and that it would be highly unlikely that vectors consisting of sub-genomic segments could pose any threat. Furthermore, most current work on viral vectors is carried out in tissue culture cells which is itself a tightly contained system.

A.M. SKALKA: I want to ask Dr. Martin what the logic of this permissible quarter is?

M.A. MARTIN: That's a very good question. The cloning of foreign DNAs using an animal viral vector was discussed at the Ascot meeting. I think everybody there agreed that this was the one area of recombinant DNA experimentation where there were many unknowns. The major concern that people had at that meeting involved the alteration of the host range of a viral vector following the insertion of a piece of foreign DNA. In many cases that we discussed, particularly those involving such vectors as SV40, polyoma and BK virus DNAs, no-one could generate a convincing scenario involving an alteration of virus host range. Our main problem was that the group at Ascot felt that they did not know enough about many of the other viruses to be absolutely certain that insertion of heterologous DNAs would not alter the host range. Since a consensus was not reached on this issue, the "25% viral DNA segment" compromise was agreed upon. It's worth pointing out that the discussions at Ascot focussed on work in tissue culture systems. As you will hear tomorrow, the administration of recombinant DNA molecules to animals has presented a multiplicity of problems.

F.S. ROLLESTON: I think as somebody who tries to administer a set of national guidelines that there are problems arising from a large number of individual decisions such as have been taken at the NIH and I would, in parenthesis there, say we are extremely grateful in Canada for the work that the NIH has done because it certainly has made our job a lot easier. They are certainly bearing the brunt of this whole discussion. What I am saying is in sympathy rather than in criticism.

But we are thinking of saying that, given that we are working in tissue culture which must be classified as having a very high biological containment factor, can one then make a generalised statement about eukaryotic viruses as vectors according to their ability to infect or transform, or whatever, a target population which presumably is human-beings. For example, if they transform or do whatever to human-beings, they would be classified as rather low in biological containment but if they are restricted to non-human and non-primate organisms they would be classified as a high biological containment. Could one then use the same table as the *E. coli* host-vector systems use based on the different sources of DNA? One can approach a form of logic of that type to try and make the guidelines at least internally consistent, though it, of course, begs the question of whether they are externally consistent.

J. SAMBROOK: You can lower the containment within any one set of guidelines by juggling the vectors. However, there is no rational basis for conjuring tricks, because there is no proven risk.

F.S. ROLLESTON: But that is a function of external consistency.

S.N. COHEN: I have a question about the host range point. As I understand it there are no regulations or restrictions in infecting a particular cell with two viruses of different host range, and as we have all heard just a short while ago, the experiment that Lewis did along these lines resulted in an alteration of host range. How did the Ascot workshop rationalise the requirement for limiting recombinant DNA experiments to a quarter of the viral genome, because of their concerns about host range alterations, while no specific limitations existed for experiments that could potentially bring about the same kind of alteration of host range by natural infection of a cell with two different viruses?

M.A. MARTIN: That's a very difficult question to answer. The people who attended the Ascot meeting didn't address experiments involving viruses themselves. We concentrated exclusively on recombinant DNA studies using either *E. coli* or animal viral vector systems. I think that many of us, myself included, were apprehensive in 1975 and 1976 about co-infecting cells with SV40 and BK or SV40 and adenoviruses. There were good scientific reasons to want to do such experiments. I think that there were other people doing molecular virology who were concerned about doing these types of experiments, even infecting cells simultaneously with a prokaryotic and a eukaryotic virus. The important point, however, was that it was important to reach agreement among the Ascot participants. A sufficient number of workshop attendees were unwilling to commit themselves on this question. Consequently, a compromise position was agreed on.

J. SAMBROOK: We said if you wanted to make novel recombinant viruses you would have to go to RAC, who would consider each proposal individually.

M.A. MARTIN: The group at Ascot had a great deal of difficulty dealing with the category of experiments where one segment of viral DNA was linked to a viral vector, so called viral-viral combinations. We had no feeling whether alterations in host range of tumorigenicity really would occur. We put this type of experiment into the category of "case-by-case" analysis because we just couldn't deal effectively with an infinite series of *ad hoc* decisions. It may turn out that those sorts of combinations will be completely innocuous. We just didn't know. Again, there was no consensus on this point among the 25 or so participants who were there and who represented a variety of disciplines.

F.S. ROLLESTON: To try and answer Dr. Cohen's suggestion or experiment, I wonder whether such an evaluation could not be done retrospectively on a case-by-case basis by testing the host range, rather than prospectively.

J. SAMBROOK: It would have to be done that way because there are no known rules which predict host range of recombinant viruses. You have to just test each one as it comes up.

F.S. ROLLESTON: It is different in Canada because we have guidelines for viruses.

S.N. COHEN: I would like to make it clear that I was not proposing an experiment - I was just pointing out what seemed to me to be an inconsistency.

CHAIRMAN's INTRODUCTION

W. J. Peacock

CSIRO Division of Plant Industry, P.O. Box 1600, Canberra City, ACT 2601, Australia

It's been a long day and by the end of the third talk it will be even longer so I thought I might give my summary now.

Initially the analyses of eukaryotic genomes by recombinant DNA methodology almost exclusively centred on members of the animal kingdom but I think you'll see tonight that plants and fungi are beginning to appear on the scene. They will do so in an ever increasing way. In fact, the most successful marriage of *E. coli* with eukaryotic genes has been with genes from yeast and *Neurospora*; Dr. Schell will discuss this. He will also discuss a naturally occurring vector system for the introduction of genes into plant cells - a field of great promise. The ability to be able to re-differentiate a whole plant from a single cell in many species opens the way to genome manipulation without the great disturbances that are caused by the normal means of plant breeding and selection in agricultural plants which usually have very highly selected genome construction.

Plants produce 70% of the protein man consumes and many of our crop plants have disease problems which drastically reduce the potential yields. For many of the diseases there are simple genetic resistances available, often single gene resistances. We sometimes know these resistances to be present in the wild species relatives of our crop plants, but they are unavailable to us because of sexual crossing barriers. I think both for microorganism-induced diseases, in a number of our agricultural plant species, there is real scope for the introduction of a single gene element with tremendous consequences for man.

Plant proteins often need amino acid adjustment for our best health and it's become clear in recent years that sometimes this could be achieved if we could manipulate the relative levels of different constituent proteins in any given seed-storage-protein system. I think that by learning something of the control of protein synthesis we might ultimately be able to achieve this sort of manipulation.

Now, you've all been aware of the long-term hopes to negate the use of fertiliser by producing plants that do their own thing. Some aspects of a nitrogen fixation system for plants may in fact be a reality relatively soon. I think, for example, it's feasible to think of changing the host range of many of the nitrogen-fixing bacteria so that they will be able to be associated with other plant species. We should be able to do something about selecting strains of those bacteria and manipulating their genome so that they can work under more extreme conditions, e.g. the acid soil conditions which are a problem in much of our agricultural land.

I think too that we might construct microorganisms which have the abilities of both cellulose digestion and nitrogen fixation, since this seems to be within the realms of plasmid manipulation within bacteria. We may thus be able to use plant stubbles and residues to great effect in putting nitrogen back in the soil where it's really needed.

Some of those problems Dr. Riley will cover in his analysis of the future of this methodology in agriculture.

The third talk tonight covers another central problem to any gene manipulation scheme, that of ensuring the faithful and controlled expression of any introduced gene in a particular host. Dr. Ehrlich will address this problem.

NEW INSIGHTS INTO THE MOLECULAR GENETICS OF FUNGI AND PLANTS

J. Schell and M. Van Montagu *

Max-Planck-Institut für Züchtungsforschung, Köln, Vogelsang, F.R.G.
***Laboratorium voor Genetika, Rijksuniversiteit, Gent, Belgium**

The main techniques used to study the molecular genetics of various organisms are *in vivo* or *in vitro* recombination and mutation and introduction of the recombinant and, or, mutant DNA molecules into appropriate host cells via transformation, transduction or conjugation. Such studies have three general purposes: first and foremost to isolate and amplify genes in order to achieve a more detailed understanding of their structural and functional organisation; secondly to try to produce important proteins by introducing and expressing genes coding for such proteins in cells that can easily be grown in cultures (mostly bacteria and fungi, possibly also some plant cell cultures) and finally to try and breed new organisms (e.g. plants) with added desirable genetic traits. Recently very important progress has been made in the molecular genetics (as defined above) of fungi and plants. It is the purpose of this short review to describe briefly this rapidly expanding field.

MOLECULAR GENETICS OF FUNGI

The most significant advances have been made in yeast (*Saccharomyces cerevisiae*) and in the mold, *Neurospora crassa*.

Yeast

S. cerevisiae is probably the best genetically studied lower eukaryote and recent developments to be described below will undoubtedly result in a knowledge as detailed and accurate as is now available for prokaryotic genes and genomes. The following steps have been decisive in bringing about this progress:

1. Yeast chromosomal genes can be isolated by *in vitro* insertion into *E. coli* phages or plasmids. Struhl, Cameron and Davis (1976) and Struhl and Davis (1977) have isolated viable molecular hybrids of phage λ and random yeast chromosomal fragments. By looking for complementation of *E. coli* auxotrophic mutations by these hybrid DNA molecules, it was possible to directly select for functional activity of the cloned yeast genes in *E. coli*. A λgt.Sc.*his* hybrid was thus isolated which was shown to contain the yeast *his*3 gene, and which was able to complement *E. coli* *his*B auxotrophs.

Ratzkin and Carbon (1977) and Carbon and others (1978) have used the same general strategy to isolate, in a ColE1 plasmid, the yeast genes coding for the synthesis of imidazole glycerol phosphate dehydratase, tryptophan synthase, β-isopropyl malate dehydrogenase and argininosuccinate lyase.

The most striking aspect of these observations is that apparently a considerable number of genes of lower eukaryotes can be correctly transcribed and translated in prokaryotes. This is in sharp contrast to most genes of higher eukaryotes that have thus far been cloned in *E. coli*.

2. Yeast cells can be transformed with *E. coli* vectors containing cloned yeast genes. The possibility of transferring DNA sequences between yeast cells was the one important missing element in the study of the molecular biology of this genetically and biochemically well characterised organism. The cloning of some yeast genes in *E. coli* vectors, providing a convenient source of defined yeast DNA sequences, opened the way to look for conditions under which yeast cells would be able to take up, maintain and express genes in the form of purified DNA sequences. Using this strategy, Hinnen, Hicks and Fin (1978) have developed such a yeast transformating system. Using spheroplasts of a non-reverting double auxotrophic *leu*2^- mutant as recipient strain and polyethylene glycol and $CaCl_2$ to make the spheroplasts competent for the uptake of ColE1 - *leu*2^+ hybrid plasmid DNA, these authors showed that $1/10^7$ of the regenerated cells had become LEU^+. It turned out that the whole ColE1 - *leu*2^+ plasmid had been stably integrated at several locations into the chromosomes of the LEU^+ transformants and that, once integrated, the bacterial plasmid sequence (ColE1) and the *leu*2^+ gene were inherited as simple Mendelian factors. In a follow-up paper Hicks, Hinnen and Fink (1978) describe further experiments showing that integrated bacterial plasmid sequences can be retrieved from the yeast genome and re-introduced, by transformation, into *E. coli* where they replicate as autonomous plasmids. Of great practical importance is the observation that co-transformation of a single cell by two independent plasmids, occurs at high frequencies. This makes it possible to screen for transformed cells harbouring both a selected and an unselected gene.

3. DNA host-vector systems can be made that can replicate both in *E.coli* and in yeast. In view of the fact that most eukaryotic genes that have thus far been cloned in bacteria, are not expressed properly, it was tempting to develop some lower eukaryote, such as yeast or *Neurospora,* as a host system to clone and amplify eukaryotic genes.

Certain strains of *S. cerevisiae* contain in their cytoplasm 50 to 100 copies of autonomous closed circular, double stranded DNA molecules with a monomeric circumference of $\pm$ 2μ (Sinclair and Others, 1967; Cameron, Philippsen and Davis, 1977). These 2μ plasmids, now called Scp, were therefore good candidates for a yeast host-vector system.

Ideally such a host-vector system should have the following properties: 1) the ability to replicate both in yeast and in *E. coli* cells, thus allowing the high solution of the genetic and molecular analysis that is available in *E. coli* to bear on the eukaryotic genes cloned and, hopefully, expressed in yeast cells. 2) genetic characteristics which are selectable in yeast and, ideally, also in *E. coli*. 3) sites of cleavage into which essentially any fragment of DNA can be inserted and 4) form covalently closed DNA circles that can easily be isolated from at least one of the host organisms.

Several laboratories have successfully produced host-vector systems that satisfy these ideal properties. Composite plasmids were made by inserting the whole of the 2μ yeast plasmid into an *E. coli* plasmid (Beggs, 1978; Struhl and others, 1979) or by inserting replicating natural deletions of Scpl (Struhl and others, 1979) or relevant restriction fragments of Scpl (Hicks, Hinnen and Fink, 1978) in an *E. coli* plasmid. By including the previously cloned yeast genes in these composite yeast-*coli* plasmids it was possible to select for their transfer (by transformation), maintenance and expression both in *E. coli* and in yeast. The technical possibility of screening libraries of 10^4 yeast DNA fragment cloned in such plasmids in a single experiment, makes it possible to identify, by complementation, virtually any yeast gene for which defective mutants can be obtained. Yes a further important development of these host-vector systems resulted from a recent observation in the group of R. Davis (Struhl and others, 1979). It was found that a particular 1.4 kb yeast chromosomal DNA fragment, comprising the *trpl* gene, when inserted in an *E. coli* plasmid,allowed the autonomous replication of this hybrid plasmid, probably in the nucleus of yeast. In essence a new yeast mino-chromosome was thus made that could be extracted, transformed and replicated in *E. coli*. This could well be the ideal vector to study the precise structures and DNA sequences that control the expression of eukaryotic genes in yeast because, in contrast to the other yeast host-vectors, these mini-chromosomes do not integrate in other yeast chromosomes and are therefore ideally suited for complementation studies and for establishing structure-function relations as detailed and as precise as are available in prokaryotes.

Recently J. Begg and C. Weissmann(personal communication) have introduced the cloned rabbit β-globin into yeast. They found that the initiation of the transcription of this mammalian gene was incorrect in yeast and that the transcript was polyadenylated but that no correct splicing took place.

Neurospora crassa

Most of the recent developments in our understanding of the molecular genetics of *N. crassa,* stem from the work of N. Giles and his collaborators (see Alton and others, 1978). These authors cloned the gene for *N. crassa* catabolic dehydroquinase (C-DHQase) and found that it was able to complement a dehydroquinase-deficient auxotroph of *E. coli*. With these clones it was possible to answer a number of questions about transcription and translation of mold-genes in *E. coli*. Two independently derived clones, harbouring the structural gene for *N. crassa* dehydroquinase, indicate that C-DHQase inserted in opposite orientation in the *E. coli* vector plasmid, is expressed equally well in *E. coli*. The results demonstrate that transcription is efficiently initiated within the cloned *N. crassa* DNA and the resulting m-RNA is translated at a level which is similar to the level of translation of other m-RNAs derived from the plasmid vector.

It is therefore clear that not only yeast but also mold-genes can be properly expressed in *E. coli*. This probably means that at least some yeast and mold transcripts do not require splicing in order to give active messengers.

To our knowledge no host-vector system has yet been developed for *N. crassa*. The recent claim of Martini, Grimaldi and Guardiola (1978) that a phytopathogenic fungus, *Fusarium oxysporum,* harbours a 60 x 10^6 M.D. plasmid DNA that can be transferred, replicated and possibly expressed in *E. coli*, would, if further substantiated, indicate that autonomous plasmids exist in molds and therefore could be developed as host-vector systems.

MOLECULAR GENETICS OF PLANTS

At least one of the prerequisites for the efficient study of plant genes by genetic engineering, i.e. the availability of a host-vector system in which defined genes can be isolated, amplified, modified and re-introduced in plants - has recently been fulfilled. The remarkable fact here is that such a host-vector system already existed in Nature and was developed by natural selection in bacteria!

Agrobacterium tumefaciens form a group of Gram-negative soil bacteria that induce so-called crown-gall tumour on most dicotyledonous plants. Work in our team (Schell, 1974; Zaenen and others, 1974; Van Larebeke and others, 1974 and 1975) and by others (Watson and others, 1975) has demonstrated that a group of large plasmids, called Ti-plasmids, are involved in this neoplastic transformation of plant cells. Since the discovery of the Ti-plasmid, several important observations have allowed a fairly precise explanation of this neoplastic transformation of plant cells. First of all it was demonstrated that Ti-plasmids carry genes that somehow determine the specificity of synthesis of so called opines by transformed plant cells (Bomhoff and others, 1976). Opines were discovered by Morel and his collaborators and their fundamental significance for the understanding of the crown-gall phenomenon was first indicated by Petit and others (1970). The Ti-plasmids were also shown to harbour genes that allow *Agrobacteria* to specifically use opines as sole carbon, nitrogen and energy source. These observations provided genetic evidence in favour of a model involving the Ti-plasmid in a DNA-transfer from bacterium to plant. That such a Ti-DNA transfer actually occurs was subsequently demonstrated first by Chilton and others (1977). The demonstration that the T-DNA (i.e. the Ti-plasmid DNA segment present in transformed plant cells) is actually responsible for the synthesis of the specific opine in the transformed plant cells and probably also for the maintenance of their tumourous condition, was reported by our group (Schell and others, 1978a, 1978b). This work involved both a genetical and a direct physical identification of the Ti-DNA segments that were involved in tumour induction and in opine synthesis and that were found to be present in transformed plant cells.

At the present level of evidence the following picture of the Ti-plasmid and of the T-DNA emerges: the Ti-plasmid is a highly evolved system with several functions necessary for the effective transfer and maintenance of its T-DNA in the transformed plant cells.

The T-DNA corresponds to one continuous segment of the Ti-plasmid with a molecular weight of about 15×10^6. This T-DNA is integrated in some plant DNA. Very recent hybridization results obtained by Dr. L. Willmitzer and M. De Beuckeleer in our laboratory, using DNA extracted from nuclei purified from crown-gall plant cells by Dr. L. Willmitzer in the G.B.F. labs in Stockheim, West Germany, have demonstrated that the T-DNA is located in the nucleus of transformed plant cells. The T-DNA thus either forms an independent nuclear replicon containing some plant DNA, or more likely, is simply integrated in some chromosomal replicon. The number of different plant DNA segments that are associated with the T-DNA both in cloned and uncloned tumour-callus cultures derived from a given tumour, appear to be limited to one or two. Whether or not these DNA segments are identical in independently derived tumour-lines has not been established yet.

By genetic means (insertion and deletion mutants) it was established that the T-DNA consists of at least three different functional units. A central segment, of about 4.5 M.D., is conserved in all types of Ti-plasmids. Mutations in this segment invariably result in the loss of capacity to produce plant tumours. On either side of this conserved DNA segment, we found DNA sequences that are different in different types of Ti-plasmids. Mutations in these non-conserved segments of the T-DNA do not result in the loss of capacity to induce tumours. Mutations in what has been called the right side of the T-DNA ($\pm$ 1.5 M.D.), do, however, result in Ti-plasmids that induce tumours in which no opine synthesis occurs, thus identifying the DNA sequence involved in opine synthesis. Possible aberrant phenotypes of plant tumours induced with Ti-plasmids with a mutation in the "left" ($\pm$ 9 M.D.) end of the T-DNA, have not yet been carefully studied. Drummond and others (1977) and Kemp (personal communication) have demonstrated that at least part of the T-DNA is actively transcribed in transformed plant cells.

With the realisation that Ti-plasmids are in fact a natural host-vector system able to promote transfer, integration and expression of "foreign" DNA in plants, we asked whether or not the properties of this system could be used to introduce, at will, foreign DNA sequences in various plants (Schell and Van Montagu, 1977). Recently (Schell and others, 1978a, 1978b) we were able to report that this was indeed the case. The general strategy used was to incorporate, by *in vivo* recombination, a bacterial transposon, Tn7 (Hernalsteens and others, 1978) into the 1.5 M.D. segment of the T-DNA that codes for opine synthesis. Subsequently we were able to demonstrate that tobacco tumour-lines induced with such Tn7 harbouring Ti-plasmids, did indeed contain a T-DNA segment consisting of the whole of the T-DNA (15 M.D.) of the wild-type plasmid together with the whole of the Tn7 DNA (9.5 M.D.) still inserted in its original site in the T-DNA. Whether or not the Tn7 DNA is transcribed in these plant cells has not yet been established.

An important question is whether or not plant cells containing a T-DNA segment can be used to breed new plants. Recent observations made by Braun and Wood (1976), have demonstrated that whole plants can be regenerated from transformed teratoma cultures. In these studies tobacco teratoma derived tumour shoots were isolated and grafted to cut stem tips of normal tobacco plants of a morphologically distinct cultivar. This way, shoots were obtained that developed quite normally and ultimately flowered and set viable seed. It was found that the leaves of these grafts were normally organized but were still transformed in the sense that, when such specialized cells were isolated and planted on a basic culture medium, they grew as crown-gall cells and synthesized nopaline. Gordon and his collaborators (personal communication) and also our group have studied these tissues and found that they still contain T-DNA from the TiT37 plasmid. In fact we were able to demonstrate that the T-DNA segment found in tissues derived from organized leaves of the teratoma grafts, was exactly the same as the T-DNA segment found in the original teratoma tissues with which the grafts were initiated. No gross rearrangement of the T-DNA had therefore occurred during the differentiation of tumour cells to organized plant cells. It was thus demonstrated that it is possible to create new plants containing a specific DNA segment using the Ti-plasmid as a host-vector.

Other Possible Plant Host-Vectors

In analogy to the use of SV40 as a host-vector for some mammalian cells, it may be possible to develop some plant DNA viruses as vectors for plants. One of the very few candidates that can be considered is CaMV (cauliflower mosaic virus). However, work with this system is very difficult and little progress has been made thus far (for a recent review, see Hull, 1978).

Cloning of Plant Genes in *E. coli* Vectors

Except for chloroplast DNA genes (see e.g. Coen and others, 1977), relatively few plant genes have thus far been cloned in *E. coli*. Recently cDNA copies of some plant storage proteins have been cloned, such as, the two subunits of the maize zein protein (Burr, 1978: Wienand, Brüschke, and Feix, 1978). Sidloi-Lumbroso and Schulman (1977) have purified the soybean leghemoglobin mRNA. One can therefore expect that cDNA copies - and possibly genes - of leghemoglobin will soon be available in *E. coli* vectors.

REFERENCES

Alton, N.K., J.A. Hautela, N.H. Giles, S.R. Kushner, and D. Vapnek (1978). Transcription and translation in *E. coli* of hybrid plasmids containing the catabolic dehydroquinase gene from *Neurospora crassa*. *Gene*, *4*, 241-259.

Beggs, J.D. (1978). Transformation of yeast by a replicating hybrid plasmid. *Nature*, *275*, 104-109.

Braun, A.C., and H.N. Wood (1976). Suppression of the neoplastic state with the acquisition of specialized functions in cells, tissues, and organs of crown gall teratomas of tobacco. *Proc. Nat. Acad. Sci.* (USA), *73*, 496-500.

Bomhoff, G., P.M. Klapwijk, H.C.M. Kester, R.A. Schilperoort, J.P. Hernalsteens, and J. Schell (1976). Octopine and nopaline synthesis and breakdown gene tically controlled by a plasmid of *Agrobacterium tumefaciens*. *Mol. gen. Genet.*, *145*, 177-181.

Burr, B. (1978). Identification of zein structural genes in the maize genome. In *Abstracts International Symposium on seed protein improvement in cereals and grain legumes*, *IAEA*, Neuherberg (BRD), in press.

Cameron, J., P. Phillipsen, and R.W. Davis (1977). Analysis of chromosomal integration and deletions of yeast plasmids. *Nuc. Acids Res.*, *4*, 1429-1448.

Carbon, J., B. Ratzkin, L. Clarke, and D. Richardson (1978). The expression of cloned DNA in prokaryotes. *Brookhaven Symposia in Biology*, *29*, 277-297.

Chilton, M.D., H.J. Drummond, D.J. Merlo, D. Sciaky, A.L. Montoya, M.P. Gordon, and E.W. Nester (1977). Stable incorporation of plasmid DNA into higher plant cells: the molecular basis of crown gall tumorigenesis. *Cell*, *11*, 263-271.

Coen, D.M., J.R. Bedbrook, L. Bogorad, and A. Rich (1977). Maize chloroplast DNA fragment encoding the large subunit of ribulosebiphosphate carboxylase. *Proc. Nat. Acad. Sci.* (USA), *74*, 5487-5491.

Drummond, M.H., M.P. Gordon, E.W. Nester, and M.D. Chilton (1977). Foreign DNA of bacterial plasmid origin is transcribed in crown gall tumours. *Nature*, *269*, 535-536.

Hernalsteens, J.P., H. De Greve, M. Van Montagu, and J. Schell (1978). Mutagenesis by insertion of the drug resistance transposon *Tn7* applied to the Ti plasmid of *Agrobacterium tumefaciens*. *Plasmid*, *1*, 218-225.

Hicks, J. B., A. Hinnen, and G.R. Fink (1978). Properties of yeast transformation. *C. S. H Symposium* .

Hinnen, A., J.B. Hicks, and A. Hinnen (1978). Transformation in yeast. *Proc. Nat. Acad. Sci.* (USA), *75*, 1929-1933.

Hull, R. (1978). The possible use of plant viral DNAs in genetic manipulations in plants. *TIBS*, *111*, 254-256.

Martini, G., G. Grimaldi, and J. Guardiola (1978). Extrachromosomal DNA in phytopathogenic fungi. In A.W. Boyer and S. Nicosia (Eds), *Genetic Engineering*, Elsevier/North Holland, Amsterdam. pp. 197-200.

Petit, A., S. Delhaye, J. Tempé, and G. Morel (1970). Recherches sur les guanidines des tissus de crown gall. Mise en évidence d'une relation biochimique spécifique entre les souches d' *Agrobacterium* et les tumeurs qu'elles induisent. *Physio. vég.*, *8*, 205-213.

Ratzkin, B., and J. Carbon (1977). Functional expression of cloned yeast DNA in *Escherichia coli*. *Proc. Nat. Acad. Sci.* (USA), *74*, 487-491.

Schell, J. (1974). The role of plasmids in crown gall formation by *A. tumefaciens*. In L. Ledoux (Ed.), *Genetic manipulations with plant material*, Plenum Press, New York, pp. 163-181.

Schell, J., and M. Van Montagu (1977). The Ti plasmid of *Agrobacterium tumefaciens*, a natural vector for the introduction of *NIF* genes in plants? In A. Hollaender (Ed.), *Genetic engineering for nitrogen fixation*, Plenum Press,

New York, 159-179.

Schell, J., M. Van Montagu, M. De Beuckeleer, M. De Block, A. Depicker, M. De Wilde, G. Engler, C. Genetello, J.P. Hernalsteens, M. Holsters, J. Seurinck, B. Silva, F. Van Vliet, and R. Villarroel (1978a). Interactions and DNA transfer between Agrobacterium tumefaciens, the Ti-plasmid and the plant host. Proc. R. Soc. Lond. B, in press.

Schell, J., and M. Van Montagu (1978b). Transfer of genes into plants via the Ti plasmid of A. tumefaciens. In S. Rapoport (Ed.), Proc. 12th F.E.B.S. Meet., Dresden, no. 3533.

Sidloi-Lumbroso, R., and H.M. Schulman (1977). Purification and properties of soybean leghemoglobin messenger RNA. Biochemica et Biophysica Acta, 476, 295-302.

Sinclair, J.H., B.J. Stevens, P. Sanghavi, and M. Rabinowitz (1967). Mitochondrial-satellite and circular DNA filaments in yeast. Science, 156, 1234-1237.

Struhl, K., J.R. Cameron, and R.W. Davis (1976). Functional genetic expression of eukaryotic DNA in Escherichia coli. Proc. Nat. Acad. Sci. (USA), 73, 1471-1475.

Struhl, K., and R.W. Davis (1977). Production of a functional eukaryotic enzyme in Escherichia coli: cloning and expression of the yeast structural gene for imidazole-glycerolphosphate dehydratase (his3). Proc. Nat. Acad. Sci. (USA), 74, 5255-5259.

Struhl, K., D.T. Stinchcomb, S. Scherer, and R.W. Davis (1979). High frequency transformation of yeast: autonomous replication of hybrid DNA molecules. Proc. Nat. Acad. Soc. (USA), 76, in press.

Van Larebeke, N., G. Engler, M. Holsters, S. Van den Elsacker, I. Zaenen, R.A. Schilperoort, and J. Schell (1974). Large plasmid in Agrobacterium tumefaciens essential for crown gall-inducing ability. Nature, 252, 169-170.

Van Larebeke, N., C. Genetello, J. Schell, R.A. Schilperoort, A.K. Hermans, J.P. Hernalsteens, and M. Van Montagu (1975). Acquisition of tumour-inducing ability by non-oncogenic agrobacteria as a result of plasmid transfer. Nature, 255, 742-743.

Watson, B., T.C. Currier, M.P. Gordon, M.D. Chilton, and E.W. Nester (1975). Plasmid required for virulence of Agrobacterium tumefaciens. J. Bacteriol., 123, 255-264.

Wienand, U., C. Brüschke, and G. Feix (1978). Cloning of double stranded DNAs derived from polysomal mRNA of maize endosperm: isolation and characterization of zein clones. In Abstracts International Symposium on seed protein improvement in cereals and grain legumes, IAEA, Neuhenberg (BRD), in press.

Zaenen, I., N. Van Larebeke, H. Teuchy, M. Van Montagu, and J. Schell (1974). Supercoiled circular DNA in crown gall inducing Agrobacterium strains. J. Mol. Biol., 86, 109-127.

DISCUSSION

J.K. SETLOW: What's the host range of this - we are not so interested in fixing tobacco.

J. SCHELL: The host range is very wide. In fact, the host range is itself a property determined by the type of Ti plasmid. There are Ti plasmids that specify a wide host range and Ti plasmids that specify a narrow host range. If you take the whole collection of Ti plasmids the host range is most, if not all, dicotyledonous plants. Monocotyledons are still a problem. There is only one that has been reported to be transformable. Possibly this is due simply to the fact that an initial conjugating-like step is needed to bring the cells together for the initial transfer of the Ti plasmid. If this is the case, then probably one can get around this problem by using protoplasts of monocotyledons and introducing pure Ti plasmid into them and transforming them that way. This is in progress but has not yet succeeded.

J. SAMBROOK: What do the other two sets of genes of the plasmid do, which are necessary for transformation but don't appear in the transformed cells?

J. SCHELL: You mean the two other groups of genes on the Ti plasmid that are involved in oncogenicity. We don't know yet. We only have indirect evidence. One group is most likely involved with this initial conjugation-type mechanism that brings plant cells and bacterial cells together. What the other one does, we have no idea. We want to analyse this with the protoplast system to see which of these groups of genes we can omit and still transform protoplasts. Certainly they are not present in a transformed cell and we think that these genes are part of the mechanism by which this whole transfer phenomenon can be so efficient in Nature.

N.D. ZINDER: Does the T-DNA segregate in a cross and thus pass through the seed?

J. SCHELL: The only experiment that has been done is the one that I illustrated at the end of my talk and in that case no segregation was observed. The seeds made on this plant, regenerated from teratomas, were grown into new plants; they were analysed and they were not transformed. There was no T-DNA and no synthesis of opines. But, I cannot really answer the question - does it segregate? All that has been done, I have illustrated here - some plants were made, from these plants a number of seeds were grown and from those seeds tissues were made, analysed and found to be untransformed.

W. SZYBALSKI: Are the genes of the Tn7 transposon, which control the streptomycin and trimethoprin resistance expressed in the plant cells?

J. SCHELL: We don't know yet. We're looking for the mRNA.

W. SZYBALSKI: Is it possible to re-design the experiments so as to study plant cell transformation without using plant tumour formation as the selective property?

J. SCHELL: Certainly, this is something which could be done. Until now we used the transformation as an easy marker to recognise the transformed cells but whether it's essential for the transfer of the DNA we don't know. Experiments are planned to answer that question.

J.R.S. FINCHAM: I have a couple of questions - the first relates to the case of cloning the *Neurospora* dehydroquinase gene in *E. coli*. I think you said that the gene was expressed in *E. coli* as well as it would have been had it been a *coli* gene. There is some doubt in my mind as to whether that is really true. As I recall, the specific activity of the enzyme in *E. coli* was somewhat lower and then there is a question of how many gene copies there were.

J. SCHELL: I can give you the paper from which I extracted this, from the Norman Giles group - it's an unpublished paper. What they have done is first to insert the cloned DNA is two orientations and to demonstrate that they obtain the same efficiency of transcription in both cases. They have also measured the rate of translation of the *N. crassa* enzyme in *E. coli* as compared to the translation of some of the *coli* vector transcripts and they found that both transcripts were translated at the same rate.

J.R.S. FINCHAM: I want to come back to the question of what's special about tobacco? It seems to me that what may be special about tobacco is not that it's the only good host for *Agrobacterium* but it's about the only plant which you can regenerate from calluses with any degree of efficiency.

J. SCHELL: Your point is well taken. It certainly is not the only host for *Agrobacterium*. As I said, all dicotyledons are sensitive. It is one of the convenient ones to do regeneration experiments with. It is a good model system, no more.

R. BREATHNACH: Did you mention what the copy number was for the yeast mini-chromosome of Cameron and Davis?

J. SCHELL: I am sure I didn't mention it because I do not remember the data.

RECOMBINANT DNA AND AGRICULTURAL RESEARCH

R. Riley and R. B. Flavell *

Agricultural Research Council, Great Portland Street, London W1N 6DT, U.K.
***Plant Breeding Institute, Cambridge, U.K.**

The ultimate function of agricultural research is to provide scientific knowledge that can be applied to improve the quantity or quality of the plant and animal products that are derived by man from his crops and domesticated animals. Much research in DNA technology for agriculture will initially be fundamental in Nature - and not different from general research on recombinant DNA. Indeed agricultural research will expect to benefit in many ways from developments in recombinant DNA research. It will apply knowledge and methods of a general nature to its particular problems. Finally, however, agricultural researchers aiming purposefully to modify nucleic acid structures must work with the organisms that provide most of the foods, hides, beverages and natural fibres, on which man depends. These organisms are rarely those of experimental choice for the quick and clear resolution of scientific phenomena. Nevertheless new technologies are only useful to agriculture in so far as they can be used on our crops and livestock.

At this stage in the history of recombinant DNA research, it is not possible to point to a list of practical achievements. There is one, but this will be left to the end of these notes. First of all reference will be made to some opportunities that may arise as DNA science improves in sophistication. Part of my purpose in indicating the concerns of agriculture is to illustrate to molecular biologists where it is that agricultural research needs their help. As the technology develops it will be increasingly important for those working in established agricultural research and technology to be aware of the opportunities presented by DNA research. A responsibility will also lie with DNA specialists to maintain an awareness of the needs of agriculture and agricultural research so that if simple answers, or methods, can be supplied by molecular biology they are speedily available. It is for this reason that in the UK the ARC programme on the genetic manipulation of plants is juxtaposed to other better established research which already has a proven track record of practical achievement. We expect each to provide stimulus and challenge to the other.

In agriculture recombinant DNA technology should make possible improvements to:-

1. Economically important micro-organisms
2. Farm livestock
3. Crop plants

and each will be dealt with separately in these notes.

MICRO-ORGANISMS

A great deal of the work on micro-organisms that will have an impact on agriculture will not be different from that which will be undertaken for the pharmaceutical industry. It may be expected that antibiotics, hormones and control peptides,analagous to those developed for use on man, will also be used on farm livestock.

The proposal to release into the environment micro-organisms that have been modified by recombinant DNA technology is not unique to agriculture, but it is rare elsewhere. Nevertheless the engineering of micro-organisms that are components of the root-soil-rhizosphere ecosystem could be of value if they can be established in the soil. For example, hints are beginning to emerge that host specificity in *Rhizobium* is plasmid determined. There are also suggestions that the *nif* systems of *Rhizobium* may also be plasmid located. If it eventually proves to be true that these two critical characters of *Rhizobium* are readily accessible to manipulation, then a potentially valuable new range of variation may be obtained. However, it is necessary to be extremely cautious about the feasibility of transferring nitrogen fixing ability to plant species which at present do not form symbiotic relationships with micro-organisms with this property. The nitrogenase information of the DNA is complex, and, moreover, the enzyme only functions in the absence of atmospheric oxygen.

As part of the ARC programme on the genetic manipulation of plants there is work using yeast, and its 2 micron plasmid, as a model for plant cells. However, no reference will be made to yeast as an agricultural organism.

FARM ANIMALS

Originally the great attraction to working on the genetic manipulation of plants, by means of DNA insertion or modification, lay in the property that an entire plant sporophyte could be developed from a single protoplast or cell. Therefore, the consequences of the manipulations applied to a single cell could be expected to be observed on an intact organism. However, manipulation of the reproduction of farm animals, particularly cattle, now makes it appear that we are close to the point where they may also become suitable objects for the practical use of genetic manipulation. Successful *in vitro* fertilisation of the egg of a cow has yet to be achieved. However, 7 to 9 day old embryos are readily available. Indeed several commercial companies have been established to harvest and transplant embryos of that age.

The system is to (i) treat the cow with prostaglandin to fix the time of oestrus, (ii) Next treat her with pregnant mares serum or follicle stimulating hormone to induce superovulation, and the release of up to 12 eggs at oestrus, (iii) Inseminate, (iv) 7 - 9 days later insert 3 coaxial flexible tubes to harvest the embryos which may then be frozen and preserved, or immediately inserted into the uterine horn of a recipient mother. This system gives a high rate of successful pregnancy.

Clearly when the embryo, or possible in the future the fertilised egg, is outside the animal it is accessible to experimental manipulation. It would then be possible to introduce information sequences to improve the adult animal. There seems little doubt before long that this will be within our capabilities.

Let us imagine that, in the way I have described, we could incorporate in the egg of a cow the information for the production of antibodies against a particularly prevalent debilitating virus, that the gene was transcribed, translated and so on. Would it be worthwhile or economic? We cannot know until the full range of opportunities have been assessed, and the fundamental research has made further progress.

CROPS

The potentiality for agricultural advance by recombinant DNA technology has always seemed more likely with plants, because of the capacity to regenerate an entire individual from a single cell or protoplast. Plant protoplasts are produced experimentally by the enzymatic degradation of cell walls. The protoplast can regenerate the cellulose/hemicellulose walls, but while in the naked state will take up viruses or plasmids. Consequently the protoplast is the route by which transformation can be envisaged.

There is still uncertainty concerning the means by which alien genetic information may be incorporated in protoplasts, and in plants derived from them. Particular attention is being paid to the use of *Agrobacterium tumefaciens* and the cauliflower mosaic virus as vectors. Much remains to be learned, but if alien DNA is to be incorporated it may go into any of three genomes, those of the nucleus, chloroplast or mitochondrion. If transformation is to be an adjunct to conventional plant breeding then the incorporated DNA sequences must be transmitted through meiosis and fertilisation, except for those few crops that are propagated vegetatively. In view of the non-transmission in this way of the tumour inducing plasmid of *A. tumefaciens*, this may be a considerable obstacle for which research solutions should be sought.

In addition, it is important to recognise that when the transformation of crop plants by DNA technology is successful it will be only the first step in the exploitation of alien genetic information in crop production. Further manipulation by conventional cytogenetics or genetics may be necessary as well as normal plant breeding technology. Consequently there will probably be a long interval between the first transformation of a crop species and the exploitation of the resulting plant material in commerce. Nevertheless, at the Plant Breeding Institute in Cambridge we are so established that plant material can flow from transformation to cytogenetic analysis and manipulation and be fed into our breeding programmes. In organising this structure we have been concerned with the reality that a successful plant variety must have the optimal expression of numerous characters, and the modification of one by DNA technology is not alone a sufficient basis for varietal advance.

Recombinant DNA research may offer the potentiality for the modification of crops in new ways, but it is not always easy to see what characters might be incorporated from alien sources. Some general principles are clear however. The introduced DNA sequences should give clear phenotypic expression, and, therefore, be easily selected. This implies that they will probably be short sequences producing simple products, although it is probably not essential to know the first product of gene action.

A good example might be the capacity to detoxify the triazine herbicide. Maize detoxifies this by enzymatically substituting a hydroxyl group for a chlorine atom. Consequently the triazine herbicides can be used with great specificity in the maize crop, but no other species has a similar property. A useful outcome of applying DNA technology to plants would be the introduction into other crops of this characteristic of maize.

Another proposal is that resistance to fungal pathogens might be developed in plants by the incorporation in them of genetic information from the pathogen.

The proposal arises because in many interactions between host and pathogen there is resistance only when the pathogen carries at a specific locus an allele for avirulence, and the host carries at a specific locus an allele for resistance. From this it has been inferred that resistance derives from the interaction of the products of these two alleles. This may be illustrated as follows :-

		HOST	
		Resistant	Susceptible
PATHOGEN	Avirulent	No disease	Disease
	Virulent	Disease	Disease

If resistance is so caused it might be possible to include both alleles in the plant by transferring the pathogen allele by recombinant DNA technology as follows :-

1. Isolate shotgun DNA from avirulent fungal pathogen.

2. Incorporate fungal DNA in bacterial plasmid and clone plasmids in bacteria.

3. Make protoplast of virulent form of pathogen and incorporate bacterial/fungal plasmid.

4. Regenerate pathogen from plasmid-containing protoplast.

5. By tests for virulence on hosts determine which plasmids contain the dominant pathogen gene for avirulence.

6. Further engineer avirulence-containing plasmids so that they also contain a biochemical marker (e.g. drug resistance).

7. Make protoplasts from cells of crop host species using genotypes carrying allele giving resistance to the avirulent form of the pathogen.

8. Incorporate biochemically marked, avirulence-containing plasmid in host protoplasts.

9. Regenerate protoplasts and select for presence of the biochemical marker (e.g. drug resistance).

10. Final outcome will be plants incorporating the dominant fungal allele for avirulence and carrying the dominant host allele for resistance.

As a final example reference will be made to the use already being made in plant breeding of a piece of recombinant DNA research. It concerns the relationship between rye and wheat and the synthetic, potentially useful synthetic crop species - called Triticale - derived from hybrids between these two parental species. In the haploid genome of rye there is about 30% more DNA than in a haploid genome of diploid wheat. Much of this extra DNA of rye is in the form of repeated sequences not present in wheat and several of these have been cloned in *E.coli* plasmids. cRNA copies of cloned rye sequences have been hybridised *in situ* to wheat and rye chromosomes and demonstrated to be specific for rye chromosomes.

One such sequence which represents about 4% of the rye genome, is about 475 bases long and is present in rye in between half and three-quarters of a million copies. This sequence is located overwhelmingly in the telomeres of rye chromosomes. These telomeres are heterochromatic and late replicating in the cell cycle compared with the interstitial regions, or to wheat chromosomes. This late replication of the rye chromosomes disrupts the development of endosperm in *Triticale* causing "poor filling" of the grain.

The advantage of the molecular probe of cloned rye DNA is that not only can rye chromosomes be recognised in a wheat background, but also rye can be selected lacking the telomeric repeat sequences. When incorporated in *Triticale* such rye chromosomes will undergo a normal cycle of replication so removing the disadvantages of grain collapse.

This example of the current use of DNA technology has been left until last because so much talk in this field, especially in agriculture, is of jam tomorrow, and this is jam today. No doubt, in the future, more exciting examples will occur, but we are already, in a small way, gaining benefit in agricultural research from recombinant DNA.

DISCUSSION

P. STARLINGER: I have a question on nitrogen fixation, concerning the idea of introducing nitrogen fixation into other crop plants. As a layman, I do not understand why this is a goal. I mean, nitrogen, if fixed by the plant, can always be substituted by nitrogen fertiliser and as the raw material for nitrogen fertiliser is abundantly present the production of nitrogen fertiliser is mainly a question of energy. If you replace nitrogen fertiliser by a plant which, probably at a cost in yield, is fixing its nitrogen itself, you use prime soil for the conversion of solar energy and as this is a goal which, as you said, will be reached only some time from now, it will probably be reached at a time when prime soil will be very scarce and when energy may no longer be a problem. Why do we want to fix nitrogen in other plants than those which do it now?

R. RILEY: The straightforward answer to that is that in a great deal of the kind of research with which I am concerned, you research for options. One option is that there may be inadequate industrial energy for the fixation of atmospheric nitrogen. What we have to do is to work for a range of scenarios and one may well be right. Certainly you make a perfectly valid point that there is an energy cost to the plant in fixing nitrogen. I don't think we are yet in a position to say that we can ignore the attempt to transfer nitrogen fixing as a potential object.

M.G.P. STOKER: For Dr. Szybalski. Under what category is this work done and why?

R. RILEY: I'll answer the first part. It's done under a Williams' II.

S.N. COHEN: Along the lines of the last question, you mentioned that in agricultural research it is necessary to work within the organism that happens to be carrying out the steps that you are interested in, and that it often is not possible to work with an organism like *E. coli* K-12. It seems to me that one of the small ironies in the recombinant DNA debate over the years has been the relationship which currently exists in the various national guidelines regarding the use of *E. coli* K-12 versus other hosts. Initially, *E. coli* K-12 was thought by some to be a villain, thought to inhabit the gut, and thought to be a poor selection as a host for recombinant DNA experiments. Subsequently it has been recognised that *E. coli* K-12 is not the villain it was once thought to be and ironically, this host is now viewed by the guidelines as being the only, or almost only, safe organism to use. Since the kind of experiments you are talking about would require genetic manipulation in non K-12 organisms, particularly microorganisms for the fermentation work, I was wondering if you can put the use of non K-12 hosts in perspective by commenting upon the other kinds of genetic manipulation that are commonly carried out in such hosts for routine fermentation and agriculture work.

R. RILEY: I am not sure I fully grasp the question. However, we have in our portfolio of research work an example of *Streptomyces* - that's a good fermenting organism. Some of the work is genetic manipulation, a good deal, however, is straightforward genetics. Our purpose at the John Innes Institute in working with

Streptomyces is that this is the group which is the prime antibiotic producer. In *Streptomyces* there seems little difference between conventional genetics and recombinant DNA research. We feel that we need to know more about the fundamental genetics of this organism, in part for a reason which Dr. Starlinger referred to during the day, namely the worries about the extent to which antibiotics are used as a supplement to the feedstuffs of livestock and the consequence that this has on the development of resistance which may affect the human therapy.

W. SZYBALSKI: Are there any regulations for managing compost piles and manure fermentation? One could always imagine some dangers.

R. RILEY: Manure is a major environmental hazard, but primarily for odour. Agriculture research does a good deal of work on this. I am not aware of any instances of infections from manure fermentation.

W.J. PEACOCK: I would like to comment on the last point Dr. Riley made about the loss of heterochromatin. Incidentally this will stress the power of selection that Walter Bodmer talked about this morning. Dr. Appels and Dr. May in our laboratory in Canberra have analysed one of these apparent losses in heterochromatin from a rye chromosome in a largely wheat genome, and shown that in fact the "loss" is a translocation of a wheat segment on to the rye chromosome. The identification has been possible with wheat-specific probes. The segment has been identified to a particular chromosome arm. This selection of a rare translocation event may have occurred in the cases of apparent heterochromatin loss.

EXPRESSION OF FOREIGN GENES

S. D. Ehrlich and A. Goze

Laboratoire d'Etude des Acides Nucléiques,
Institut de Recherche en Biologie Moléculaire,
Université Paris VII, 2, Place Jussieu, Paris 75005, France

In vitro recombinant DNA technology has been used, in numerous studies, to introduce foreign genetic material into a new intra-cellular environment. The purpose of such interventions has often been isolation, and subsequent structural study of a defined DNA segment. Frequently, however, the main goal was to obtain expression of foreign genes in the new host cells.

Some of the foreign genes which confer upon their new host, at least qualitatively, the same phenotype as they did on their original hosts, are listed in Table 1. Such results were obtained relatively early, and were interpreted to mean that, in general, prokaryote and lower eukaryote genes will be expressed in a new bacterial host. This notion was, however, contradicted by the finding that some *E. coli* genes are not expressed in another prokaryote, *Bacillus subtilis* (Table 2).

TABLE 1 Foreign Genes Expressed in New Hosts

Phenotype	Original host	New host(s)	References
Ap^R	*S. aureus*	*E. coli*	(1)
Cm^R	*S. aureus*	*E. coli*, *B. subtilis*	(2,3)
Tc^R	*B. cereus*	*E. coli*, *B. subtilis*	(4)
Tc^R	*S. aureus*	*E. coli*, *B. subtilis*	(2,3)
Km^R	*S. aureus*	*E. coli*, *B. subtilis*	(5,6)
Sm^R	*S. aureus*	*B. subtilis*	(7)
thy^+	*B. subtilis*	*E. coli*	(8)
thy^+	*E. coli*	*B. subtilis*	(9)
leu^+	*B. subtilis*	*E. coli*	(10,11)
pyr^+	*B. subtilis*	*E. coli*	(11)
his^+	*S. cerevisiae*	*E. coli*	(12)
trp^+	*S. cerevisiae*	*E. coli*	(13)
arg^+	*S. cerevisiae*	*E. coli*	(13)
leu^+	*S. cerevisiae*	*E. coli*	(14)
aro^+	*N. crassa*	*E. coli*	(15)

Since the lack of expression was not due to alterations of *E. coli* genes in *B. subtilis* (they functioned when isolated from the new and re-introduced into the original host), these results meant that some barriers to heterospecific gene expression were encountered. What were these barriers, and more generally, what kind of barriers to expression of genes stably maintained in a foreign host one can expect to encounter? An exhaustive treatment of this vast problem is beyond the scope of the present discussion. We limit ourselves here to presenting a few illustrative examples.

TABLE 2 *E. coli Genes not Expressed in B. subtilis*

Phenotype	Reference
Ap^R	(3,4)
Km^R	(3,4)
Cm^R	(4)
Tc^R	(17)
trp^+	(18)

BARRIERS TO HETEROSPECIFIC GENE EXPRESSION

Expression of genetic information involves : (a) transcription of DNA; (b) processing of RNA; (c) translation of RNA; (d) processing of proteins. The correct execution of each of these steps depends on the correct interpretation by the host cell of the different signals carried by implicated macromolecules. Consequently, the barriers to heterospecific gene expression can be related to divergence of such signals between organisms. An exception deserves, however, to be mentioned, concerning the possible toxicity of a foreign gene product for its new host, where in spite of the correct interpretation of all the signals, the expression may not be observed.

TRANSCRIPTION OF DNA

Transcription of DNA into RNA is a complex biochemical reaction. It is performed by an extensively studied enzyme RNA polymerase (19,20). Two signal-dependent steps of this process, initiation and termination of transcription are schematically represented in Fig. 1. Two regions are of particular importance for initiation of transcription in *E. coli*. About 10 nucleotides before the site where the RNA synthesis starts, a sequence related to a motive described by Pribnow (21) is found. Some 25 nucleotides further another typical sequence exists (cf. 22). The efficiency of *E. coli* promoters seems to be related to their conformity to these two sequences, weaker promoters deviating more. A foreign gene may be devoid of these signals, and consequently not be transcribed in *E. coli*.

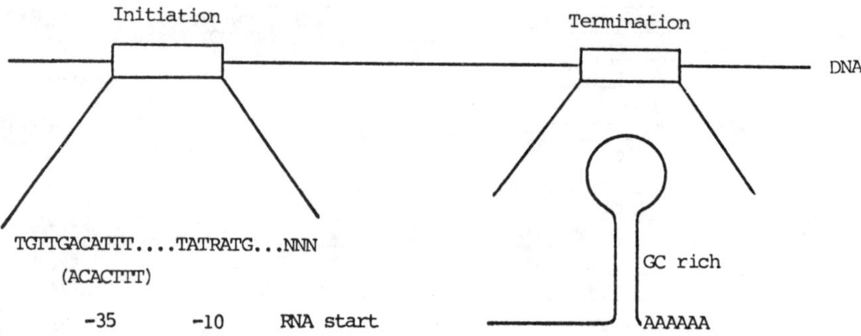

Fig. 1. *E. coli* transcription signals.

Different transcription initiation signals are possibly present in other organisms. The inability of B. subtilis RNA polymerase to recognize E. coli signals may be the reason for the lack of expression of E. coli genes in B. subtilis (23).

Termination of transcription is also a signal-dependent process (reviewed recently, 24). E. coli RNA polymerase terminates transcription at sites like the one presented schematically in Fig. 1.Less GC rich structures lacking the stretch of As can also be used as termination signals, but only in the presence of the termination protein Rho (24). Since the terminators are found not only at the end, but also within operons (24) their correct interpretation is important for the transcription of promoter-distal genes. Lack of heterospecific gene expression could be due to over-interpretation of such signals by a foreign RNA polymerase.

RNA PROCESSING

In both prokaryotes and eukaryotes long transcription products are found, which are subsequently processed to give the corresponding mature mRNAs. A well studied prokaryote example represented in Fig. 2 (25-27) is that of T7 mRNA. The long transcript is cleaved by RNase III at specific sites, which seem to be recognized due to their specific primary and secondary structure. Lack of processing may impair the translatability of mRNA (28).

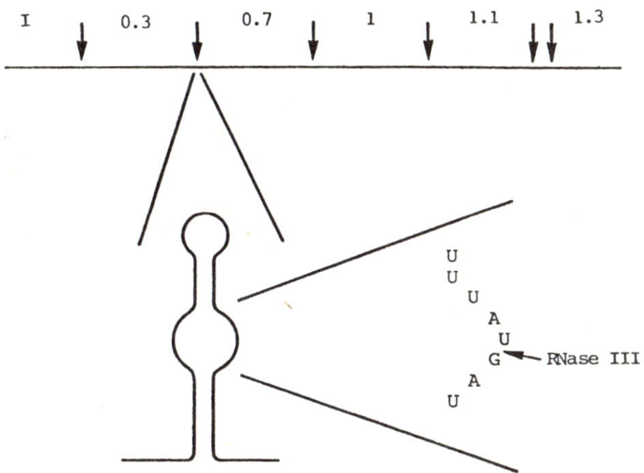

Fig. 2. T7 RNA processing.

RNA processing in eukaryotic systems, schematically shown in Fig. 3, is of paramount importance. It involves both the elimination of the region preceding the 5' end of the mature mRNA (29) and the excision, by "splicing" of the internal regions not coding for the gene product (30-35). Little is known at present about signals governing these processes, except that a sequence AGGU exists in several systems on the boundary of the excised and the preserved RNA (35). Correct processing of yeast tRNA genes in Xenopus laevis oocytes indicates evolutionary conservation of signal sequences among eukaryotes (29). The lack of splicing machinery in prokaryotes represents probably the most important single barrier to expression of eukaryote genes in such hosts.

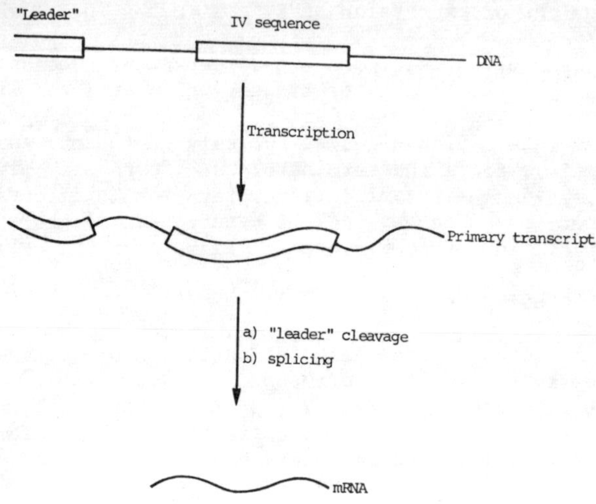

Fig. 3. RNA processing in eukaryotes

TRANSLATION OF RNA

Once the genetic information has been transcribed from DNA to RNA, and processed, it is translated into proteins through a cellular machinery composed of ribosomes, tRNAs and a variety of protein factors (36). In different organisms some of these components can diverge to a point where heterologous mRNA is no more efficiently translated. The best studied example is the B. stearotermophilus derived system, which did not translate E. coli phage f2 RNA (37,38). The phenomenon was correlated to the specificity of the 30S ribosome subunit of B. stearotermophilus, and possibly its lack of the counterpart of E. coli S1 ribosomal protein, which is essential for ribosome binding and translation of phage mRNAs (38a). Another 30S ribosomal protein, S12, has also been implicated in the specificity of translation (39). The signals which the 30S subunit proteins presumably recognize on mRNA are not known at present.

Sequences found about 10 nucleotides in front of the translation initiation codon RUG in E. coli mRNA, are complementary to a part of the 3' end of 16S rRNA (Fig. 4, ref. 40). The complementarity seems to contribute to the initial binding of ribosome to mRNA (38a), and its absence may impair the translatability of a foreign mRNA.

PROTEIN PROCESSING

Numerous proteins need to be modified by proteolytic cleavage in order to assume their physiological role. A recent review (41) lists many cases, examples being the conversion of proinsulin to insulin or trypsinogen to trypsin (42). The latter illustrates extreme specificity of the reactions of this type, since the activation is due to a single peptide bond cleavage (out of 228) by enterokinase, which is not known to cleave any other protein (43). It is not likely that the functional trypsin could be synthesized in very many foreign hosts into which the corresponding gene could be introduced.

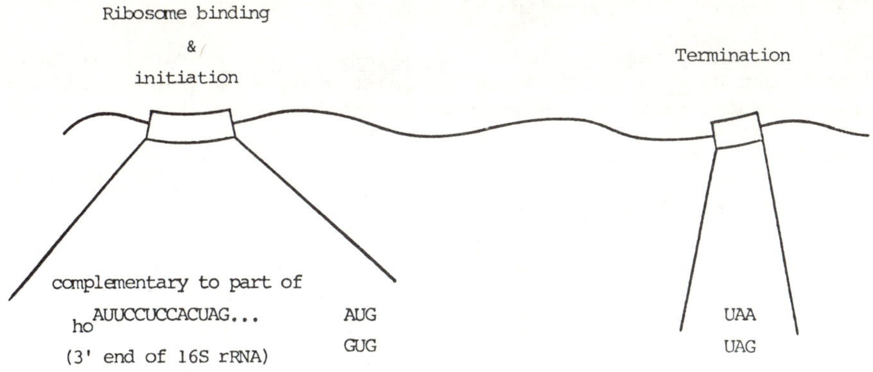

Fig. 4. E. coli translation signals

Another type of modification by proteolytic cleavage, which occurs concomitantly with the protein synthesis, is observed for proteins transported across membranes. These have been generally found to carry an amino-terminal signal sequence, thought to be responsible for initiating the passage of protein through the membrane (44). The comparison of signal sequence of a number of excreted proteins has revealed that, although none was identical, they all contained non-polar and hydrophobic amino acid residues (45). E. coli mutants unable to export maltose binding protein to the periplasm, probably because of a defective signal sequence of the protein, were recently isolated (46). This finding points to the sequence specificity of the excreting mechanism, and indicates that a sufficiently divergent signal sequence may be unrecognized in a heterospecific system. This explanation was put forward in the case of lack of expression of E. coli β-lactamase encoding gene in B. subtilis (W. Goebel, personal communication).

A phenomenon possibly related to the processing of proteins is their degradation. Short polypeptides resulting from nonsense mutations are rapidly degraded in E. coli, while the wild-type parental proteins are stable (47,48). It can be envisaged that the foreign proteins will be rapidly degraded if their configurations and/or aminoacid sequence does not protect them from intra-cellular nucleases, a possibility invoked to explain the initial lack of expression of somatostatin gene in E. coli (49).

OVERCOMING THE BARRIERS

Expression of a gene in a foreign host can be impaired for numerous reasons, which were discussed briefly above. It cannot therefore be assured by applying some generally valid strategy. Several useful rules can nevertheless be put forward.

1) Obtaining genetic information which encodes no more than the active protein. This obviates the need for in vivo processing of the gene product, either at the RNA or the protein level. Two approaches were successfully used to achieve this goal. The first consists of in vitro DNA synthesis, using mRNA as a template. In this way, genetic information processed correctly at the RNA level by a natural host is exploited. Recent examples include synthesis of mouse tetrahydrofolate reductase (50) and chicken ovalbumine (51) genes.

The second approach, applicable at present only to relatively short genes, is to synthesize de novo the desired gene, corresponding to the known protein sequence. Somatostatin (49) and human insulin (52) genes were created in this way.

2) Providing adequate transcription and translation signals. These can be added to the coding sequence in vitro, prior to joining the hybrid to the convenient cloning vector. Lac promoter region was often used for that purpose (e.g. 51). Alternatively, the signals carried by the cloning vector can be exploited by inserting the coding sequence in an appropriate position (50).

3) Assuring protection of foreign protein from degradation. This seems to be important in the case of small polypeptides synthesized in E. coli (49). Fusion between the host gene coding for the large protein, β-galactosidase, and the somatostatin gene, was successfully used for that purpose. The desired polypeptide, somatostatin, could be released from the chimeric protein by in vitro chemical cleavage (49). 9). The gene encoding β-lactamase, a protein endowed with signals for excretion from the host cell, was used similarly in the case of rat preproinsulin synthesis in E. coli (53,54).

An example of expression of a foreign gene in a new eukaryote host has recently been described (55). In this case, the coding sequence derived from rabbit globin mRNA was inserted in an SV40 vector, in a way which did not interfere with the SV40 RNA processing signals. When the chimera was introduced into cultured mouse kidney cells, the rabbit globin was synthesized.

HETEROSPECIFIC REPLICATION

Signals involved in the initiation of replication (56-64) represent a class of genetic information different from that which encodes RNA and protein products. The ability of various hosts to recognize such foreign signals is of a particular interest, since it may be necessary for heterospecific gene exchange in Nature.

Various plasmids are able to replicate in phylogenetically distant prokaryote hosts. Examples abound for Gram-negative bacteria (16) and exist also for Gram-positive ones (2). This indicates that such promiscuous plasmids may carry widely recognized replication initiation signals. In order to test the limits of heterospecific replication, we have examined the capacity of the small plasmid pC194, which functions in Gram-positive hosts S. aureus and B. subtilis, to replicate in a Gram-negative host, E. coli. For that purpose,we have introduced into an E. coli polAts strain the plasmid pHV14 (3), which is a hybrid between pC194 and E. coli plasmid pBR322 (65). The latter needs high amounts of DNA polymerase I for its replication, and is therefore rapidly lost from the polAts host at the restrictive temperature (Fig. 5, ref. 66). The hybrid plasmid pHV14 is 1000 times more stable under the same conditions. This is not due to the hypothetical synthesis of a pC194-encoded DNA polymerase I, since a) the strain carrying hybrid plasmid is still polAts; b) hybrid plasmid displayed no trans-effect (data not shown). The pBR322 moiety of the hybrid has not become polA-independent, as tested with the pBR322 reconstituted by the excision of pC194 part from the hybrid (Fig. 5).

Hybrid plasmid pHV14 could be introduced into the polA1 strain of E. coli by DNA transformation while pBR322 plasmid could not (Table 1). A deletion mutant of pHV14 which has lost the replication origin of pC194 (17, 67) did not transform polA1 strain either (Table 3) and was rapidly lost from the polAts host at the restrictive temperature (not shown).

These results indicate that at least some of the replication functions of pC194 are expressed in E. coli, in addition to being expressed in S. aureus and B. subtilis. Determination of nucleotide sequence of its replication origin may therefore give us some information relative to the replication initiation signals recognized in very distant hosts. The data also point to the fact that the potential for natural genetic exchange may be much greater than usually expected.

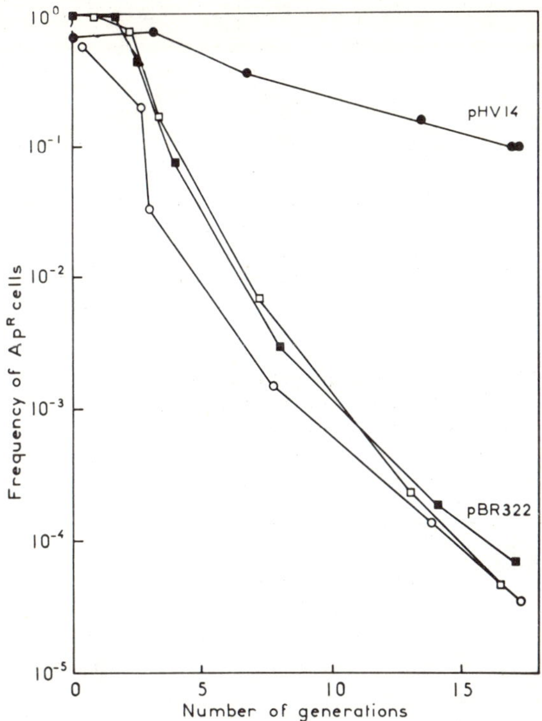

Fig. 5. Plasmid loss at the restrictive temperature from a polAts strain of E. coli.
Open circles refer to authentic, squares to re-constituted pBR322.

TABLE 3 Transformation of E. coli Cells with ColE1-derived Replicons

	pol^+	pol^-
pBR322	10^3	0
pHV14	10^3	50
pHV32	10^3	0

About 1 ng of plasmid DNA was used. Clones resistant to ampicillin (10 μg/ml) were selected. Average values from several experiments are shown.

CONCLUSION

Some genes are expressed in foreign hosts, while many need to be first adapted to the requirement of the gene-to-phenotype converting machinery of the new host. Successful adaptations, already reported, raise hope that the heterospecific expression will be attainable for almost any gene-host combination.

ACKNOWLEDGEMENTS

The authors are on the INSERM research staff. This work was supported, in part, by a grant from Université Paris VII, and a DGRST grant 78-7-0346.

REFERENCES

1) Chang, A.C. and Cohen, S.N. (1974) Proc. Nat. Acad. Sci. USA 72, 1030.
2) Ehrlich, S.D. (1977) Proc. Nat. Acad. Sci. USA 74, 1680.
3) Ehrlich, S.D. (1978) Proc. Nat. Acad. Sci. USA 75, 1433.
4) Kreft, K., Bernhardt, W. and Goebel, W. (1978) Mol. Gen. Gen. 162, 59.
5) Gryczan, T.J. and Dubnau, D. (1978) Proc. Nat. Acad. Sci. USA 75, 1428.
6) Grandi, G. Ehrlich, S.D. and Sgaramella, V. (1979) Atti. Assoc. Gen. It., in press.
7) Löfdahl, S., Sjöström, J.E. and Philipson, L. (1978) Gene 3, 149.
8) Ehrlich, S.D. et al. (1976) Proc. Nat. Acad. Sci. USA 73, 4145.
9) Rubin, E.M., Wilson, G.E. and Young, F.E. (1978) Proc. IV European Meeting on Transformations and Transfection, in press.
10) Mahler, J. and Halvorson, H.O. (1977) J. Bact. 131, 374.
11) Chi, N.Y., Ehrlich, S.D. and Lederberg, J. (1978) J. Bact. 132, 816.
12) Struhl, K., Cameron, J.R. and Davis, R.W. (1976) Proc. Nat. Acad. Sci. USA 73 1471.
13) Clarke, L. and Carbon, J. (1978) J. Mol. Biol. 120, 517.
14) Ratzkin, B. and Carbon, J. (1977) Proc. Nat. Acad. Sci. USA 74, 487.
15) Vapnek et al. (1977) Proc. Nat. Acad. Sci. USA 74, 3508.
16) Reaney, D.C. (1976) Bact. rev. 40, 552.
17) Primrose, S.B. and Ehrlich, S.D.,unpublished.
18) Goze, A. and Ehrlich, S.D., unpublished.
19) Losick, R. and Chamberlin, M. (eds.) (1976) RNA polymerase, Cold Spring Harbor Laboratory.
20) Lathe, R. (1978) Curr. Top. Microbiol. Imm. 83, 37.
21) Pribnow, D. (1975) Proc. Nat. Acad. Sci. USA 72, 784.
22) Calos, M.P. (1978) Nature 274, 762.
23) Ehrlich, S.D. and Sgaramella, V. (1978) Trends Biochem. Sci. 3, 259.
24) Adhya, S. and Gottesman, M. (1978) Ann. Rev. Biochem. 47, 967.
25) Dunn, J.J. and Studier, F.W. (1973) Proc. Nat. Acad. Sci. USA 70, 1559.
26) Robertson, M.D., Dickson, E. and Dunn, J.J. (1977) Proc. Nat. Acad. Sci. USA 74, 822.
27) Rosenberg, M. and Kramer, R. (1977) Proc. Nat. Acad. Sci. USA 74, 984.
28) Cannistraro, V.J. and Kennell, D. (1979) Nature 277, 407.
29) De Robertis, E.M. and Olson, M. (1979) Nature 278, 137.
30) Brack, C. and Tonegawa, S. (1977) Proc. Nat. Acad. Sci. USA 75, 5652.
31) Breathnach, R., Mandel, J.L. and Chambon, P. (1977) Nature 270, 314.
32) Aloni, Y. et al. (1977) Proc. Nat. Acad. Sci. USA 74, 3686?
33) Berget, S.M., Moore, C. and Sharp, P.A. (1977) Proc. Nat. Acad. Sci. USA 74, 3171.
34) Chow, L.T. et al. (1977) Cell 12, 1.
35) Sakano, et al (1979) Nature 277, 627.
36) Haselkorn, R. and Rothman-Denes, L.B. (1972) Ann Rev. Biochem. 42, 397.
37) Lodish, H.F. (1970) Nature, 226, 705.
38) Steitz, J.A. (1973) J. Mol. Biol. 73, 1.
38a) Steitz, J.A. et al. (1977) in "Nucleic Acid-Protein Interaction", ed. Vogel, H.J., Academic Press.
39) Held, N.A., Gelter, W.R. and Nomura, M. (1974) Biochemistry, 13, 2115.
40) Shine, L. and Delgarno, L. (1974) Proc. Nat. Acad. Sci. 71, 1342.
41) Neurath, H. and Walsh, K.A. (1976) Proc. Nat. Acad. Sci. USA 73, 3825.
42) Steiner, D.F. et al (1975) in "Proteases and Biological Control", eds. Reich, E. Rifkins, D.B. and Shaw, E. Cold Spring Harbor, Vol. 2, pp. 51-64.

43) Maroux, S.,Baratti, J. and Desnuelle, P. (1971) J. Biol. Chem.246, 5031.
44) Blobel, G. and Dobberstein, B. (1975) J. Cell. Biol. 67, 835.
45) Habener, J. et al. (1978) Proc. Nat. Acad. Sci. USA 75, 2616.
46) Bassford, P. and Beckwith, J. (1979) Nature 277, 538.
47) Goldschmidt, R. (1970) Nature 228, 1151.
48) Platt, T., Miller, J.H. and Weber, K. (1970) Nature 228, 1154.
49) Itakura, K. et al. (1977) Science 198, 1056.
50) Chang, A.C.Y. et al (1978) Nature 275, 617.
51) Mercereau-Puijalon, O. et al (1978) Nature 275, 505.
52) Crea, R. et al. (1978) Proc. Nat. Acad. Sci. USA 75, 5765.
53) Ulrich, A. et al. (1977) Science 196, 1313.
54) Villa-Komaroff, L. et al. (1978) Proc. Nat. Acad. Sci. USA 75, 3727.
55) Mulligan, R.C., Howard, B.H. and Berg, P. (1979) Nature 277, 108.
56) Ravetch, J.V., Horiuchi, K. and Zinder, N. (1977) Proc. Nat. Acad. Sci. USA 74, 4219.
57) Tomizawa, J.I., Ohmori, H. and Bird, R. (1977) Proc. Nat. Acad. Sci. USA 74, 1865.
58) Denniston-Thompson, K. et al. (1978) Science 198, 1051.
59) Grosschedl, R. and Hobom, G. (1979) Nature 621.
60) Fiddes, J.C., Borell, B.G. and Godson, G.N. (1978) Proc. Nat. Acad. Sci. USA 75, 1081.
61) Tijan, R. (1978) Cell 13, 165.
62) Crews, S. et al.(1979) Nature 277, 192.
63) Sugimoto, K. et al.(1979) Proc. Nat. Acad. Sci. USA 76, 575.
64) Meijer, M. et al (1979) Proc. Nat. Acad. Sci. USA 76, 580.
65) Bolivar, F. et al. (1977) Gene 2, 75.
66) Ehrlich, S.D. et al. (1978) in "Genetic Engineering", eds. Boyer, H.W. and Nicosia, S. pp. 25-38.
67) Niaudet, B. and Ehrlich, S.D. (1979) Plasmid 2, 48.

DISCUSSION

J.K. SETLOW: What did you say your explanation was for the fact that in the last slide the pHV14 in the *pol*$^-$ was a factor of 20 lower?

S.D. EHRLICH: Well, I don't really know - the question of transformation efficiency is complicated and we don't understand it really well. If I were to formulate a hypothesis I would think it may be related to the fact that the hybrid in the *polA* strain is not as stable as the parental plasmid is in the *pol*$^+$ plasmid because the replication is not really as efficient. This is indicated by the fact that under restrictive conditions the hybrid plasmid is still lost from the *PolAts* strain. It is much more stable than the parental one but is still lost. We have indirect evidence that the copy number is much lower under restrictive than under permissive conditions, where ColE1 replication functions are expressed. *PolAts* strain under restrictive conditions, or the *polA* strain carrying pHV14 is sensitive to chloramphenicol and that can be most easily explained by the difference in the plasmid copy number in the *pol*$^+$ relative to *pol*$^-$ strain. For the ampicillin it's the same thing. Ampicillin resistance in a *pol*$^+$ strain is 100 μg/ml, but about 10 μg only in a *PolAts* strain under restrictive conditions or in the *pol*$^-$ strain. Both results indicate there is a copy number question, and it may influence the transforming efficiency of the hybrid plasmid.

W.F. BODMER: On the assumption that mice and humans are foreign to each other, it might be worth pointing out, of course, that somatic cell hybrids prove that you in general get expression of virtually all foreign genes in those combinations under situations where one might have anticipated many of the problems you mention. In the early days of this sort of work, we did have those worries. For example, the human major histocompatability antigens, the HLA system, can be expressed very well in what essentially is a mouse cell, for gene products which presumably involve all sorts of processing, including signal sequences and so on. As far as I am aware, it's only perhaps with respect to mitochondrial function and, of course, differential chromosome segregation where there is some evidence of this sort of incompatability. There is a remarkable lack of problems, as far as I am aware, at least in hybrids between mammalian cells.

S.D. EHRLICH: One could conclude that *Bacillus subtilis* is much further away from *E. coli* than mouse is from man, I think. I have in my summary an example taken from Berg's group using SV40 - globin gene hybrid expressed in mouse cells.

W.J. PEACOCK: I think it's clear that the quality of the talks must have been high to have kept you alert for so long. I wish you to join me in thanking the three speakers.

PRACTICAL BENEFITS OF RECOMBINANT DNA RESEARCH

CHAIRMAN'S INTRODUCTION

W. J. Whelan

Department of Biochemistry, University of Miami School of Medicine, P.O. Box 016129, Miami, Florida 33101, U.S.A.

As we move into the third day, we come towards the end of the purely scientific presentations with the papers now to be given by Bill Rutter and Charles Weissmann.

Last night, Dr. Riley referred to remarks that I made on Sunday, quoting Thomas Sprat's history of The Royal Society, but I think Dr. Riley put the wrong emphasis on what Sprat was trying to say. Sprat was criticising those people who have time only for experiments which will bring immediate gain, and criticised those people who deride basic research. I do not think Sprat was deriding applied research *per se*, only the attitude that this is the only kind of research that should be done.

This morning we move into the area of applied aspects of recombinant DNA technology. Two years ago at a U.S. National Academy of Sciences meeting on recombinant DNA, one of the speakers was someone who is in the audience today, Irving Johnson. He talked about recombinant DNA technology in relation to its potential for producing valuable proteins and he spent a good deal of his time on the topic of insulin, demonstrating that the demands were approaching the total available supply. He then went on to indicate seven methods by which the pharmaceutical industry was considering meeting the demand for more insulin or devising alternative forms of therapy for diabetics. The last of these was insulin production by genetically manipulated cells. That was March 1977. Of course, it was known that people were already engaged in the possibility of using recombinant DNA technology to synthesise insulin. I do not suppose many people, certainly myself, appreciated how quickly this would be realised because we are talking about only two years ago. Within six months Herb Boyer and his colleagues had synthesised somatostatin and within eighteen months reports appeared of the synthesis of insulin by this technique, something that will surely loom large in the talk by Bill Rutter on the production of valuable proteins.

PRODUCTION OF "VALUABLE" PROTEINS IN ALTERNATE BIOLOGICAL HOSTS

W. J. Rutter

Department of Biochemistry and Biophysics, University of California, San Francisco, California 94143, U.S.A.

THE PROSPECTS

Recombinant DNA technology and the procedures of modern molecular biology provide in principle a means for producing proteins in alternate hosts. The methods are general, thus they offer the potential for producing any desired protein in any quantity. This paper focuses on the potential production of proteins that have some value to society outside of their intrinsic scientific interest. For the most part, the problems in their production are generic and the issues dealt with elsewhere in this symposium are therefore applicable here.

Currently a number of proteins are produced commercially for various purposes. These include, besides food products and additives, various enzymes and proteins, viral proteins used as vaccines and various protein hormones, and antibodies used for passive immunization. In many of these instances the protein represents some compromise between the most desired molecule and the one(s) for which there was an adequate (convenient) source. If the new technology is implemented, then an increase in the quality and quantity of available products can be anticipated. Perhaps the most exciting prospect is the range of novel compounds that may be specifically isolated or constructed for a variety of new purposes. In biomedical science, for example:

- Proteins which control growth and differentiation are frequently present in very low quantities in tissues. Thus they have been largely unavailable for research purposes let alone general biomedical use.

- Systematic variations in the structures of biologically active molecules such as hormones or enzymes may produce a more efficacious product. For example, a hormone analogue that binds to its receptor longer and thus has a more extended biological effectiveness (a superhormone) may be produced; or alternatively one that binds effectively to the receptor without biological activity (an anti-hormone) and hence can counteract natural overproduction.

Molecules with novel properties can be created such as proteins that transport a desired drug to an effective target site and specifically constructed attenuated viruses that will allow the production of desired antigens without the consequent production of pathological proteins. These "neutral" viruses then may help to immunize against a variety of viral infections for which there is no currently effective remedy.

In the chemical industry, many processes will employ enzymatic catalysts with an inherent increase in specificity, increased yields and lower energy demands. Eventually synthetic pathways providing starting materials and intermediates for industrial chemical processes will be developed within cells;thus, the nature of the processes may be substantially modified. More chemistry may be done via genetics in natural systems and less done in the classical chemical laboratory. Ultimately, efficient processes for conversion of light energy to hydrocarbons may aid substantially in the resolution of world-wide energy problems.

While these spectacular vistas seem on the relatively near horizon, many technical and theoretical problems must be resolved before the feasibility and the limits of this approach become fully evident.

THE PROCESS

The production of proteins by recombinant DNA technology requires a DNA molecule encoding the protein. This DNA is then integrated appropriately in a vector so that the expression can occur in the biological system to be employed. In bacteria the vehicles are typically plasmids or bacteriophages which replicate independently of the host chromosome; they frequently have been modified so that they can be maintained in the cells by selection procedures.

Coding Sequence

The coding sequence for the gene can be obtained by reverse transcription of the mRNA, by chemical synthesis or by isolation of a fragment genomic DNA containing the gene or a combination of procedures. Quite frequently proteins of interest, for example, hormones are present in very low concentrations in biological tissues and hence the problem of obtaining significant amounts of mRNA is not a trivial one. Strategies have typically involved the development of procedures for isolating particular cell types which produce or can be induced to produce relatively high levels of hormone (and presumably have high levels of the mRNA). Alternatively, tumours which continue to produce high levels of hormones and hence are enriched in mRNAs may be used. Methods for selecting a particular mRNA of interest from a crude mixture of mRNAs by specific priming by synthesized oligonucleotides is a significant new development (Noyes and co-workers, 1979).

To obtain a genomic DNA fragment containing a desired gene is usually a difficult problem because of the high molecular complexity of most genomes; it requires either a specific (usually cDNA) probe or a selection technique. Furthermore, many genes in eukaryotic organisms contain untranslated sequences which interrupt the coding sequences. These "intervening sequences" must be removed before a functional mRNA can be obtained. Bacteria apparently do not contain these cleavage-ligase systems. Usually genomic DNA is a practical source for the coding sequence when the gene does not contain intervening sequences and when the genome is relatively small (e.g. in viruses).

New methods in chemical synthesis coupled with enzymatic ligation allow the facile construction of any given polynucleotide sequence. Of course, a knowledge of the primary structure of the protein is required for the appropriate choice of codons. Although the methods in principle allow the synthesis of molecules of any size, synthetic procedures are obviously more practical for the smaller proteins (<100 amino acids).

Expressing Vehicle

For expression the DNA encoding the gene must be integrated into the DNA in an appropriate relationship to a promoter such that transcription proceeds along the correct strand and in phase with the appropriate initiator codon. In addition the sequence must contain a ribosomal binding site, a sequence complementary to a region of the smaller rRNA (16S RNA in bacteria or its equivalent in other systems). The extent of expression depends then upon the copy number of the DNA in the cells, the strength of the promoter and of the inducer (if the promoter is regulated) and the ribosomal binding sites. In the future artificial promoters and ribosomal binding sites may be constructed synthetically from knowledge of their general potency. These can then be employed on superexpressing vehicles. At present, the promoters present on certain bacteriophages (e.g. lambda, T5) and in other well studied operons (β-galactosidase, tryptophan synthetase) as well as the β-lactamase (penicillinase) found on certain bacterial plasmids are being used to advantage.

A different approach involves the construction of a polycistronic gene in which there is independent initiation of translation of the inserted gene. This was achieved fortuitously by Cohen, Schimke and colleagues (see Cohen's article in this volume). It may provide a general avenue for the expression of foreign proteins.

The construction of an expressing plasmid seems relatively straightforward; most problems can be handled with current technology and more sophisticated approaches will be developed in time.

The host cell. The expression of a particular gene in a given host cell may also be greatly affected by other factors. If any of these become significant problems, remedies appear available.

Codon Abundance. Because of the redundancy of the genetic code and because different organisms have different relative concentrations of the various tRNAs related to each of the codons, the tRNAs may become limiting for translation of a foreign protein. Then a phenomenon called "stuttering" can occur in which there is higher frequency of tRNA misreading, or premature termination (Parker and co-workers, 1978; O'Farrell, 1978). To solve the problem a synthetic gene may be constructed using the codon abundances of the host.

Protein conformation. There is good reason to believe that in a *particular system* the amino acid sequence of a molecule determines its conformation. The ability to form the functional configuration, however, may be dependent critically upon a number of environmental factors such as pH, ionic strength, oxidation-reduction potential, enzymes that catalyze the formation of SS bonds from SH bonds, and auxiliary proteins that may interact with the molecule to aid in the formation of a particular configuration. Alternate hosts may have substantially different internal environments, and the formation of certain foreign proteins might be prohibited (this seems likely in extreme cases such as in halophilic and thermophilic organisms, it may be important in others as well). Should this problem arise it should be able to shift to another organism with a more compatible internal environment. Alternatively, certain molecules may be unfolded and refolded accurately under *in vitro* conditions to form the native structure.

Degradation. There are several systems in bacteria that are active in the destruction of incomplete foreign peptides. Unfolded structures may be selectively destroyed in certain organisms; thus the rate of folding into a native configuration may influence the yield. Varying the culture conditions to change the synthetic rate or internal environment or selecting mutants deficient in degrading pathways (some are known in bacteria) represent approaches to these problems.

Selection. The utility of random selection is not to be underestimated. Selections for biological function, and for quantities of the desired molecule via scoring for immunological crossreactivity, for example, have all been used to advantage.

THE PRODUCT

Fused Proteins

Although it is in principle possible to produce precisely the desired polypeptide, in practice the formation of molecules in which this polypeptide is linked to one or more other amino acids is likely. In fact it appears that the degree of expression sometimes depends upon the molecule to which the polypeptide is fused. For example, large molecules frequently express more effectively when fused with a smaller peptide (ovalbumin, growth hormone) whereas the reverse also appears to be true (somatostatin, insulin). For these reasons the fused polypeptide as well as the amino acid sequences at the junction are of considerable interest. It should be possible to make the fused polypeptide advantageous, for example, it could be useful in some aspect of the preparative procedure prior to its cleavage. Thus if it were selected to bind to a particular affinity matrix the protein of interest might be purified with greater facility.

In most instances it will be necessary to cleave the peptides. In restricted cases a specific chemical cleavage reaction can be utilized. For example, when methionine residues are not present in the desired polypeptide, and are present at the junction, they can be cleaved with cyanogen bromide to yield a homoserine C-terminus, and the desired polypeptide with a free NH_2-terminus. Specific chemical cleavage is not a general solution, since the specificity of the reactions depend largely upon residues immediately adjacent to the bond being cleaved. More general solutions depend upon additional requirements for cleavage specificity. There are endopeptidases with complex specificities, e.g. plasmin, thrombin, collagenase, etc. It should be possible to synthesize nucleotide linkers that code for the amino acids required for cleavage and join them to the gene such that the fused protein produced can be cleaved precisely at the N-terminus.

PROCESSING

Many functional proteins are produced by selective processing of a precursor. When a non-chemically synthesized cDNA or gene is used to code the desired protein then the sequence can be modified to eliminate the undesired peptide if it cannot be eliminated by other means.

Products of hormone mRNA invariably include an additional amino acid sequence on the N-terminus. This "prepeptide" is apparently a signal for secretion, and is naturally removed during the vectorial transport of the synthesized molecule into the endoplas reticulum prior to the processes of packaging and secretion. Such "signal peptides" exist in bacterial cells and are cleaved similarly by enzymes present in membranes. It is possible, though perhaps not likely, that the signal peptides present in non-bacterial genes can direct the transport of synthesized proteins into the periplasmic or extracellular space concomitant with the cleavage of the peptide.

We have considered primarily the synthesis of simple proteins. In instances where modification of the protein is required for activity and does not occur in the alternate host, then a secondary process *in vivo* or *in vitro* must be developed.

EXPERIMENTAL PARADIGMS

Although questions of efficiency become a major concern in any practical process, the problems now being faced are qualitative ones. The currently favoured experimental systems involve the production of hormones that are in short supply. Thus production of even small quantities of these molecules represents a substantial advance. In addition certain genes of viruses (e.g. hepatitis B) that cannot be cultivated *in vitro* may be cloned and expressed in alternate hosts. The protein in turn may be used to produce an effective vaccine. Furthermore the potent general antiviral agent interferon (normally found in cells in infinitesimal quantities) is the subject of other investigations. The first reported successful studies on expression carried out by Itakura, Riggs, Boyer and colleagues (1977) involved the chemical synthesis of the coding sequence, the formation of a chimera near the C-terminus of the β-galactosidase gene with a methionine codon at the junction (somatostatin contains no methionine), synthesis in bacteria and subsequent cleavage by cyanogen bromide. A product with somatostatin immunoreactivity and biological activity was detected.

Insulin

Experiments on the expression of insulin have been carried out by several groups (Rutter, Goodman and colleagues; Itakura, Riggs and colleagues; Gilbert and colleagues). The insulin molecule is composed of two chains, A and B. The immediate precursor, proinsulin, however, is a single polypeptide in which the two chains are joined by a connecting peptide which is removed by processing. The ultimate precursor, however, contains an additional prepeptide joined at the N-terminus which is removed during transport into the endoplasmic reticulum. The cloned rat preproinsulin cDNA fragment was introduced into a position in the β-galactosidase gene near the C-terminus of the β-galactosidase. A polypeptide larger than β-galactosidase and crossreacting with anti-insulin antibodies was produced (Ullrich and co-workers, 1978).

Insulin cDNA has also been inserted into the β-lactamase gene of a bacterial plasmid (Villa-Komaroff and co-workers, 1978). Selection of clones using a sensitive immunological screening technique resulted in the discovery of a number of separate clones producing molecules crossreacting with anti-insulin. More recently cloned human preproinsulin cDNA has been obtained (Bell and co-workers, 1979) so that experiments on its expression can proceed. Efforts to produce human insulin using the native gene naturally must deal with the problem of elimination of the prepeptide sequence (yielding proinsulin) and the subsequent conversion of proinsulin to insulin itself (a process that can occur by known enzymatic procedures).

Itakura, Riggs and their colleagues (Goeddel and co-workers, 1979) have employed another strategy. Nucleotides coding for the B and A chains were synthesized and were each linked separately to the end of the β-glactosidase gene (as in somatostatin) to form a chimera. Independent synthesis of the sequences of the B and A chains were obtained. Subsequent cleavage of the fused proteins by cyanogen bromide yielded the B and A chains which were then combined by chemical methods to produce a molecule with insulin immunoreactivity. This very promising result may be the first successful synthesis of a "valuable" molecule in an alternate host.

Growth Hormone

Attempts to synthesize growth hormone in bacteria have been made by Goodman, Baxter and colleagues. Rat growth hormone and human growth hormone cDNAs have been cloned in bacteria. The rat cDNA has been inserted into the β-lactamase gene. This resulted in the formation of a fused peptide that crossreacts with anti-growth hormone antibodies (Seeburg and co-workers, 1978). The human growth hormone cDNA has been inserted in the appropriate phase into the *trp* gene. Expression of a linked protein has been achieved which crossreacts with anti-human growth hormone (Martial and co-workers, 1979).

These results suggest that molecules resembling natural hormones can be produced in bacteria. Encouraging as they are, however, additional experiments are required before it can be concluded that the molecules are synthesized with fidelity. In most instances the assay for the protein product has been only by immunological procedures. It is well known that immunological crossreactivity is no guarantee of molecular homogeneity or biological activity. Additional experiments to confirm the fidelity of the translation process are in progress in several laboratories.

The general view is optimistic: in the near future human hormones and other useful proteins will be produced in bacteria.

REFERENCES

Bell, G.I., W.F. Swain, R. Pictet, H.M. Goodman, W.J. Rutter (1979). Submitted to *Nature*.

Goeddel, D.V., D.G. Kleid, F. Bolivar, H.L. Heyneker, D.G. Yansura, R. Crea, T. Hirose, A. Kraszewski, K. Itakura and A.D. Riggs (1979). *Proc. Natl. Acad. Sci. USA*, 76, 106-110.

Itakura, K., T. Hirose, R. Crea, A.D. Riggs, H.L. Heyneker, F. Bolivar and H.W. Boyer (1977). *Science*, 198, 1056-1063.

Martial, J.A., R.A. Hallewell, J.D. Baxter and H.M. Goodman (1979). *Science*, in press.

Noyes, B.E., M. Mevarech, R. Stein and K.L. Agarwal (1979). *Proc. Natl. Acad. Sci. USA*, 76, 1770-1774.

O'Farrell, P.H. (1978). *Cell*, 14, 545-557.

Parker, J., J.W. Pollard, J.D. Friesen and C.P. Stanners (1978). *Proc. Natl. Acad. Sci. USA*, 75, 1091-1095.

Seeburg, P.H., J. Shine, J.A. Martial, R.D. Ivarie, J.A. Morris, A. Ullrich, J.D. Baxter and H.M. Goodman (1978). *Nature*, 276, 795-799.

Ullrich, A., J. Shine, P.H. Seeburg, E. Tischer, R. Pictet, B. Cordell, G.I. Bell, J.M. Chirgwin, W.J. Rutter and H.M. Goodman (1978). In: *Proc. Sym. Proinsulin, Insulin, C-Peptide, Japan*, 1978, in press.

Villa-Komaroff, L., A. Efstratiadis, S. Broome, P. Lomedico, R. Tizard, S.P. Naber, W.L. Chick, W. Gilbert (1978). *Proc. Natl. Acad. Sci. USA*, 75, 3727-3731.

THE EXPRESSION OF A CLONED RABBIT CHROMOSOMAL ß-GLOBIN GENE IN MOUSE L CELLS AND YEAST

A. van Ooyen, N. Mantei, J. van den Berg *,
J. D. Beggs * *, W. Boll, R. F. Weaver and C. Weissmann

Institut für Molekularbiologie I, Universität Zürich, 8093, Switzerland
***Present address: Gist-Brocades, Delft, The Netherlands**
****Present address: Plant Breeding Institute, Cambridge, U.K.**

A promising strategy for elucidating the mechanism of expression and its control in eukaryotic cells involves the cloning and analysis of a gene and its surrounding regions, *in vitro* modification of the cloned DNA and its re-introduction into a eukaryotic cell (Weissmann, 1978). In this paper we show that a cloned chromosomal rabbit β-globin DNA fragment can be introduced and maintained in mouse L cells and that it is transcribed to yield a β-globin mRNA indistinguishable from the rabbit reticulocyte mRNA. In contrast, yeast cells transformed with the rabbit β-globin DNA yield β-globin-specific RNA which is not spliced and has abnormal 5' and 3' termini.

TRANSFORMATION OF MOUSE L CELLS WITH RABBIT β-GLOBIN DNA

Fig. 1 shows the nucleotide sequence of the cloned rabbit β-globin gene contained in the hybrid plasmid Z-pCRI/RchrβG-1 (RβG for short) (van den Berg and co-workers, 1978), as well as 223 and 109 nucleotides preceding and following it, respectively (van Ooyen and co-workers, 1979).

The plasmid RβG and a hybrid plasmid containing the thymidine kinase (TK) gene of Herpes simplex I virus (Boll and co-workers, 1979; Weissmann and colleagues, 1979) were both cleaved at their single *Sal*I sites and concatenated with DNA ligase. TK negative (TK^-) mouse L cells were transformed (Wigler and co-workers, 1977) with the concatenates, and TK positive (TK^+) clones were selected (Szybalski, Szybalska and Ragni, 1962). DNA extracted from 21 clones was cleaved with *Pst*I and *Eco*RI, the fragments were separated by agarose gel electrophoresis and transferred to a Millipore membrane by the Southern (1975) technique. Hybridization with a rabbit β-globin-specific ^{32}P-DNA probe demonstrated the presence of the expected 1200 and 900 base pair β-globin DNA fragments in 19 of 21 TK^+ cell lines. We estimate that the different cell lines contained about 0.25 to 20 copies of rabbit β-globin DNA per cell (data not shown).

IDENTIFICATION AND CHARACTERIZATION OF RABBIT β-GLOBIN SPECIFIC TRANSCRIPTS FROM TRANSFORMED MOUSE L CELLS

In vivo transcription of the extraneous rabbit β-globin gene was monitored by a modification of the Berk-Sharp procedure (Berk and Sharp, 1977) developed by Weaver and his colleagues (1979), as exemplified in Fig. 2. A specific probe was prepared as follows. The hybrid plasmid RβG contains a *Bgl*II site immediately following the 3' end of the β-globin coding sequence. The hybrid was cleaved with *Bgl*II and labelled with ^{32}P at the 5' termini. After digestion with *Pst*I, the *Pst*I-*Bgl*II (1291/1299) fragment was isolated. The labelled probe was hybridized to

TAGCAATTAGTACTGCTGGTATGGGTCTGGGAGATACATAGAAGGAAGGCTGAGTCTGTCAGACTCCTAAGCCATTGCCATAACTGCCAA
-220 -210 -200 -190 -180 -170 -160 -150 -140

PstI BspI
GGACAGGGGTGCTGTCATCACCCAGACCTCACCCTGCAGAGCCACACCCTGGTGTTGGCCAATCTACACACGGGGTAGGGATTACATAGT
-130 -120 -110 -100 -90 -80 -70 -60 -50

PvuII Cap
TCAGGACTTGGGCATAAAAGGCAGAGCAGGGCAGCTGCTGCTTACACTTGCTTTTGACACAACTGTGTTTACTTGCAATCCCCCAAAACA
-40 -30 -20 -10 0 10 20 30 40

MboII BspI
GACAGAATGGTGCATCTGTCCAGTGAGGAGAAGTCTGCGGTCACTGCCCTGTGGGGCAAGGTGAATGTGGAAGAAGTTGGTGGTGAGGCC
Met Val His Leu Ser Ser Glu Glu Lys Ser Ala Val Thr Ala Leu Trp Gly Lys Val Asn Val Glu Glu Val Gly Gly Glu Ala
50 60 70 80 90 100 110 120 130

CTGGGCAGGTTGGTATCCTTTTTACAGCACAACTTAATGAGACAGATAGAAACTGGTCTTGTAGAAACAGAGTAGTCGCCTGCTTTTCTG
Leu Gly Ar
140 150 160 170 180 190 200 210 220

TaqI
MboII
CCAGGTGCTGACTTCTCTCCCCTGGGCTGTTTTCATTTTCTCAGGCTGCTGGTTGTCTACCCATGGACCCAGAGGTTCTTCGAGTCCTTT
gLeu Leu Val Val Tyr Pro Trp Thr Gln Arg Phe Phe Glu Ser Phe
230 240 250 260 270 280 290 300 310

GGGGACCTGTCCTCTGCAAATGCTGTTATGAACAATCCTAAGGTGAAGGCTCATGGCAAGAAGGTGCTGGCTGCCTTCAGTGAGGGTCTG
Gly Asp Leu Ser Ser Ala Asn Ala Val Met Asn Asn Pro Lys Val Lys Ala His Gly Lys Lys Val Leu Ala Ala Phe Ser Glu Gly Leu
320 330 340 350 360 370 380 390 400

AluI AluI BamHI
AGTCACCTGGACAACCTCAAAGGCACCTTTGCTAAGCTGAGTGAACTGCACTGTGACAAGCTGCACGTGGATCCTGAGAACTTCAGGGTG
Ser His Leu Asp Asn Leu Lys Gly Thr Phe Ala Lys Leu Ser Glu Leu His Cys Asp Lys Leu His Val Asp Pro Glu Asn Phe Arg
410 420 430 440 450 460 470 480 490

AGTTTGGGGACCCTTGATTGTTCTTTCTTTTTCGCTATTGTAAAATTCATGTTATATGGAGGGGGCAAAGTTTTCAGGGTGTTGTTTAGA
500 510 520 530 540 550 560 570 580

MboII HinII
ATGGGAAGATGTCCCTTGTATCACCATGGACCCTCATGATAATTTTGTTTCTTTCACTTTCTACTCTGTTGACAACCATTGTCTCCTCTT
590 600 610 620 630 640 650 660 670

AluI
ATTTTCTTTTCATTTTCTGTAACTTTTTCGTTAAACTTTAGCTTGCATTTGTAACGAATTTTTAAATTCACTTTTGTTTATTTGTCAGAT
680 690 700 710 720 730 740 750 760

TGTAAGTACTTTCTCTAATCACTTTTTTTTCAAGGCAATCAGGGTATATTATATTGTACTTCAGCACAGTTTTAGAGAACAATTGTTATA
770 780 790 800 810 820 830 840 850

ATTAAATGATAAGGTAGAATATTTCTGCATATAAATTCTGGCTGGCGTGGAAATATTCTTATTGGTAGAAACAACTACATCCTGGTCATC
860 870 880 890 900 910 920 930 940

BspI
HpaII
ATCCTGCCTTTCTCTTTATGGTTACAATGATATACACTGTTTGAGATGAGGATAAAATACTCTGAGTCCAAACCGGGCCCCTCTGCTAAC
950 960 970 980 990 1000 1010 1020 1030

MboII AluI EcoRI
CATGTTCATGCCTTCTTCTTTTTCCTACAGCTCCTGGGCAACGTGCTGGTTATTGTGCTGTCTCATCATTTTGGCAAAGAATTCACTCCT
Leu Leu Gly Asn Val Leu Val Ile Val Leu Ser His His Phe Gly Lys Glu Phe Thr Pro
1040 1050 1060 1070 1080 1090 1100 1110 1120

BspI BglII
CAGGTGCAGGCTGCCTATCAGAAGGTGGTGGCTGGTGTGGCCAATGCCCTGGCTCACAAATACCACTGAGATCTTTTTCCCTCTGCCAAA
Gln Val Gln Ala Ala Tyr Gln Lys Val Val Ala Gly Val Ala Asn Ala Leu Ala His Lys Tyr His
1130 1140 1150 1160 1170 1180 1190 1200 1210

pA
AATTATGGGGACATCATGAAGCCCCTTGAGCATCTGACTTCTGGCTAATAAAGGAAATTTATTTTCATTGCAATAGTGTGTTGGAATTTT
1220 1230 1240 1250 1260 1270 1280 1290 1300

TTGTGTCTCTCACTCGGAAGGACATATGGGAGGGCAAATCATTTAAAACATCAGAATGAGTATTTGGTTTAGAGTTTGGCAACATATGCC
1310 1320 1330 1340 1350 1360 1370 1380 1390

Fig. 1. The nucleotide sequence of a rabbit β-globin gene and its flanking regions.

Position 1 corresponds to the capped nucleotide of the β-globin mRNA. The five nucleotides marked by an asterisk have not been reliably determined. "Cap" and "pA" designate the positions of the capped and the polyadenylated nucleotides, respectively.

to the RNA sample, treated with nuclease S_1 (Berk and Sharp, 1977) and analysed by polyacrylamide gel electrophoresis (cf. Fig. 2). Mature rabbit β-globin mRNA protects a fragment of 134 nucleotides length, from the *Bgl*II site to the position where the sequences of β-globin cDNA and β-globin chromosomal DNA no longer match. A precursor containing both introns protects 1200 nucleotides and a precursor containing only the large intron, 929 nucleotides. The probe discriminates between mouse and rabbit globin mRNA because the rabbit mRNA has a *Bgl*II recognition sequence (Efstratiadis, Kafatos and Maniatis, 1977) while mouse mRNA does not (N. Mantei, unpublished results and Konkel, Tilghman and Leder, 1978). RNA from transformed cell lines P1/R-3, P1/R-4, P1/R-5 and M2/R-1 yielded as major product the 134-nucleotide fragment. Thus, the majority of globin-specific RNAs are partly or entirely devoid of the large intron. RNA from cells transformed with only a TK plasmid gave no detectable bands. We estimate that cell line M2/R-1 contains about 2200 rabbit β-globin RNA molecules per cell. The absence of the small intron was determined by a similar approach, using a probe labelled at the *Bam* cleavage site (data not shown).

The 5' terminus of the β-globin-specific transcripts was determined as shown in Fig. 3. Hybridization of the appropriate probe to either authentic rabbit β-globin mRNA or to transformed cell RNA gave rise to a labelled fragment about 128 nucleotides long. Thus, the only detectable 5' terminus of the β-globin transcripts mapped in the same position as that of authentic β-globin mRNA. About 80% of the β-globin transcripts had a poly$(A)^+$ tail (data not shown).

The mobility of the rabbit β-globin-specific RNA present in the cell line M2/R-1 was compared with that of natural rabbit globin mRNA using the "Northern transfer" procedure of Alwine and co-workers (1977). Poly$(A)^+$ RNA derived from m2/R-1 RNA gave an autoradiographic band at the same position and with about the same intensity as 8ng of authentic rabbit globin mRNA (Fig. 4A).

To determine whether or not transformed L cells produce mouse β-globin mRNA, we again used the Weaver-Berk-Sharp assay, with the 5'-^{32}P-labelled 123-nucleotide *Bsp-Bsp* minus strand fragment of cloned mouse β-globin cDNA (Curtis, Mantei and Weissmann, 1977) as probe. 0.1ng of mouse β-globin mRNA gave a strong band while 10ng rabbit globin mRNA gave no signal. 50μg total RNA from L cell lines transformed with TK DNA only or with rabbit β-globin-TK DNA concatenates gave no detectable signal (data not shown). From this experiment we estimate that the transformed L cell lines we tested, in particular the one producing >2000 molecules rabbit β-globin mRNA (M2/R-1), contained less than 50 strands per cell of mouse β-globin mRNA.

TRANSFORMATION OF YEAST WITH CLONED RABBIT CHROMOSOMAL β-GLOBIN DNA

A hybrid DNA, pJDB219, consisting of pMB9, the 2 μ yeast plasmid and the *leu*2 gene of yeast was constructed by Beggs and shown to be maintained predominantly in plasmid form in yeast (Beggs, 1978). The rabbit β-globin DNA sequence of RβG was excised with *Hha*I (van den Berg and colleagues, 1978) and linked into the *Hpa*I site of the pMB9 moiety of pJDB219 by the dG-dC tailing procedure (cf. Lobban and Kaiser, 1973; Jackson, Symons and Berg, 1972). Two hybrid plasmids, Z-pJDB219/RchrβG-1 and Z-pJDB219/RchrβG-3 (μRβG-1 and μRβG-3 for short), containing the β-globin *Hha*I fragment were used for the transformation experiments.

S. cerevisiae MC16 *leu*$^-$ was transformed with both plasmids and leucine autotrophic colonies were isolated. One clone from each transformation was examined in greater detail. Total DNA from the transformed yeast clones, as well as the original hybrid plasmid DNAs were digested with *Pst*I and *Bgl*II and subjected to agarose gel electrophoresis. The DNA fragments were transferred to nitrocellulose

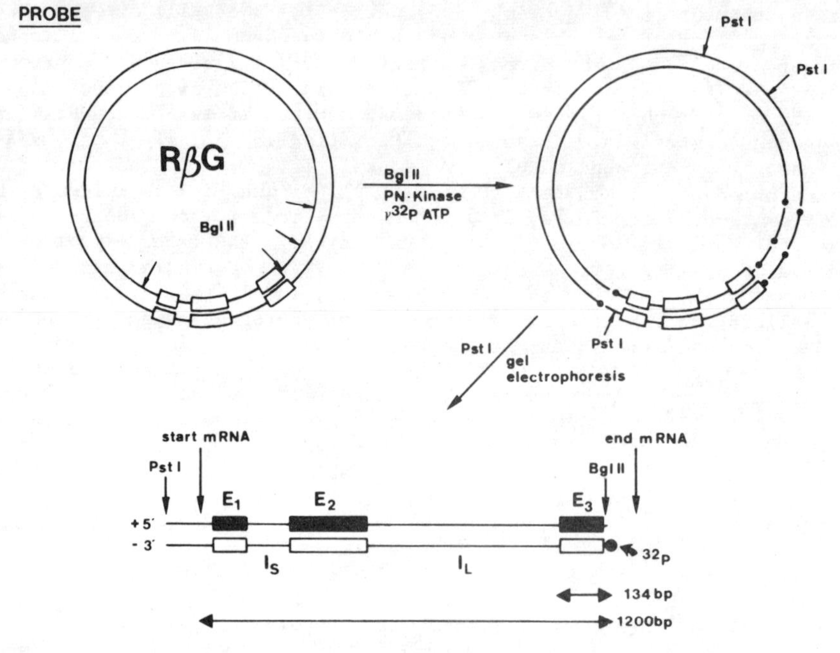

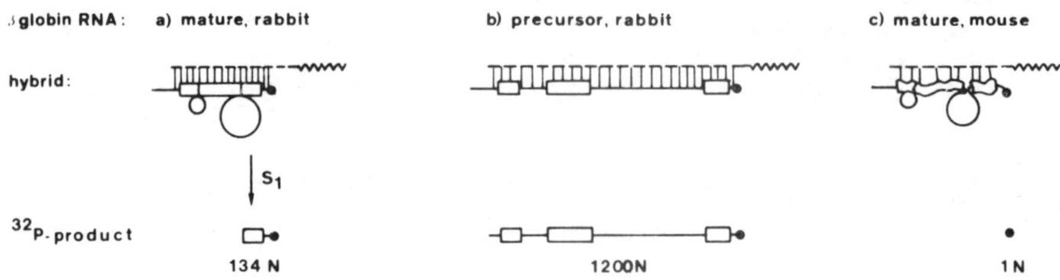

Fig. 2. Detection and characterization by the Weaver-Berk-Sharp assay of rabbit β-globin RNA in transformed mouse L cells.

(A) The procedure is a modification of the method of Berk and Sharp (1977). The probe is a DNA restriction fragment ^{32}P-labelled at one terminus (rather than being uniformly labelled). The denatured DNA is hybridized to the RNA sample in 80% formamide, digested with S_1 nuclease, denatured, and analyzed by polyacrylamide gel electrophoresis in 7M urea. Only if the labelled terminus of the probe was hybridized to RNA is a labelled DNA fragment recovered. The *Pst*-*Bgl*II (1291/1299) fragment derived from the rabbit chromosomal β-globin plasmid (cf. Fig. 1), labelled at the *Bgl*II site, is diagnostic for plus strand β-globin sequences and allows the distinction between rabbit precursor RNA, rabbit mature RNA, and mouse RNA. E, exon; I_S and I_L, small and large intron, respectively.

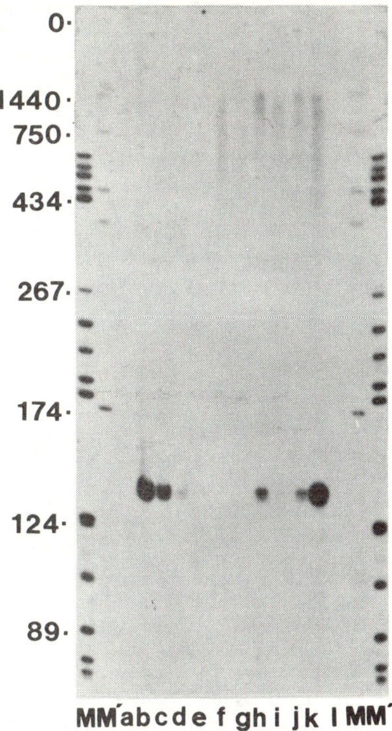

Fig. 2. Detection and characterization by the Weaver-Berk-Sharp assay of rabbit β-globin RNA in transformed mouse L Cells.

(B) Total cell RNA was prepared from about 10^8 cells as described (Curtis and Weissmann, 1976). The assay was carried out as outlined in (A).

Lanes a to e, 0, 3, 1, 0.3 and 0.1ng rabbit (α+β) globin mRNA. Lanes f and g, RNA from cell lines P1/S-1 and P1/S-2 (transformed only with TK-DNA plasmid). Lanes h to k, RNA from cell lines P1/R-3, P1/R-4, P1/R-5 and M2/R-1 (transformed with rabbit β-globin DNA linked to TK DNA. Lane 1, 50 ng mouse globin mRNA. Lanes M and M', 5'- ^{32}P-labelled fragments of plasmid pBR322 (Sutcliffe, 1978) as size markers.

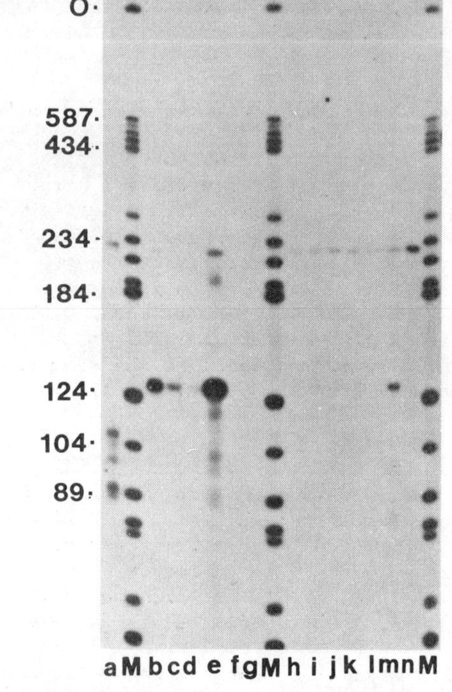

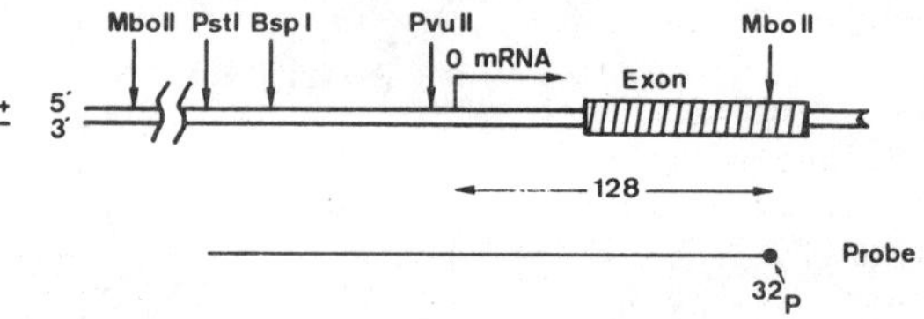

Fig. 3. Mapping of the 5' terminus of rabbit β-globin-specific RNA from transformed mouse L cells and transformed yeast.

The assay was as described in Fig. 2. The probe for mapping the 5' end (cf. scheme) was prepared as follows. Plasmid RβG was digested with *Hha*I, the 6.4 kilobase fragment isolated, cleaved with *Mbo*II, and 5'-terminally labelled with ^{32}P (specific activity, 3200 Ci/mmole). After cleavage with *Pst*I, the 223 bp fragment was isolated by agarose gel electrophoresis. Lane a, RNA from yeast cells transformed with Z-pJDB219/RchrβG-3. Lanes b to d, 3, 1, and 0.3 ng rabbit (α+β) globin mRNA. Lane e, 20 μg rabbit bone marrow RNA (a gift from R. Flavell). Lane f, no RNA. Lane g, 50 ng mouse globin mRNA. Lanes h and i, RNA from cell lines P1/S-1 and P1/S-2, transformed with a *Sal*I-cleaved TK-gene plasmid only. Lanes j to m, RNA from cell lines P1/R-3, P1/R-4, P1/R-5 and M2/R-1, transformed with the rabbit β-globin DNA-containing hybrid linked to a TK-gene plasmid. Lane n, DNA probe not digested with S_1 nuclease. M. size marker: pBR322 digested with *Bsp*I and 5'-terminally ^{32}P-labelled.

filters (Southern, 1975) and hybridized to ^{32}P-labelled, cloned rabbit β-globin cDNA. The pattern of globin-specific bands given by the transformed yeast DNA was the same as that of the original β-globin DNA-containing plasmid RβG (data not shown).

CHARACTERIZATION OF β-GLOBIN-SPECIFIC TRANSCRIPTS IN YEAST

To determine whether globin-specific RNA had been synthesized in the transformed yeast cells, total RNA was subjected to agarose gel electrophoresis under denaturing conditions, transferred to diazobenzyloxymethyl paper (Alwine, Kemp and Stark, 1977), and hybridized with ^{32}P-labelled, cloned β-globin cDNA. As shown in Fig. 4B, a strong β-globin-specific band with a mobility corresponding to an 850-nucleotide RNA was found in yeast cells transformed with either μRβG-1 or μRβG-3. This RNA is about 100 nucleotides longer than mature rabbit β-globin mRNA, but 600 nucleotides shorter than the 15S β-globin precursor RNA (about 1450 nucleotides). By comparing the intensity of the globin-specific RNA band with that of known amounts of β-globin mRNA, we estimate that about 0.01% of the total yeast RNA is β-globin-specific.

To map the 5' terminus of the transcript, the *Mbo*II-labelled probe described above cf. Fig. 3) was used in the Weaver-Berk-Sharp assay. As shown in Fig. 3, natural rabbit β-globin mRNA protected a DNA fragment about 128 nucleotides in length; the DNA protected by RNA from β-globin-transformed yeast cells gave five major bands, from 88 to 110 nucleotides. The β-globin-specific transcripts from yeast apparently have several 5' termini, all of which map downstream from the position corresponding to the 5' terminus of natural β-globin mRNA.

A hybridization experiment similar to the one above showed that the small intron was present in the β-globin-specific RNA (data not shown).

Finally, the 3' termini of the transcripts were mapped using a *Bam-Sal* 14,000 base pair fragment labelled at the 3' (minus strand) terminus of the *Bam* site. Using this probe, hybridization with RNA from transformed yeast yielded 4 major bands from 270 to 350 nucleotides (Fig. 5). Since the distance from the *Bam* site to the 5' border of the large intron is 18 nucleotides, many or all of the β-globin transcripts contain part of the large intron. The distance from the *Bam* site to the 3' end of the large intron is 591 nucleotides. Since the largest fragment protected in this experiment is only 350 nucleotides, it appears that none of the yeast transcripts includes the entire intron. This experiment by itself does not exclude that some β-globin RNA lacking the large intron was present, since an 18 nucleotide fragment could not have been detected under the conditions of the experiment. However, since the "Northern transfer" had not revealed much RNA smaller than 850 nucleotides, the proportion of mature β-globin mRNA could not have been significant.

DISCUSSION

Most mouse L cell lines transformed with rabbit β-globin DNA produced rabbit β-globin-specific RNA. Cell line M2/R-1, which contained the highest number of gene copies, about 20, also contained the highest number of RNA molecules, namely about 2200 per cell; however, this correlation was not general.

The rabbit β-globin-specific RNA of the line M2/R-1 was indistinguishable from natural mRNA in regard to its electrophoretic mobility, its 5' terminus, the absence of the large and the small (Mantei, Boll and Weissmann, 1979) intron and the presence of a 3' terminal poly(A) tail. The β-globin RNAs of the other cell lines, when tested for one or the other of these criteria, had similar properties.

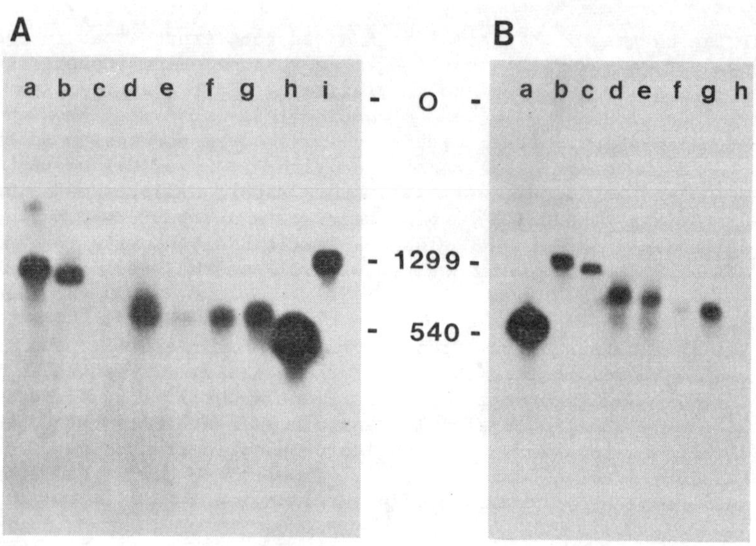

Fig. 4. The electrophoretic mobility of rabbit β-globin RNA from transformed mouse L cells and yeast cells.

Samples were denatured with glyoxal (McMaster and Carmichael, 1977), subjected to electrophoresis through a 1.5% agarose gel, transferred to diazobenzyloxymethyl paper (Alwine, Kemp and Stark, 1977) and hybridized with a ^{32}P-labelled rabbit β-globin cDNA probe. Denatured restriction fragments of rabbit β-globin-specific DNA and β-globin mRNA were used as size markers. A. (a) RβG DNA digested with *Pst*I and *Bgl*II (1299 base pairs.) (b) RβG DNA digested with *Pvu*II and *Bgl*II (1209 base pairs). (c) Poly(A)$^+$ RNA from P1/S-1 cells RNA (transformed with TK DNA only). (d) Poly(A)$^+$ RNA from M2/R-1 cell RNA (transformed with rabbit β-globin DNA and TK DNA). (e-g) Rabbit (α+β) globin mRNA 1, 4 and 8ng, respectively. (h) Cloned rabbit β-globin cDNA (Meyer and co-workers, 1979); ca. 540 base pairs. (i) RβG cleaved with *Pst*I and *Bgl*II (1299 base pairs). B. (a) same as A(h). (b) same as A(a). (c) same as A(b). (d) and (e) RNA from *S. cerevisiae* Mc16 transformed with Z-pJDB219/RβG-1 (d) or with Z-pJDB219/RβG-3 (e). (f and g) Rabbit globin mRNA 2ng (f) and 10ng (g). (h) RNA from *S. cerevisiae* Mc16 transformed with pJDB219.

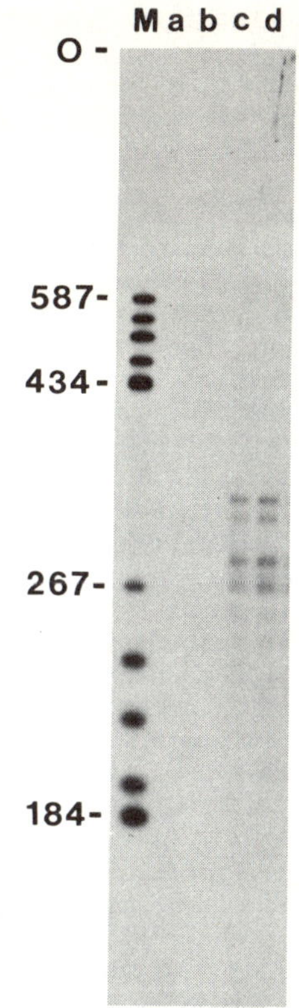

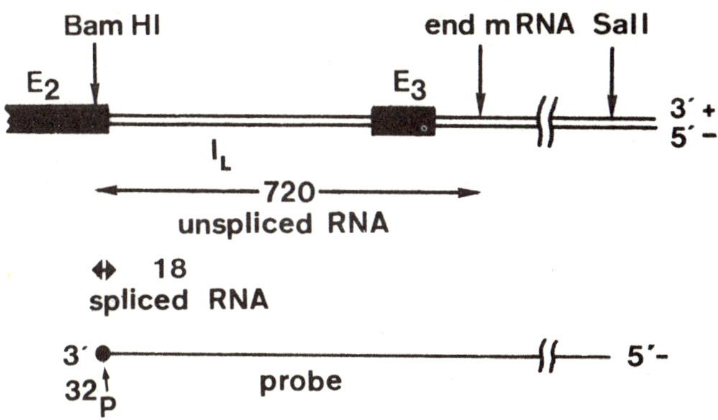

Fig. 5. Mapping of the 3' terminus of β-globin-specific RNA from transformed yeast cells.

The assay was carried out by the method explained in Fig. 2. However, the probe was labelled at the 3'-terminus of the *Bam*-cleavage site. The probe was prepared by digesting RβG with *Bam*HI, filling in the cohesive ends with DNA polymerase I and cleaving with *Sal*I. M, marker (PBR322 cleaved with *Bsp*I and 5'-terminally ^{32}P-labelled). (a) and (b), rabbit (α+β) globin mRNA 4 ng and 1 ng, respectively. (c) and (d), RNA from *S. cerevisiae* Mc16 transformed with Z-pJDB219/RchrβG-1 and Z-pJDB219/RchrβG-3, respectively. The rabbit β-globin RNA should give rise to a 18-nucleotide band which could not be detected on this gel.

Although the β-globin DNA-transformed yeast cells also produced substantial quantities of β-globin-specific RNAs, these transcripts differed considerably from natural β-globin mRNA: they were shorter by 18-40 nucleotides at the 5' end, they contained the small intron, part of the large intron, and lacked the normal 3' terminus.

In the case of mouse β-globin mRNA, the 15S precursor and the mature mRNA have the same 5' terminal sequence and the same cap structures (Weaver, Boll and Weissman, 1979; Curtis, Mantei and Weissmann, 1978). If the 15S β-globin precursor is the primary transcript then we may conclude that the rabbit β-globin RNA produced in transformed mouse L cells originates at the true promotor. We cannot yet, however, exclude the possibility that the transcript is initiated upstream at a plasmid or a host DNA promotor, and that the 5' terminus arises by processing. In the β-globin DNA-transformed yeast cells transcription either starts downstream from the correct position or a nuclease cleaves a longer transcript.

Why do transformed mouse L cells synthesize rabbit but no, or very little, mouse β-globin mRNA? One explanation is that transcription starts at an unphysiological promotor. If, on the other hand, the physiological promotor is being utilized, then the mechanism that prevents the transcription of the mouse gene may not be active on the rabbit gene, for example, because of sequence differences in the control regions, or because of a position effect. The experimental results clearly argue against a simple dosage effect, since several transformed cell lines containing only one or few copies of the rabbit β-globin gene produce detectable amounts of rabbit but not of mouse mRNA.

There is little doubt that the introns are accurately excised from the rabbit transcript in mouse L cells. This is not surprising in view of the similarity of the intron-exon junction regions in the rabbit and mouse β-globin gene (van den Berg and colleagues, 1978) and of the finding that splicing appears to occur normally in very heterologous systems, such as SV40 transcripts in *Xenopus* oocytes (De Robertis and Mertz, 1977). The fact that β-globin transcripts are not (or incorrectly) spliced in yeast could have one or more causes. First, yeast may not contain the appropriate splicing enzymes for dealing with the β-globin transcript, or the enzymes may not be in the same cellular compartment as the hybrid DNA. Second, the splicing enzymes may be present, but incapable of processing the incomplete transcript formed in the yeast cell.

At this stage it would seem that thymidine kinase-linked transformation of mouse L cells will be useful for the study of transcription and splicing, whereas the applicability of the yeast system to the study of transcription and translation of genes of higher organisms remains to be demonstrated.

ACKNOWLEDGEMENTS

This work was supported by the Schweizerische Nationalfonds (No. 3.114.77) and the Kanton of Zürich. A.v.O. was supported by grants from EMBO and the Netherlands Organisation for the Advancement of Pure Research. J.v.d.B. received grants from EMBO and Koningin Wilhelmina Fonds and J.D.B. had a short-term EMBO Fellowship and R.F.W. was supported by an American Cancer Society Research Scholarship.

REFERENCES

Alwine, J.C., D.H. Kemp and G.R. Stark (1977). Method for detection of specific RNAs in agarose gels by transfer to diazobenzyloxymethyl-paper and hybridization with DNA probes. *Proc. Natl. Acad. Sci. USA*, 74, 5350-5354.

Beggs, J.D. (1978). Transformation of yeast by a replicating hybrid plasmid. *Nature*, 275, 104-109.

van den Berg, J., A. van Ooyen, N. Mantei, A. Schamböck, G. Grosveld, R.A. Flavell and C. Weissmann (1978). Comparison of cloned rabbit and mouse β-globin genes showing strong evolutionary divergence of two homologous pairs of introns. *Nature*, 275, 37-44.

Berk, A.J. and P.A. Sharp (1977). Sizing and mapping of early adenovirus mRNAs by gel electrophoresis of S_1 endonuclease-digested hybrids. *Cell*, 12, 721-732.

Boll, W., N. Mantei, N. Wilkie, B. Clements, P. Greenaway and C. Weissmann (1979) Cloning of the thymidine kinase gene of Herpes simplex I in pBR322. *Experientia*, in press.

Curtis, P.J., N. Mantei and C. Weissmann (1978). Characterization and kinetics of synthesis of 15S β-globin RNA, a putative precursor of β-globin mRNA. *Cold Spring Harbor Symp. Quant. Biol.*, 42, 971-984.

Curtis, P.J. and C. Weissmann (1976). Purification of globin messenger RNA from dimethylsulfoxide-induced Friend cells and detection of a putative globin messenger RNA precursor. *J. Mol. Biol.*, 106, 1061-1075.

De Robertis, E.M. and J.E. Mertz (1977). Coupled transcription-translation of DNA injected into *Xenopus* oocytes. *Cell*, 12, 175-182.

Efstratiadis, A., F.C. Kafatos and T. Maniatis (1977). The primary structure of rabbit β-globin mRNA as determined from cloned DNA. *Cell*, 10, 571-585.

Jackson, D.A., R.H. Symons and P. Berg (1972). Biochemical method for inserting new genetic information into DNA of Simian virus 40: Circular SV40 DNA molecules containing lambda phage genes and the galactose operon of *Escherichia coli*. *Proc. Natl. Acad. Sci. USA*, 69, 2904-2909.

Konkel, D.A., S.M. Tilghman and P. Leder (1978). The sequence of the chromosomal mouse β-globin major gene: Homologies in capping, splicing and poly(A) sites. *Cell*, 15, 1125-1132.

Lobban, P.E. and A.D. Kaiser (1973). Enzymatic end-to-end joining of DNA molecules. *J. Mol. Biol.*, 78, 453-471.

Mantei, N., W. Boll and C. Weissmann(1979). Rabbit β-globin mRNA production in mouse L cells transformed with cloned rabbit β-globin chromosomal DNA. Submitted for publication.

McMaster, G.K. and G.G. Carmichael (1977). Analysis of single and double-stranded nucleic acids on polyacrylamide and agarose gels by using glyoxal and acridine orange. *Proc. Natl. Acad. Sci. USA*, 74, 4835-4838.

Meyer, F., H. Heijneker, H. Weber and C. Weissmann (1979). Transposition of AT-linked, cloned DNA from one vector to another. *Experientia*, in press.

van Ooyen, A., J. van den Berg, N. Mantei and C. Weissmann (1979). The complete sequence of a cloned rabbit DNA segment containing a β-globin gene and its flanking regions; comparison with the homologous mouse gene. Submitted for publication.

Southern, E.M. (1975). Detection of specific sequences among DNA fragments separated by gel electrophoresis. *J. Mol. Biol.*, 98, 503-517.

Sutcliffe, J.G. (1978). pBR322 restriction map derived from the DNA sequence: accurate DNA size markers up to 4361 nucleotide pairs long. *Nucl. Acids Res.*, 5, 2721-2728.

Szybalski, W., E.H. Szybalska and G. Ragni (1962). Genetic studies with human cell lines. *Natl. Cancer Institute Monograph*, 7, 75.

Weaver, R., W. Boll and C. Weissmann(1979). A hybridization procedure allowing full discrimination between closely related RNAs. *Experientia*, in press.

Weissmann, C. (1978). Reversed genetics. *TIBS*, 3, N109-N111.

Weissmann, C., N. Mantei, W. Boll, R.F. Weaver, N. Wilkie, B. Clements, T. Taniguchi, A. van Ooyen, J. van den Berg, M. Fried and K. Murray (1979). Expression of cloned viral and chromosomal plasmid-linked DNA in cognate host cells. *11th Miami Winter Symposia*, in press.

Wigler, M., S. Silverstein, L-S. Lee, A. Pellicer, Y-C. Cheng and R. Axel (1977). Transfer of purified Herpes virus thymidine kinase gene to cultured mouse cells. *Cell*, 11, 223-232.

DISCUSSION

I.S. JOHNSON: I do not have a question as much as an observation on both the introduction and Dr. Rutter's paper. I thought I heard Bill Whelan say on the second slide that we were considering these seven points as alternatives to insulin and we really weren't. I only meant to indicate in the National Academy Forum that these were factors which might influence the supply of pancreases and the requirement for insulin. The problem, of course, was and is that the diabetic population, at least in the United States, is rising at about a 4% rate while the normal population is somewhere between zero and 2%. Commenting on Dr. Rutter's paper, I'd like to indicate that we have an interest in both the natural and the synthetic approaches and we collaborate with Dr. Rutter on his approach as well. Both approaches in a practical sense have advantages and disadvantages. One of the disadvantages of the natural route, besides the problem of getting mRNA - and purifying the messenger, of course - is the connecting peptide of the proinsulin. While it is true that you can cleave it, the enzyme which does it normally has not really been identified. It's probably a cathepsin D type molecule, but with trypsin and other enzymes, you actually end up with desthreonine-insulin and while it's just as biologically active as normal insulin, no one really wants it, they want the real thing.

In the synthetic approach, I think it *is* possible to define impurities with a great deal of resolution. It's certainly possible to get materials that react in a radio-immuno assay and that will react in a radio receptor assay and yet will show almost no target protein by high pressure-liquid chromatography analysis. But if all three of these assays correlate quantitatively along with sequencing and amino-acid analysis I believe you can be sure of the configuration of your target protein and its purity.

C. WEISSMANN: But, I mean, the case of tetrahydrofolate reductase. Has an analysis there shown that in fact all the codons are compatible for *E. coli*?

W.J. RUTTER: In the case of tetrahydrofolate reductase the enzyme is selected only for activity. Not all the enzyme molecules need have the same sequence.

C. WEISSMANN: Are the codons compatible with the *E. coli* pattern?

W.J. RUTTER: I don't know that. I don't know the amino-acid sequence. Can you answer that, Stanley?

S.N. COHEN: To get enough enzyme for detectable phenotypic activity probably doesn't require complete compatibility of codon selection patterns. I think it's hard to say whether they were more compatible. The question is "What is compatibility?", and if one were to use codons that were optimally designed whether there would be more enzyme made than is made under the present circumstances. I really can't answer that question at the present time.

C. WEISSMANN: It's really a very essential question for the people involved in these problems because if one makes the statement that you have to have compatibility in codons, it would change the approach quite a bit. On the other hand, the evidence we are seeing in the case of tetrahydrofolate reductase would suggest one would not have to worry so much. I was interested to know whether any hard facts are known on this matter. Can you comment?

W.J. RUTTER: We do not really have hard facts. We have the one single case in tetrahydrofolate reductase where there is no question the enzyme is active. What the population of molecules are that comprise the activity is the point at issue. In the case of insulin we hope to get the sequences to provide direct information on this matter.

S.N. COHEN: In the case of the introduction of mouse dihydrofolate reductase coding sequences in *E. coli* K-12, we calculated that about 0.01% of the total protein was active enzyme. I think the point that Bill is making is that, while we clearly showed that a substantial amount of active enzyme was being made in bacteria, theoretically one might propose that with a different codon mix there might be an even greater amount of active enzyme and I can't dispute that.

P. CLARKE: I'd like to make a comment and put a question to Dr. Rutter about enzymes of industrial importance. Most of the enzymes of industrial importance come from microorganisms, but not usually from *E. coli*. The strains that are used have been selected to give a good yield of enzyme and also for robust growth in fermentors. It's quite clear that it's possible to clone genes from such organisms in *E. coli* and to get high expression in laboratory experiments. Perhaps, in the long run the most useful thing would be to clone the genes in *E. coli* and then put them back, perhaps having replaced the natural promoters by better promoters, into the original host where they would grow better and survive in large-scale fermentor growth conditions. One of the things we don't know very much about is how we can get genes cloned in *E. coli* and then put them back and get them to survive well in alternative hosts, in fact, in the original host. I wondered if Dr. Rutter would comment on that.

W.J. RUTTER: I would agree that your strategy sounds excellent. I would only add that instances may occur where an enzyme in a host which isn't at present practical to get in commercial quantities might now be easily obtained by the cloning method as well.

W. FIERS: I would like to make a comment regarding this codon usage. As regards *E. coli* there are mainly two cases in which a concentration of the isoaccepting tRNA is very low, and this is AUA for isoleucine, and AGA and AGG for arginine. There are several cases known of proteins which contain many of these codons yet still are made in quite appreciable amounts. If you think in terms of evolution, a protein being made at a rate of 90% rather than 100% may have quite an effect, but in terms of these kinds of experiments, synthesis of useful proteins by "engineered" bacteria , I don't think this selective codon usage is that important.

However, there may be other problems. As far as I understand it, e.g. on the basis of studies which have been done by John Atkins, you can get a certain derailment - probably of the ribosome when specific codons follow each other. So the problem may be much more subtle and unpredictable at the moment. Each codon can be followed by any other one, but in certain cases you may have a defined probability of derailment, and if these instances accumulate then your total yield of protein may decrease.

W.J. RUTTER: Yes, what is detected in these experiments are a set of molecules of both different charge and different lengths. The point I'd like to make is that in those experiments with β-glactosidase-insulin one sees a number of molecules accumulating of different charge and apparent size.

GUIDELINES AND LEGISLATION

CHAIRMAN'S INTRODUCTION

W. Szybalski

McArdle Laboratory for Cancer Research, University of Wisconsin, Madison, Wisconsin 53706, U.S.A.

During the previous sessions we heard about many triumphs and beneficial applications of genetic techniques, including the recombinant DNA technique. We also heard that natural evolution employs tools analogous or identical to those used in the laboratory by geneticists. Now we will hear about the restrictions placed on this basic research when one particular genetic technique, the recombinant DNA method, is employed. Regulation of basic research is a very serious matter, because of several scientific, social and legal considerations, which I am planning to discuss very briefly.

First, comprehensive regulations and "a prior restraint", as represented by the NIH Guidelines (1978), are "antithetical to creativity, and creativity is the most important component of scientific advance" (Stetten, 1978). Complex regulations are usually inflexible and, since they cannot anticipate frequent new discoveries, they cannot be applied to rapidly developing research. Thus, as could have been predicted and based on the present experience, many novel experiments are currently blocked by regulations. This problem cannot be solved by making the present regulations (Guidelines, 1978) still more complex, but only by abolishing them or replacing them by a single sentence as proposed by Stetten (1978).

Second, the present regulations of the recombinant DNA technique are based on hypothetical, imaginary or conjectural risks and on fears of the unknown, rather than on any evidence of existent or predictable dangers. On such "non-grounds", based on the "heavy-question and metaphor arguments" (Cohen, 1978), almost any human activity could be prohibited or regulated.

Third, all human activities, including imposition of regulations on the recombinant DNA technique, could create new risks and could be damaging to society. For example, novel laboratory hazards are being created and the existing ones aggravated. There could also be political and legal dangers, including loss of freedoms, designing laws which invite disrespect, misleading the public, and creating an atmosphere of mistrust and witch hunting. Valuable resources could be depleted and misdirected, unnecessary bureaucracy fostered, benefits to society denied or delayed, and environmental and other laudable public concerns misdirected.

Fourth, basic research in biology has enjoyed traditional freedoms during the past hundreds of years. Although there were attempts to restrict such freedoms, with genetics and evolution often the favoured targets, no one at present takes pride in such infamous examples as medieval restrictions on research in anatomy or the Scopes trial.

Fifth, the regulations based on the NIH Guidelines require a prior approval for most of the experiments employing the recombinant DNA technique. In legal terms, this constitutes "a prior restraint", which in analogy to the freedom of the press corresponds to "censorship". Many arguments for or against censorship could be advanced and it would be difficult to reach unanimous agreement whether one should permit a reporter to print his article in the newspaper before it was evaluated and approved by a censor or to let a scientist test his idea in the laboratory before it was evaluated and approved by a special agency. However, censorship of the press is generally not popular, unless under very special and compelling circumstances, and censorship of basic science was hardly ever practised. The experience teaches us that both journalists and scientists could generally be trusted and that censorship is counterproductive. In some countries the freedom of the press and of basic research is under the protection of the Constitution (e.g. in the FRG), which generally means that the Government is not allowed to institute a censorship and regulations until presenting a convincing proof that very compelling reasons exist. Such reasons could not be just some general apprehensions, quasi-scientific scenarios, and political, sociological or religious doctrines, advocated by a vocal minority or even majority (Emerson, 1977).

Finally, although the freedom of basic research is not explicitly mentioned in the United States Constitution, and only the freedom of speech, press and assembly are specified in the First Amendment, several rulings of the U.S. Supreme Court indicate that the freedom of enquiry, including basic research, is also protected. In the famous case of Grisvold v. Connecticut, the U.S. Supreme Court (1965) ruled that "The right of freedom of speech and press includes not only the right to utter or to print, but the right to distribute, the right to receive, the right to read ... and freedom of enquiry, freedom of thought, and freedom to teach ... indeed the freedom of the entire university community ...".

Also, several Constitutional scholars, including Emerson (1977), Delgado (Delgado, 1979; Delgado and Millen, 1978), and Robertson (1978), have advanced convincing reasons that basic research is protected by the United States Constitution. Such protection, which includes both ideas and experiments, is not absolute, but it means that basic research cannot be censored, regulated or prohibited, unless there are *compelling* and *well-documented* reasons. Moreover, the proof that compelling reasons exist has to be provided by the agency that plans to institute regulations, and according to the Administrative Procedures Act. The eventual ruling cannot be arbitrary or capricious. The general apprehensions, political doctrines of pressure groups, gut feelings, scare scenarios, and bureaucratic inertia in view of pre-existing but unnecessary regulations, among others, do not constitute compelling and well-documented reasons for regulations. Also, the cost-benefit analysis of the proposed regulations has to be provided.

I mailed the following questions to the speakers a few weeks before the meeting and I will be interested to hear their comments.

1. Which, if any, compelling reasons are officially cited in those countries which have promulgated mandatory guidelines, regulations or legislation?

2. Was there an evaluation made on the validity of such officially cited *compelling* reasons and has the due process been followed?

3. Was there a risk-benefit or cost-benefit evaluation made *before* adoption of any recombinant DNA regulations? It is my understanding that a recent United States Federal Court ruling in Louisiana has mandated the cost-benefit analysis before issuance of health-related regulations.

4. Could suggestions be made of any expedient amendments, in case the present regulations are *not* based on compelling and well-documented reasons? Would it be logical to suggest immediate suspension of all regulations until evidence for compelling reasons is agreed to, and accepted for specific experiments?

5. Could one predict the answer of a granting or censoring agency to the following request? The grantee has an already active grant valid for the next three years for studying the controls of the *lac* operon in *E. coli*, and he plans to test the activity of *lac* controls in several species for which so-called HV1 systems were not certified. He notifies the granting agency that he plans to clone the *E. coli lac* operon or its control elements in an actinophage for *Streptomyces lavendulae* or *coelicolor*, and in bacteriophages or plasmids for *Thiobacillus thiooxidans*, *Nitrosocystis oceanus*, *Rhodospirillum rubrum*, *Caulobacter crescentus*, *Gallionella ferruginea*, *Leptothrix ochracea*, *Anabaena catanula* and in few more saprophitic and innocuous bacteria or algae. He is not planning to undertake any developmental and bureaucratic work necessary for certification of all HV1 host-vector systems, and he has no P3 facility or plans to build one. He states that his proposed experiments are within the scope of his present grant, are scientifically sound and beneficial, free of any conceivable practical risks, and that under the principle of the freedom of academic basic research (or Constitutional protection under the First Amendment and the "doctrine of unconstitutional conditions", in the case of the USA) he would seek protection in the courts in case his funding should be discontinued as a penalty for performing his experiments. He considers any prohibitions to his proposed experiments as capricious and unfounded infringement on the freedom of scientific enquiry, unless the granting agency could provide any scientifically compelling reasons to censor his research and impose "a prior restraint".

I am looking forward to this session in an open-minded fashion, especially since I am not involved in any recombinant DNA research regulated by the NIH Guidelines (1978). I hope to be convinced by the speakers that the present regulations should be either strengthened, or left unchanged, weakened or abolished.

Unfortunately, our first speaker, Dr. Fredrickson, is unable to be here, but his paper will be read by Dr. W. Gartland.

REFERENCES

Cohen, C. (1978). Restriction of research with recombinant DNA: the dangers of inquiry and the burden of proof. *Southern Calif. Law Rev.* 51, 1081-1113.

Delgado, R. and D.R. Millen (1978). God, Galileo and Government: toward constitutional protection for scientific inquiry. *Washington Law Rev.* 53, 349-404.

Delgado, R. (1979). Science and the First Amendment. *Bull. Atomic Scient.* 35, 60-62.

Emerson, T.I. (1977). Testimony before the House Subcommittee on Science, Research and Technology. *In* "Science Policy Implications of DNA Recombinant Molecule Research". Hearings before the Subcommittee on Science, Research and

Technology of the Committee on Science and Technology, US House of Representatives, 59th Congress, March 29th - September 8th, 1977, Report No. 24, US Government Printing Office, Washington, D.C. pp. 874-890, 905-915.

Guidelines for Research Involving Recombinant DNA Molecules. (1978). *Federal Reg.* 43, 60108-60131.

Robertson, J.A. (1978). The scientist's right to research: a constitutional analysis. *Southern Calif. Law Rev.* 51, 1203-1279.

Stetten, DeW. (1978). Valedictory by the chairman of the NIH Recombinant DNA Molecule Program Advisory Committee. *Gene* 3, 265-268.

US Supreme Court (1965). Grisvold v. Connecticut. *US Reports*. 381, p. 482.

A HISTORY OF THE RECOMBINANT DNA GUIDELINES IN THE UNITED STATES

D. S. Fredrickson

National Institutes of Health, U.S. Department of Health, Education and Welfare, Bethesda, Maryland 20205, U.S.A.

On December 16, 1978, a telegram purporting to be from the Vatican was hand-delivered to the office of Joseph A. Califano, Jr., Secretary of Health, Education, and Welfare. "Habemus regimen recombinatum," it proclaimed, in celebration of the end of a long struggle to revise the NIH Guidelines for Research Involving Recombinant DNA Molecules. It was not the first telegram the Secretary had received on this subject. From Peking the preceding June, I had responded to his cabled instructions concerning proposed revisions of the Guidelines which I had taken to China. The U.S. liaison officers found my reply inappropriate for transmission, but I delivered it on my return, imprinted on a rice paper temple-rubbing.

These sophomoric tricks were moments of comic relief in a three-year period of coping with the scientific, political, and legal problems created by the advent of the "new biology". The following pages summarize my impressions of this turbulent experience.

My personal history of the DNA Guidelines in the United States recognizes three phases to date. Phase I is the period between the early concern about possible hazards of recombinant DNA technology and the delivery to NIH of proposed rules for conducting research. Phase II covers the promulgation of the NIH Guidelines in 1976 and the thirty months before their official revision. Phase III began on January 2, 1979, with a new set of rules painfully formulated during this unprecedented curtailment of experimentation in biology.

THE END OF THE BEGINNING

In this collection are reminiscences of the first apprehensions (1973), the decision to develop guidelines (1974), the Asilomar agreements (1975), and the exhausting constructions of the NIH Recombinant DNA Advisory Committee (RAC). Three versions of guidelines had emerged from RAC meetings after Asilomar. In La Jolla, California, on December 5, 1975, the Committee, with the "variorum edition before it, finally succeeded in scaling conjectural hazards by parliamentary procedure. Chairman Hans Stetten went to the telephone to inform the NIH Director in Bethesda that the nation had acquired rules for recombinant DNA research. Much later I was told how he had returned to the conferees, shoulders drooping, success drained from his face. "He wants to have a public hearing on them", he mumbled.

PUBLIC AIRING BEGINS

From the beginning the decision to "go public" was variously understood and was resented by many. After all, the Director, NIH, has long had authority to promulgate guidelines for investigators the agency supports. There is no requirement for hearings or public comment. I became aware of new responsibilities heading my way sometime in the autumn of 1975, when I had been Director for only two or three months. At that time, I had barely heard of restriction enzymes and could not even have explained the crucial distinctions between Federal guidelines and regulations. From the first I was inclined--and after a little study and consultation, quite determined--to air in an open and public manner the scientific and social issues. This was the only way to decompress rising tensions and to prepare to defend whatever actions would be taken against certain criticism.

The Director's Advisory Committee (DAC) was convened in February 1976 for public discussion of the Guidelines. The transcript, like all the other relevant documents on the subject, is available in the "public record" published by NIH.[1] The hearing demonstrated the difficulties of holding a town meeting on molecular biology and exposed the full range of opinions on the risks of the new technology. It was apparent that our decisions would have to run a gamut of adversarial reactions and that some, in the end, might well be tested in the courts. After the hearing, the voice of Judge Bazelon lingered longest in my mind: " . . . the healthiest thing that can happen is to let it all hang out, warts and all, because if the public doesn't accept it, it just isn't worth a God-damn."

We made some changes in the proposed Guidelines after the DAC meeting, mainly adding administrative structure. We then set out to acquaint key people and agencies with the details, for NIH supported most but by no means all of the affected research. The widening circle included the National Science Foundation, the Department of Agriculture, other Federal agencies whose authorities were crossed by the NIH Guidelines, the staffs of Congressional committees with jurisdiction over biomedical research, and representatives of industry doing what private research of this sort there was at the time.

ISSUANCE OF THE GUIDELINES

The NIH Guidelines were issued on June 23, 1976. It was front page news, but the reactions were muted. We also established the Office of Recombinant DNA Activities (ORDA), under the direction of William Gartland.

The NIH Guidelines were just that--guidelines, not regulations, which have more of the force of law. The verbs tended to be "shoulds," though some "shalls" had been substituted after the February hearing. It was stated that the Guidelines would be frequently revised, but no special procedures for doing so were laid out. They were expected to evolve as understanding of the subject grew. As it turned out, it was not the subject of the Guidelines, but "due process" for changing and administering

[1]Office of the Director, NIH (1976-78). Recombinant DNA research, Vols. 1-4 (4,015 pp. in all); for sale by Superintendent of Documents, U.S. Govt. Printing Office, Washington, D.C. 20402, and available in about 600 public libraries of the GPO depository system. (GPO stock no. for Vol. 1, 017-040-00398-6; Vol. 2, 017-040-00422-2; Vol. 3, 017-040-00429-0, and appendix, 017-040-00430-3; and Vol. 4, 017-040-00443-5, and appendix, 017-040-00442-7.) The environmental impact statement, cited in footnote 2, was published as a supplement to Vol. 2 (not included in above page-count).

them, which became the focus for opposition to the research. For the next two and a half years, the Guidelines were to be practically frozen, while the science expanded impatiently within.

EXTENSION OF THE GUIDELINES BEYOND NIH

NIH had no illusions that it was creating guidelines for all the recombinant research in the world. Scientists are citizens of different nations whose laws can supersede intellectual accord. Even within the United States, extension of the same rules to all laboratories could not be achieved by any simple move.

Two different kinds of protest about this incompleteness were brewing in 1976-78. One encouraged extension of the jurisdiction of state and local communities to regulation of laboratory research, a legal area hitherto unexploited. The other sought to persuade DHEW, its Food and Drug Administration (FDA), and other regulatory agencies to use certain narrow authorities to force compliance with common rules. If that failed, the Department was to seek a Federal law to that end. Some fervent advocates of legislation fought for preempting local jurisdictions from enacting more stringent standards if they wished. Others just as vehemently opposed Federal preemption. A Balkanization of recombinant DNA research was one of the most serious and extraordinary threats of this period.

In May 1976 we informed our Department superiors about our intention to issue guidelines, and urged then-Secretary David Mathews to ask the President to direct all relevant Federal agencies to coordinate recombinant DNA activities through an interagency committee. Mathews agreed, but no words emanated from the White House. In July, Senators Kennedy and Javits addressed a letter to President Ford advocating the extension of NIH Guidelines to all Federal and private research. Local hearings in Cambridge, Mass.; Ann Arbor, Mich.; San Diego, Calif.; and New York added to a sense of urgency.

The President's letters were finally dispatched in September, and the Federal Interagency Committee on Recombinant DNA Research was promptly convened in Bethesda. The research agencies readily agreed to use the NIH Guidelines for the research they supported or conducted. The committee then undertook to examine the regulatory authorities of each of the member agencies and to develop recommendations for possible new legislation. Later the committee would examine patent policy and the international aspects of regulating DNA research.

NEPA AND THE FRIENDS OF THE EARTH

A full discussion of the National Environmental Policy Act (NEPA) with reference to laboratory research in general and to how the NIH Guidelines for recombinant DNA research became involved would fill a volume. NEPA, a law passed in 1969, requires the Federal agencies to determine whether contemplated actions will significantly affect the environment. If so, the action must be heralded by an Environmental Impact Statement (EIS). In the spring of 1976 we were made aware that if we released the Guidelines before issuing an EIS, we could be charged with violating NEPA.

Although an EIS had become common in proposals to level mountains or build dams, the adaptation of NEPA to conjectural hazards of laboratory research was a nightmare. The situation was aggravated by ambiguous and arbitary procedures for implementing NEPA within DHEW. The tortoise-like march from draft to final EIS could take years.

But delaying the issuance of the Guidelines pending completion of the EIS process was never an alternative. The voluntary agreements made at Asilomar were losing their hold on the scientists, confusion was mounting, and dissidents in various communities threatened to obtain either local regulation or prohibition of the research if Federal standards were not quickly forthcoming. It was obvious that the public interest would be better served--and the opportunity of scientists to continue experiments, better protected--with guidelines than without them, even if an EIS were not published until after their issuance.

We therefore released the Guidelines in June with an announcement that an EIS was to follow. The draft EIS was filed in September 1976, the final one in October 1977.[2] In May 1977 two suits against NIH were launched in separate Federal courts. One, brought by an organization called The Friends of the Earth, sought to enjoin all recombinant research. The other sought to block the Rowe-Martin risk assessment study. The final EIS was entered as part of the Government's defense in the latter suit (Mack v. Califano). In finding for the Government in March 1978, the Court concluded that NIH, in its EIS, had indeed "taken a hard look" at the consequences of experiments with recombinant DNA. The plaintiff was denied an injunction.

THRUSTS TOWARD LEGISLATION

A bill was introduced in the Senate (S. 621) in February 1977 by Senator Bumpers (D., Ark.), with a companion bill in the House by Rep. Ottinger (D., N.Y.). They were the first of 12 bills to regulate DNA research submitted to the 95th Congress. On February 23, representative scientific leaders were invited to NIH to read selected passages of the proposed new legislation, including heavy penalty provisions. Two weeks later, the last traces of their indifference were dispelled by the acrimonious tone of a forum on "genetic engineering" at the National Academy of Sciences.

The following May the Interagency Committee conveyed to the new HEW Secretary, Joseph A. Califano, Jr., its conclusions that a Federal statute would be required if the Guidelines were to be extended to all recombinant research in the country. It also offered the elements of what it considered an "ideal" law--elements that were quickly converted to an Administration bill introduced by Senator Kennedy (S. 1217) on April 1, 1977. Kennedy then revised the bill radically. In an intensive reaction to this and other proposed laws, scientists and their organizations soon made strong appeals to the Congress. The ardor of the legislators for statutory regulation cooled progressively during 1977-78.

REVISION IS NEEDED

Within six months of their appearance, the NIH Guidelines clearly needed revision. The molecular biologists who had constructed them, if given that chance again, would surely have engaged other disciplines on the route from Asilomar to Bethesda. Especially lacking had been the counsel of experts on infections, who had a better perspective of the improbability that *E. coli* K-12 could be converted into an epidemic pathogen. And more thought should have been given to the containment levels for

[2]Office of the Director, NIH (1977). National Institutes of Health Environmental Impact Statement on NIH Guidelines for Research Involving Recombinant DNA Molecules, Part One (147 pp.) and Part Two (appendices, 438 pp.); for sale by Superintendent of Documents, U.S. Govt. Printing Office, Washington, D.C. 20402, and available in about 600 public libraries of the GPO depository system. (GPO stock no. 017-040-001413-3.)

dealing with viral DNA, to the prohibitions, and to the coverage of organisms known to exchange DNA in Nature.

It is also notable that the guidelines constructed at about the same time in the U.S. and the U.K. were quite different in form. The Americans wrote an extensive codification and the British opted for common-law evolution of minimum rules. Both sets, however, were meant to be interpreted and administered centrally, by a GMAG or an ORDA. The U.S. scientists did not want local committees to second-guess their experimental protocols. More comfortable with central decision-making by study sections in Bethesda, they preferred ORDA's interpretations and administration.

But delays in administrative actions were inevitable. The requirement for prior NIH approval of all changes in ongoing projects particularly irked investigators. In rejecting at the start nearly all suggestion of control by their institutions, the scientists had made it difficult to regain a proper balance between local application and national standards.

As a broker between the molecular biologists and the various public interests, NIH also failed to perfect the Guidelines before issuing them. We did not incorporate mechanisms for revision. Discretionary powers to make interpretative judgments and minor changes, essential in so complex and fast-moving a subject, were lacking. There were reasons, however, for avoiding imitation of the formal and formidable procedures of the regulatory agencies. The more we embedded the Guidelines in inflexible administrative molds, the less chance there would be for timely accommodation to the tide of new information that was already rising.

REVISION BEGINS

In January 1977 the RAC commenced to prepare a revised set of Guidelines. A workshop was held in June at Falmouth, Mass., to synthesize old and new information about bacterial host-vector systems. In September the proposed revision--a complete rewriting of the text--was formally presented to me and published in the Federal Register. A two-day hearing was held in December 1977 at which the RAC members defended their proposed changes. Most critics raised questions of process; but some containment levels were severely challenged, and additional meetings of experts on viruses and plant pathogens led to further alterations.

In Departmental clearance, the revised Guidelines encountered more difficulties than the original. Recombinant DNA research had emerged as a scientific issue with immense appeal to laymen, and Secretary Califano's staff took a strong and sophisticated interest in how all relevant law and administrative practices pertained to the new draft. By now, some dissidents and a militant fraction of the environmental movement had also launched a concerted campaign to exact, if science wished to proceed, a more generous tax in procedure.

On July 28, 1978, the proposed revision, accompanied by our environmental impact assessment and a Director's decision paper, was published in the Federal Register. An introductory memorandum from the Secretary invited the public to comment and announced that, after a 60-day period, there would be another hearing chaired by Peter Libassi, the HEW General Counsel.

REVISION COMPLETED

The Libassi hearing took place on September 15 at the HEW headquarters in Washington. NIH staff and I--the "Kitchen RAC"--then dissected the comments received in testimony and 170 letters, and joined in numerous discussions with Mr. Libassi and his committee. Special meetings were also held with a group of environmentalists who wished to reinforce some of their earlier demands. We also met with representatives of pharmaceutical firms, other Federal research agencies, and members of institutional biosafety committees to discuss their problems with the proposed revisions. A culmination of the Libassi hearings was the reconstitution of the RAC to broaden its public (nonscientific) membership and to combine the technical and policy reviews, usually carried out at NIH in a two-tiered process. The appointment of RAC members was shifted from the NIH Director to the HEW Secretary. The revised Guidelines were released late in December, to become effective on January 2, 1979.

The detailed analysis led by the HEW General Counsel had added a few weeks to the long period of revision. One result was a modest additional burden of procedural safeguards, but this was offset by the removal of any grounds for complaint from the most fervent dissident that the public had not been exhaustively consulted.

There were important achievements in the revision. The new Guidelines contain provisions for continuous and orderly evolution of the rules--even to their eventual elimination when the need passes. Many experiments now judged to be harmless are exempt, and containment for other kinds of experiments has been reduced. Also, the discretion and responsibility for observing the rules are beginning to return to the research institutions, where I believe they belong.

Attempts to enact statutory regulation of recombinant DNA experimentation in the United States need not be revived soon. One hopes they will not, for some of the medieval features of the first bills tended to reappear as later ones passed through the committees. A problem remains, however, in the limits on NIH's ability to protect proprietary data submitted to the RAC. Actions taken by Secretary Califano upon release of the Guidelines, to have regulatory agencies (the Food and Drug Administration, the Environmental Protection Agency, etc.) use their existing authorities to extend the Guidelines over research in the private sector, have been helpful in exploring an alternative to a new law.

MORAL

It is possible that the "recombinant DNA affair" will someday be regarded as a social aberration, with the Guidelines preserved under glass. Even so, we can say the beginnings were honorable. Faced with real questions of theoretical risks, the scientists paused and then decided to proceed with caution. That decision gave rise to dangerous overreaction and exploitation, which gravely obstructed the subsequent course. Uncertainty of risk, however, is a compelling reason for caution. It will occur again in some areas of scientific research, and the initial response must be the same. After that, the lessons learned here should help us through the turbulence that is sure to come.

(This paper was given by W.J. Gartland, Director, Office of Recombinant DNA Activities, NIGMS, Bethesda, Maryland, USA.)

THE EARLY HISTORY OF THE RECOMBINANT DNA MOLECULE PROGRAM ADVISORY COMMITTEE OF NIH

H. DeWitt Stetten, Jr.

National Institutes of Health, Bethesda, Maryland 20205, U.S.A.

It was in October of 1974 that the Director, NIH, Dr. Robert Stone, responded to an invitation from the National Academy of Sciences to establish a committee to investigate and give guidance to the developing field of recombinant DNA molecule research. As Deputy Director for Science, I was invited by Dr. Stone to chair this Committee, with Dr. Leon Jacobs as co-chairman, and Dr. William Gartland and Dr. S. Stephen Schiaffino both acting as executive secretary. Other members of the Committee were: Drs. Adelberg, Chu, Curtiss, Darnell, Falkow, Helinski, Hogness, Littlefield, Setlow, Szybalski, and Thomas.

Our Committee was finally authorized by the Secretary, HEW, shortly before the meeting at Asilomar Conference Center in February 1975. This meeting, it will be recalled, was sponsored by the National Academy of Sciences and supported by the National Science Foundation and the NIH, and was designed to air the anxieties of the many scientists and to consider appropriate containment procedures to reduce the presumed hazard to acceptable levels. It was during this meeting that most of us heard for the first time about biological containment, which was to supplement the far better recognized physical containment that had long been familiar to the microbiologist. The Asilomar meeting was 2 1/2 days of unalloyed excitement. The mood was missionary, relatively little data were presented, and most of the participants were definitely set on fire by the procedure. Clearly the 15 invited representatives of the press were also excited by what they heard, and the matter was abundantly reported in newspapers and in popular magazines.

The Committee held its first meeting in San Francisco on the day immediately following Asilomar (February 28, 1975). This was largely an introductory meeting in which the several members expressed views as to how we should proceed and what would be required of us. A wide diversity of viewpoints was apparent among the several Committee members, ranging all the way from those who felt that recombinant DNA experiments entailed little or no increased risks to those who were deeply concerned about the possibility of massive adverse impact on our environment and our ecosystem. The Committee recommended that the Asilomar document, still in preparation at that time, should form the basis of actions by the NIH and the scientific fraternity pending the development of its own set of guidelines.

On May 12 and 13, 1975, the Committee met in Bethesda, and at that time authorized one of its members, Dr. David Hogness, to chair a subcommittee to draft a set of guidelines which would then serve as a subject for discussion at the following meeting. Temporary compliance with the Asilomar recommendations was reaffirmed.

On July 18 and 19, 1975, when the Committee reconvened in Woods Hole, Massachusetts, it had before it the Hogness draft which served as the working agenda. Sentence by sentence, paragraph by paragraph, this draft was reviewed, discussed, and modified. After two days of extremely hard work, a completed set of guidelines was established which, taken as a whole, appeared to be acceptable to all members of the Committee. There were, of course, many individual areas of disagreement, and certain of the specific recommendations were argued with considerable heat. However, I sensed that there was general satisfaction with the guidelines considered in their entirety.

This draft of the guidelines was fairly widely circulated among interested scientists in all portions of the country, and it elicited a large amount of correspondence. Many of the initial letters received in my office expressed distress at the laxity of the guidelines, and indeed our Committee was charged with being disloyal to the "spirit of Asilomar." An attempt was made to respond to every single letter. Curiously, this first wave of correspondence was followed by a second wave, possibly somewhat less numerous but equally heated, criticizing the Committee with being too restrictive and too severe. A determination was made in view of the many and diverse opinions received to try once again, and Dr. Elizabeth Kutter was asked to chair a new subcommittee to try rewriting the guidelines in the light of the several critical comments. The subcommittee consulted many scientists, both on the Committee and off, and produced its draft well in advance of the meeting at La Jolla (December 4-6, 1975). In preparation for that meeting, Dr. Bernard Talbot assembled what I have called the "Variorum Edition," which displayed the Hogness version, the Woods Hole version, and the Kutter version, showing approximately 250 variant readings among these three drafts. This permitted the Committee in an expeditious manner to consider each point of variation, to select from among the three readings the one which seemed preferable, or if none was satisfactory to draft yet another replacement. The entire exercise took two long days of work. It appeared to me that for the first time the Committee was working with an extreme unity of purpose and it worked very well. By the end of the second day, the first edition of the guidelines was essentially completed.

The work of the Committee was substantially forwarded by the fact that, immediately prior to its meeting at La Jolla, in a scientific session, announcement was made by Dr. Roy Curtiss III that after many months of labor he and his colleagues had succeeded in producing a host-vector system which appeared to meet the criteria of what was then called EK2. Up to that time, whereas such a host-vector system had been postulated in the guidelines, there was no assurance that it actually could be attained. The members of the Committee were now quite reassured and therefore placed new and high confidence in the practicality of the concept of biological containment. This set of guidelines was delivered to Dr. Donald S. Fredrickson, now Director of NIH, for his consideration.

The judgment was made that a far more public airing of the guidelines was indicated. Whereas there had been two non-scientist members added to our Committee--Dr. Emmette Redford, a political scientist, and Dr. LeRoy Walters, a theologian ethicist--extensive review by the lay public had not been provided. It was therefore determined to submit the guidelines for review by the Advisory Committee to the Director, NIH, which met on February 9 and 10, 1976. For the purposes of this review, the Committee was supplemented by a number of *ad hoc* members drawn from consumer groups, the law, and other sectors of the population. The history and background were presented, and the guidelines were reviewed in a fairly thorough fashion. Many verbal comments were collected from the audience at this time, and over the ensuing weeks written comments were also collected. These were reviewed in considerable detail by the Director, NIH, who selected from among these those alterations which he deemed acceptable. All of these suggested alterations were, in turn, delivered to the Recombinant DNA Molecule Program Advisory Committee which, at its meeting on

April 1 and 2, 1976, considered each one seriously and recommended its acceptance, modification, or rejection. From the synthesis of all this information, the final publishable version of the guidelines was drafted in Dr. Fredrickson's office and ultimately was published in the *Federal Register* of Wednesday, July 7, 1976.

During all of these months, collateral events were occurring. The Committee had sponsored the initiation of certain risk-assessment studies, and these were undertaken by Drs. Wallace Rowe and Malcolm Martin. They were slow in getting started because they required access to a P4 laboratory which at that time did not exist in the United States. Also, it was becoming increasingly apparent that whereas a number of positive and useful results were surfacing from recombinant DNA experiments, none of the anticipated hazards had as yet been encountered. In the minds of an increasing number of scientists, there was a growing concern that the hazards which had been aired at Asilomar had probably been vastly overstated. Furthermore, by the time the guidelines had been in effect for awhile, it became apparent that they contained a substantial number of unclear passages, a limited number of inconsistencies, and a very large number of gaps--namely, experimental situations for which no provision had been made. To some of us it had been evident from the start that no set of guidelines could be drafted which would cover all possible situations. It was certainly my belief, and I feel it was shared by most of the Committee, that since we had drafted the guidelines we would at any time have the authority to redraft the guidelines, to amend them or to alter them as the current situation demanded. It turned out, however, that this was not the case. We had failed to provide in the guidelines the machinery necessary for amendation or revision. This defect became exaggerated as the guidelines, which we had construed as merely giving guidance, were progressively more and more frequently interpreted as if they were regulations which could not be breached under threat of sanction. Whereas interpretations of the guidelines were appropriate business of the Committee, it was ruled that exemptions from the guidelines were unacceptable. Somewhat frustrated, the committee almost immediately turned its attention to the drafting of a revised set of guidelines. As chairman, I asked Dr. John Littlefield to head a subcommittee which prepared a revised draft. This subcommittee met on March 17 and April 15 and 16, 1977, the responsibilities for the several sections of the guidelines being assigned to Dr. Emmett Barkley, Dr. Susan Gottesman, Dr. LeRoy Walters, Dr. Donald Helinski, and Dr. Wallace Rowe. Meanwhile, one other event of considerable importance occurred when a meeting was convened in Falmouth, Massachusetts, in June 1977 to consider specifically the role of *Escherichia coli* K-12 as a host organism. The reason for doing this was bitter criticism of this organism on the part particularly of a group of Boston microbiologists, who felt that the close association between *E. coli* and the mammalian intestinal tract rendered this an unsuitable choice of organism. The experts assembled in Falmouth, however, concluded quite firmly that *E. coli* K-12 was an excellent choice of host.

Yet another important finding came from the laboratories of Wallace Rowe and Malcolm Martin when they concluded the initial risk-assessment experiments first conceived during the La Jolla meeting of the Recombinant DNA Molecule Program Advisory Committee. When these results were assembled, it turned out that the infectivity to the mouse of polyoma virus DNA was reduced by a factor of approximately a billion when the DNA was inserted in a plasmid into *E. coli* K-12. In other words, insertion of this particular DNA into *E. coli* K-12 not only did not enhance the hazard but, on the contrary, provided highly effective containment. In the opinion of some investigators, these results can be generalized to mean that insertion of DNA from a pathogenic virus into a prokaryote host may be expected to reduce virulence.

The final preparation of the revised version of the guidelines was conducted predominantly in the Office of the Director and will, I am certain, be covered by Dr. Fredrickson. Meanwhile, the Committee continued to meet and to consider problems presented to it.

The Committee ruled upon such matters as were construed to be interpretations of the guidelines, and these rulings were reviewed by Dr. Fredrickson. The Committee was instructed to abstain from making any rulings which might be considered exemptions from the guidelines. Such matters were deferred pending completion of the revision. The notion developed and became an intrinsic portion of the revised guidelines that the introduction of DNA from a microorganism into that same species of microorganism constituted no burden of risk. Likewise, the introduction of DNA from one microorganism into another constituted no significant increment of risk, provided these two microorganisms exchanged genetic information in their natural environment. The agreement developed and was incorporated in the revised edition that experiments of these sorts, whereas technically they were recombinant DNA experiments, could be excluded from the surveillance of the guidelines, and this action has been taken. Many other changes in the revised guidelines resulted in reduction of containment levels, and indeed P4 containment is not assigned as a containment level to any experiment. A substantial number of experiments, particularly those involving cells of higher organisms including primates as hosts, were assigned to a category of case-by-case consideration.

By April of 1978, I felt that my potential contribution to the work of this Committee was past. My own thinking had moved further and further in the direction that the increment of hazard which resulted from the recombinant DNA manipulations was sufficiently low that it could be in effect disregarded, and that the hazards in such experiments stemmed from the microorganisms which were used rather than from the recombinant DNA activities. For these reasons, I had come to regard the guidelines and the machinery which derived from the guidelines, including the several committees--both NIH and institutional--as no longer cost effective. For all these and other reasons, I asked Dr. Fredrickson to name my successor as chairman of the Recombinant DNA Molecule Program Advisory Committee and I was greatly relieved when this was accomplished. The choice of Dr. Jane Setlow as the second chairman ensures that the Committee will continue to function scrupulously and effectively.

RECOMBINANT DNA MOLECULE PROGRAM ADVISORY COMMITTEE AND TERMS OF MEMBERSHIP

Dr. Edward A. Adelberg, 1974-78
Dr. Ernest H. Chu, 1974-76
Dr. Roy Curtiss III, 1974-77
Dr. James E. Darnell, 1974-77
Dr. Stanley Falkow, 1974-76
Dr. Donald R. Helinski, 1974-78
Dr. David S. Hogness, 1974-77
Dr. John Littlefield, 1974-78
Dr. Jane Setlow, 1974-78
(extended to 1980 as chairman)
Dr. Waclaw Szybalski, 1974-77
Dr. Charles A. Thomas, Jr., 1974-76
Dr. Elizabeth Kutter, 1975-79
Dr. Wallace P. Rowe, 1975-79
Dr. John Spizizen, 1975-79
Dr. Emmette S. Redford, 1976-79
Dr. Peter R. Day, 1976-80
Dr. LeRoy Walters, 1976-80
Dr. Allan M. Campbell, 1977-81
Dr. Susan K. Gottesman, 1977-81
Dr. Richard B. Hornick, 1977-81
Dr. Milton Zaitlin, 1977-80
Dr. A. Karim Ahmed, 1979-82
Dr. David Baltimore, 1979-82
Dr. Francis Broadbent, 1979-80
Ms. Zelma Cason, 1979-81
Dr. Richard Goldstein, 1979-82
Ms. Patricia King, 1979-82
Dr. Sheldon Krimsky, 1979-81
Dr. Richard Novick, 1979-81
Dr. David Parkinson, 1979-81
Dr. Ramon Pinon, 1979-80
Dr. Samuel Proctor, 1979-80
Mr. Ray Thornton, 1979-82
Dr. Luther Williams, 1979-81
Dr. Frank Young, 1979-80

HOW THE NIH RECOMBINANT DNA MOLECULE COMMITTEE WORKS IN 1979

Jane K. Setlow

Biology Department, Brookhaven National Laboratory, Upton, New York 11973, U.S.A.

Many members of the Recombinant DNA Molecule Advisory Committee (RAC) currently think that to a large extent what were earlier considered potential hazards of recombinant molecule research do not exist in reality. Thus the task before us is to attempt to assemble documentation and justification for any proposed relaxation or elimination of all or parts of the Guidelines. This is in some ways a rather peculiar task, as well as a difficult one. In other areas of human endeavor we are not permitted to stop our neighbor from engaging in a certain activity unless there is some evidence that this activity is harmful. In the case of the Guidelines, such a requirement was not made when the rules prohibiting certain activities were drawn up. However, to undo the rules, we must by some kind of reverse transcriptase produce evidence, much harder to acquire, that performing many recombinant molecule experiments is safe. This becomes a complex and rather weird intellectual exercise, especially since many of us who would like less strict guidelines have come to this view not because of proof of safety of recombinant DNA experiments, but rather because we now believe that the horror scenarios earlier suggested, upon which the NIH Guidelines are to some extent founded, are so improbable that we can ignore them.

Thus a large part of the future energies of the RAC will be devoted to generating vast amounts of paper containing arguments as to why various types of experiments should now be considered safe. It has become clear that the RAC will not be allowed to die until it has spewed forth enough of this kind of paper.

An example of the type of data that the RAC must produce is provided in connection with the recent approval of the use of the yeast, *Saccharomyces cerevisiae* and certain modified strains of the bread mold, *Neurospora crassa*, as host-vector systems for carriers of foreign DNA. The documentation necessary to back up this approval involved, among many other things, evidence that these organisms do not occur in the intestinal tract or sewage (considered to be escape routes for organisms carrying recombinant DNA). Fortunately, we acquired a member of the Working Group on Lower Eukaryote Host-Vector Systems (not a RAC member) who was acquainted with such published literature as the 1959 *Proceedings of the Thirteenth Purdue Industrial Waste Conference*. I would like to emphasize that it is not easy to acquire documentation like this. Molecular biologists usually do not even know anybody who reads industrial waste conference proceedings.

A few months after I became Chairman, the composition of the RAC was changed rather drastically. Whereas the number of members was down to only eleven in 1978, with two nonscientist members, the Secretary of Health, Education and Welfare announced that there would be a total of forty members, twenty-five of them voting members. Among the voting members, there were to be seven nonscientists. The general idea seemed to be that the new appointments would produce a proper balance between divergent points of view on the hazards of recombinant molecule research. Thus among the new appointments were some individuals who in the past had expressed desire for much more stringent guidelines than those finally adopted. At that time, I had my doubts that the Committee would be able to function at all, especially since our task was to provide scientific-technical advice to the Director of NIH.

We have had the first meeting of this group, and it turns out that I was wrong to think that such a committee could not get anything done. The first meeting was particularly important, in that first meetings tend to set the tone for subsequent meetings. For example, if the group divided in two warring camps, then these would likely be perpetuated in future. In spite of the fact that there were some dangerous moments, in which cries of pain and rage were heard, no polarization occurred. It was possible to have rational discussions, in the course of which the nonscientists were apparently persuaded that the scientists were trying hard to explain the technical aspects to them, and the scientists discovered that lay members could be rational.

Since the RAC is so large, preliminary reports are all written by subcomittees and working groups, of which we now have seven. There is a subcommittee for bacterial viruses and one for plasmids (these two subcommittees mostly make recommendations for approval or disapproval of applications for EK2 systems, considered to provide a high level of biological containment). These groups are also in the process of drawing up criteria for what is a minor modification of a previously approved EK2 system, such that testing does not have to be done as required for the usual new EK2 application. Each of these two subcomittees contains a nonscientist member from the RAC, and approximately one-third of each subcommittee are not RAC members, but persons appointed for their expertise. Gigantic attempts are being made to make it possible for the nonscientists to follow the conversations in these subcommittee meetings. This is not an easy task, since for example as we all know, lambda experts do talk a strange brand of gobbledegook. The new nonscientist members are more than willing to read what we tell them is essential for background, and considerable time at RAC meetings has been devoted to descriptions in lay language of the biological systems to be discussed.

Another important subcommittee, very recently appointed, is for risk assessment. Its function is: 1) to provide input from the scientific and nonscientific community to NIH concerning what risk assessment experiments should be done, and 2) to examine critically proposed risk assessment experiments and report conclusions about such proposals to the RAC. This subcommittee consists of five RAC members, two of whom, including the Chairman, Ray Thornton, are nonscientists.

There are working groups for lower eukaryote host-vectors, for prokaryote host vectors other than *E. coli*, and a small group to consider containment problems involving plants.

Because the recombinant DNA field is moving so rapidly, some of the wording in the Guidelines has become increasingly difficult to interpret. For example, the Guidelines say, "When a cloned DNA recombinant has been rigorously characterized and the absence of harmful sequences has been established, experiments involving this recombinant DNA may be carried out under lower containment conditions". In the first version of the Guidelines, "free of harmful genes" meant free of things like genes coding for botulinum toxin. Now, however, we have to worry about recombinant DNA that contains genes for mammalian hormones, or that contains intervening sequences that presumably prevent expression of genes in prokaryote hosts, or complementary DNAs made from messenger RNAs of which the intervening sequences have been processed out. Since decisions concerning the criteria for rigorous characterization and freedom from harmful genes can, under the revised Guidelines, be made by institutional biosafety committees, it is obviously important for the RAC to make some generalizations upon which the local committees can base their decisions. Thus we have a working group on criteria for characterized clones that is busy attempting to make a clear interpretation of the Guidelines to enable local groups to make more uniform decisions. The RAC apparently differs from some of the European Committees in that it has to make many more generalizations because of the responsibility delegated to the local institutional committees in the United States. For example, the local committee may now permit certain projects involving recombinant DNA to begin without prior approval from NIH.

The administrative arm of the Guidelines is the Office of Recombinant DNA Activities (ORDA) at NIH. There is a very special relationship between this organization and the RAC. Since all but two of the RAC members are not otherwise associated with NIH, and since most of the people who wish to complain about the iniquities of the Guidelines do not know our telephone numbers, one of the nicest things that ORDA does for the RAC is to take much of the heat off, so that so far being a member of the RAC has not become an occupational hazard. Thus ORDA takes the blame day after day.

ORDA processes per year around one thousand applications for permission to perform recombinant DNA molecule experiments, and most of these can be decided upon without recourse to the RAC. The Director of ORDA, Dr. William Gartland, has been enormously helpful to me personally and to the RAC as a whole, providing all kinds of advice when asked, laundering my overly straightforward prose to make it suitable for public consumption or for Dr. Fredrickson's office, and in general making life on the RAC as comfortable as possible. Perhaps the most impressive thing about ORDA is that while a number of RAC members have worried about the effects of an entrenched and expanding bureaucracy on further attempts to eliminate parts of the Guidelines, ORDA does not give any appearance of pushing to prevent its own demise.

DISCUSSION

F.S. ROLLESTON: This is a general question to all three speakers. It seems as if every time somebody invents a new scenario for recombinant DNA the NIH is forced to develop another three inches of paper for each nullification of that possible scenario. Since I fully support the obvious feeling of this meeting that the current guidelines are grossly over-stated and the feeling expressed by Dr. Stetten about the way guidelines should be, and since it is not possible in a political sense to state simply that we do not need guidelines except as described by Dr. Stetten, I wonder whether some other means of assessing risks that require the proof of harmful nature rather than proof of freedom from harm that Dr. Szybalski is referring to should not be instituted? In other words, should we not try and go to a different means of risk assessment rather than the apparently collective gut feeling approach that seems to have gone on so far?

W. SZYBALSKI: Who will answer it? Maybe we should start with Jane Setlow.

J.K. SETLOW: There is no doubt that there is going to be a great deal of money spent on risk assessment experiments and I am very anxious for the NIH RAC Sub-committee to have a part in this so that there will be at least a little shred of science that could come out of it. I am not sure that is at all an answer.

F.S. ROLLESTON: The problem with risk assessment experiments is that we will be in this mess for another three or four years at least until such experiments are done, and anyway a risk assessment experiment is trying to prove a negative which is, of course, almost impossible. I think that the GMAG in the November 9th *Nature* made a start in the direction of being able to eliminate a large number of risk assessment experiments by perhaps a different approach using numbers that one can bring together quite readily simply from past experience. One might then leave the risk assessment experiments only in those very, very few cases where they are absolutely necessary. I am not even sure that there are any needs for risk assessment experiments.

J.K. SETLOW: I think that some of this discussion can be deferred because we are going to have an account of some very illuminating ones a little bit later.

H.D. STETTEN: Certainly it would be attractive if we could say there should be no regulations until a hazard has been shown. The question has been raised at one point about the risk-benefit analyses. You heard some of the benefits, present and in the foreseeable future, of this technique. There is no actuarial data whatsoever of the risks; therefore the risk-benefit rating cannot be evaluated at this time. I don't believe that we can talk about it in any meaningful way.

W. SZYBALSKI: To pose a more relevant, if partially hypothetical question, we could imagine that we are at ground zero, from the regulatory point of view. That is, we should assume that there are no mandatory NIH Guidelines and no bureaucratic regulations. If our present scientific information on recombinant DNA techniques is just as we heard it presented during this meeting, I wonder whether there would be any letters of warning, any proposals of moratorium, any Recombinant DNA Advisory Committees, any guidelines, any mandatory regulations for the recombinant DNA technique and any legislative proposals or actual legislation? Dear Hans, could you comment on it?

H.D. STETTEN: If the risk had never been dreamed of it would not have been there because it is only a dream at this time - a fantasy. It has been pointed out by Dr. Bernard Talbot, who has been an important cog in this machine, a very important one, that when the search was originally initiated by Albert Sabin for less virulent variants of the three strains of poliomyelitis virus, someone might have raised the question: "You'd better not let him go looking for variant strains of polio virus, he may turn up a more virulent one, and if he does we are in trouble". Had that ghost been raised at that time, I can see that it is entirely possible that Sabin's work would have been put into a holding position and we might not have poliomyelitis vaccine of the oral type at this time.

I.S. JOHNSON: To continue the question which was first raised by Dr. Rolleston, I would address this to Dr. Setlow, instead of trying to re-define what is meant by "free of harmful genes", couldn't we assess what is harmful? I think that is the question that Dr. Rolleston was really trying to raise. You are continually asking us to prove a negative which has already been mentioned as a very difficult thing to do. But injecting an animal, doing pathology or developing some test system which, if nothing harmful happens, might suggest that you go ahead and do something is a positive approach. But to try and re-define "free of harmful genes", I think, is going to be a very difficult thing and continues and compounds a process which most of us are finding very difficult and tiresome.

E.B. WEISS: There was a brief presentation at the beginning of this session by a scientist on the constitutional and legal issues involved in basic scientific research, and as a lawyer I would like to comment briefly. In considering constitutional protection for basic research under the First Amendment of the U.S. Constitution, we need to distinguish between restraints on content of the research and restraints related to the time and manner in which the research is carried out. The question of whether First Amendment protection extends to the content of basic research has not yet come up before the U.S. Supreme Court, and, indeed, the very authorities which were cited in the presentation make that point. Many of us believe that basic research would be protected were the issue to arise before the Court. By contrast, it is generally agreed that time and manner restraints on the conduct of research - which may take the form of guidelines - are constitutional. The fact that such restraints may make it more difficult to conduct the research does not affect the constitutionality as long as they are not explicit content restraints. Indeed, that is the conclusion of the articles which were cited. If you are involved with the U.S. Federal Government, there is a separate body of law which may apply, and that is known as administrative law, as embodied in the Administrative Procedures Act. Federal regulations governing the time and manner of conducting research are subject to the requirement for informal rule making that they not be "arbitrary and capricious", or in cases of formal rule making, that they be supported by "substantial evidence". The question was raised whether a cost-benefit analysis must accompany regulations concerned with research.

Let me just alert you to a recent development in U.S. administrative law, which may be relevant to this question. There is a pending case in the U.S. courts which raises the issue of whether the health standard for benzene issued under the Occupational Safety and Hazards Act, is subject to a cost-benefit analysis. The initial decision in the lower court reversed established practice and said that such an analysis is required in issuing a health regulation under OSHA. That decision is now on appeal. The decision, if upheld, could have broader implications for other environmental statutes, and possibly for research guidelines concerned with public health and safety. So I would only urge you that in considering these issues, Shakespeare may have said "Out damned spot!" but the spot doesn't go away that easily, and I think it is in the interests of all, for scientists and lawyers to work together.

D. HABER: I would like to ask Dr. Setlow in the absence, as we have heard, of hard experimental data on risks, what information the NIH does require or has required in the past in order to revise the guidelines as has been done and, as we have heard today, is being done again? What kind of information does the NIH receive, discuss and accept as criteria for lowering these guidelines?

J.K. SETLOW: The answer is I wish I knew in detail because we would have it all done already. We are blundering along and attempting to collect what documentation we think will be reasonable for those above us and hoping that it will work and so I really can't answer that question. The whole situation has changed so drastical with the new RAC and the first set of decisions. I think we did remarkably well on these. The way that I would put it is that all but about 2% of the RAC decisions were accepted by the Director of NIH and so we have gone from approximate zero to 98% in one jump. I am sorry about the other 2%, but I hope we will fix those a little later. The answer to your question is that we simply don't know exactly what kind of documentation we need. We know that we need a whale of a lot of it for each step.

A.M. SKALKA: I think it is fair at this point to say that there is not an absence of hard evidence. Some of that will be presented this afternoon and we should wait until that time to discuss it.

GUIDELINES, LEGISLATION AND LABOUR RELATIONS

Sir Gordon Wolstenholme
Former Chairman, Genetic Manipulation Advisory Group, United Kingdom

10 Wimpole Mews, London W1M 7TF, U. K.

This paper does not deal with scientific problems as such; I speak wholly within my experience in the United Kingdom as Chairman, for the first two years of its existence, of the Genetic Manipulation Advisory Group (GMAG).

One of the brief recommendations of the Ashby Report in January 1975 was that "subject to rigorous safeguards ... " these techniques of genetic manipulation "should continue to be used because of the great benefits to which they may lead". The safeguards proposed by Ashby and his colleagues included: a knowledge of the techniques of handling pathogens; equipment necessary for the containment of pathogens; funding bodies to enquire, in relation to any new research proposals, whether they involve any biological hazard; the designation of a biological safety officer within each institution to advise on the appropriate level of precautions; an active search for improvements in biological containment; the performance of tests of pathogenicity on animals, and of health monitoring of workers involved in the research; and long-term epidemiological studies on the health records of those involved in the work.

The Williams Working Party in its Report, published in August 1976, elaborated in detail the practical implementation of these safeguards, and advocated a central code of practice. Because of "the present state of knowledge of the field" Williams stated that "containment measures should allow a suitable margin of safety until any areas of doubt can be clarified by further experimental evidence". Williams saw "a need for a flexible approach ... by requiring those who plan to work in the field of genetic manipulation to submit their proposals to a central advisory group ... rather than by imposing rigid guidelines". The Williams Party considered that their proposals for categorisation were generally in line with those then being developed in the United States, but believed that for the United Kingdom a central advisory body - GMAG - was an important addition, since it could advise on all genetic manipulation experiments in universities, institutions and industry. GMAG was expected to establish a dialogue with the scientists concerned, and to be ready to respond quickly to requests for advice in regard to new scientific developments, health monitoring, training and other safety-related matters.

The Department of Education and Science of the British government acted with exceptional rapidity to constitute GMAG on a very unusual, if not unique basis. Primarily it had to contain enough scientists to be able to assess expertly the many proposals submitted to it, to be in a position to offer advice from a varied background of scientific experience, and to assess competently the adequacy of laboratory conditions and procedures. In the first two years, the burden carried by the scientific members of GMAG was out of all proportion to the responsibilities of the other members, but our scientific colleagues never failed in patient explanation of the principles under discussion. The remaining members represented the universities, industry, the health services, the trade unions and the public, and a remarkable effort was made by every individual not only to understand the issues at stake but also to become themselves expert on some facet of our work, such as the constitution of safety committees, medical supervision, psychological factors, protective cabinets etc. Representatives of government concerned with science, education, health, safety, agriculture, fisheries, and food attend GMAG's meetings as assessors, and the secretariat has been provided expertly by the Medical Research Council. If the going was sometimes rough it was as much due to having an apprentice jockey as to the unusual nature of the animal and the course.

GMAG was to act as the expert advisory body for the Health and Safety Executive. In the United Kingdom the Health and Safety at Work Act, 1974, imposes a duty on the employer to protect his employees and avoid hazard to the public - precautions wholly in respect of the human race. Under the Act, the Health and Safety Executive has powers to inspect premises and laboratories and if necessary to oblige work to be stopped until appropriate safety measures are taken and approved. An employer is expected "to take all reasonably practical steps" for safety at work, and seeking and adopting advice given by GMAG would almost certainly be an adequate defence in the courts in case of legal action. Nevertheless, Williams thought that a legal regulation, requiring all laboratories to notify GMAG of their experimental procedures in genetic manipulation would be an important extra safeguard - presumably an assurance to the public that no-one would escape or defy the regulations. In the event, notification to both GMAG and the Health and Safety Executive was on a satisfactory voluntary basis for the first 20 months, until the Health and Safety (Genetic Manipulation) Regulations 1978 came into operation in August 1978 and made notification obligatory.

The public may not have known it, but other safeguards have existed all along. Although some editors of scientific journals declared their opposition, there has been a tacit understanding that papers should not be published which described work done in neglect or defiance of GMAG's approval. And the funding bodies - research councils and private foundations - could hardly risk being associated with attempts to jump the gun or cut the corners. But the main source of reassurance (in so far as it has been desirable) for the public, the Health Executive and GMAG, has been and remains the local Safety Committees. Regulations of the Health and Safety Commission on Safety Representatives and Safety Committees (SI 1977 No. 500) came into legal effect on 1 October 1978, and compels employers and recognised trade unions to come to agreements on the establishment of properly constituted and representative safety committees. The unions are thereby in a position of widespread power and responsibility in regard to all people at work, which is defined in such a way as to include almost everyone who is active in anything. This applied to the main safety committees in all industries and institutions; GMAG itself has attached great importance to the extra formation of special safety committees or sub-committees in each centre where genetic manipulation is carried out. I regard these as far and away the best guarantee that work will be carried out properly. "Properly" implies that the work is done after full discussion with all relevant staff or their representatives; with their full understanding of the approved procedures; with their adequate prior training; with their knowledge of health risks, including counter-indications to personal involvement; with their full compliance in the appropriate containment and other safety procedures; and above all, their mutual acceptance and observance of the requisite laboratory discipline. Any significant departure from the detailed experimental proposals - including any change in the people doing the work - as approved firstly by the local safety committee and then agreed by GMAG, should be immediately discernible and made a matter for record and report.

It is at this point, rather than in dealing with the original proposals, that the bureaucratic disadvantages of GMAG are likely to be most evident and resented. If GMAG is to continue at all, two essential modifications of its procedure appear to be needed: the secretariat must continue to be available on the telephone but with greater powers of quick decision; and second, where all low category work is concerned approval must be given to institutions and staff rather than to experimental procedures - provided that full records are kept and that regular reports are made retrospectively to GMAG. Such a "licensing" procedure would put even more responsibility on the local safety committee for day-to-day monitoring of experimental production and storage techniques.

Probably most of the people at this conference will feel - I use this emotional word deliberately - that the risks of genetic manipulation are so remote, so much less significant than so many other activities in life, that GMAG should now be disbanded and discarded; it had a value in the early days of inflammatory alarm and scientific ignorance, but no longer serves any good purpose. For industry in particular, it constitutes a delay, a threat to the important possession of patents, and a disincentive to investment. Further, the trade unions have used GMAG to put forward two proposals which are seen as sinister threats to the freedom of scientific research: the first, that membership of safety committees should be restricted to union members (though the unions demanding this would seem unlikely to accept the implicit responsibility); and the second, that GMAG and safety committees should have the power to assess the scientific merit, as well as the health and safety implications, of the proposed projects - a power extending to the use of a veto.

It seems to me a nonsense that GMAG as the national safety committee, or any local or institutional safety committee, is so constituted as to be able to fulfil any such function. And it is nonsense to think that ethical or social views which are held in total ignorance of the basic science involved would be preferable from the point of view of the public interest to the elaborate systems of peer judgment operated by research-funding agencies. Demands for GMAG, or any similar body, to extend its powers to cover scientific assessment would destroy GMAG, and much more important, fatally injure the scientific innovation in genetic engineering on which many hopes for the future depend.

Something of value has been achieved in the GMAG type of discussion between science and society, and I "feel" that those who call for total freedom of scientific research and of industrial development, equally with those who call for political control of research, would both have reason to regret their destruction of this useful forum for mutual understanding, compromise, and public reassurance.

There is another justification for the continuation of GMAG and similar bodies elsewhere, namely in the international affairs so well illustrated by this conference at Wye. The genetic manipulation liaison committee of the European Science Foundation has already made a major contribution to international cooperation in this field of work. There have been arguments within that committee over whether the UK GMAG or NIH Guidelines should be used as a model - before either pattern of work was reasonably settled. The UK GMAG's first chairman, at least, has been in favour of minimal agreed recommendations within the European Community, to discourage the genetic equivalent of a "tax haven". But in the international sphere as a whole I should like to see the International Council of Scientific Unions (ICSU), in the form of COGENE, exercising a world-wide monitoring and advisory role. Just as the UK GMAG has governmental department representatives present as assessors at its meetings, so COGENE should probably have attendant on it people nominated by such bodies as the World Health Organisation, the Food and Agriculture Organisation and the International Court of Justice.

Whatever is achieved on the international scene, if genetic manipulation is to progress to the ultimate benefit of mankind, then I believe that some form of GMAG, properly representative of both science and society, will be required, continuing where it exists and being created where up to now it does not. Within each nation it should be the focus of public confidence in research and innovation in regard to genetic manipulation. And internationally it is needed to serve as the recognisable identity of each country's work in this field, in relation to whatever can be achieved in the way of worldwide cooperation by the International Council of Scientific Unions.

DISCUSSION

C. WEISSMANN: I would just like to comment that in Switzerland we do not have to apply to any committees or authorities. The Academy has recommended that we follow the American guidelines but the person responsible for the work is the group leader or the Director of the Institute. I am glad to say this has worked out very well. We haven't had a single casualty since the guidelines were instituted.

RECOMBINANT DNA GUIDELINES AND LEGISLATION

J. Tooze

European Molecular Biology Organization, Postfach 1022.40, 69 Heidelberg, F.R.G.

Concern about the conjectural hazards of recombinant DNA research was first voiced in the USA and was quickly taken up in the UK where in January 1975 the Ashby Report[1] was published, the first governmental publication on the subject. Several contributors to this Conference have discussed the history of events in the USA and the UK. Elsewhere the recombinant DNA debate and the seemingly inevitably concomitant development of national guidelines and discussion of legislative measures was altogether slower to develop.

Taken by surprise by the "Moratorium" letter published by the US National Academy's Committee in June 1974[2], and having played a relatively passive role in the Asilomar Conference, many molecular biologists outside the USA and UK acquiesced in the adoption of the proposals made by the organizing committee of the Asilomar Conference[3]. During the next two years, slowly but surely, working parties and committees, briefed to develop national guidelines and recommend appropriate administrative or legislative procedures, were established under the auspices of a wide variety of agencies ranging from government ministries through research councils and other public or semi-public agencies to national academies and purely private organisations.

European international organizations, the European Molecular Biology Organization, the European Science Foundation, and the Commission of the European Economic Community in that sequence also took up the subject, as, of course, did the World Health Organization and the International Council of Scientific Unions (ICSU). In 1977 ICSU formally established COGENE, the initiator of this meeting at Wye.

As the pace of events in the USA accelerated many European countries found themselves in a position resembling the tailenders in a race about to be lapped by the front runners. For in the autumn of 1977 the NIH was already discussing first drafts of revised Guidelines[4], which, significantly less stringent and more reasonable than the initial NIH Guidelines[5], officially came into force on 2nd January 1979. (The months between the late autumns of 1977 and 1978 were well spent in the USA removing from early drafts of the revised Guidelines anomalously stringent precautions for certain classes of experiments - those involving viral nucleic acids, for example.)

The discussion of revised NIH Guidelines that began in the autumn of 1977 certainly influenced some of the European countries still in the process of deciding their first national guidelines. At that time France was reaching such a decision and, to say the least of it, knowledge of the trends in the USA certainly helped the French national committee win official acceptance for the French Guidelines[6], which in their containment categorisations are close to the revised NIH Guidelines and enviable in their simplicity, brevity, and freedom from bureaucratic clutter.

On the other hand several countries, lapped by the USA, stuck to their initial course. While the NIH was busy preparing and clearly destined to adopt the less stringent revised version, Federal Germany, for example, continued to draft and, in 1978, finally adopted national guidelines[7] closely resembling the original and stringent NIH Guidelines. Revised Federal German Guidelines are now being drafted; allegedly they are more relaxed than the existing guidelines but having missed the opportunity, which France seized, to begin with reasonable guidelines our Federal German colleagues must negociate constitutional, legal, and bureaucratic obstacles to change. Is it perhaps an ominous sign that the Federal Ministry of Research and Technology is still considering new legislation to provide a legal framework for national guidelines even though German research organisations last year advised against rushing through any new law? For constitutional reasons, it is fortunately not possible to enact a new law before 1981, by which time it may well be recognized as unnecessary by even its present proponents.

The situation in the Netherlands has been nothing less than desperate for anyone wishing to use the recombinant DNA technique. A series of governmental crises, consequent ministerial changes, and the existence of a vocal minority of scientists calling for an "alternative" science policy of greater social relevance, have resulted in confusion and great obstruction. A report published in March 1977 by a Committee of the Royal Netherlands Academy of Arts and Sciences[8], drawing inspiration from the initial NIH Guidelines and the William's Report[9] published in the UK in August 1976, recommended very stringent safety precautions. It was never officially adopted and later that year there seemed to be a possibility that the then government might totally prohibit recombination DNA research. There are still no national guidelines, but, with the minister who considered a total ban no longer in office, the Academy's Committee last year urged some relaxation, and experiments involving the lowest two levels of physical containment have gone ahead. Individual universities are at present applying their own regulations. Requests to construct C3 laboratories have led to heated debates and the situation remains confused. Some believe legislation would be advantageous to pre-empt crippling local regulations.

Amongst the Scandinavian countries, Finland has recently established a committee under the Central Medical Board to propose how recombinant DNA should be regulated, at the same time the Ministry of Health is preparing a revision of the law on the control of infectious diseases. The two may well be linked. There is, of course, the danger that such a link will strengthen the suspicion in the public mind that recombinant DNA research is inherently dangerous. On the other hand, by comparison with the proven hazard of pathogens, the conjectural hazard of recombinant DNA should pale into insignificance. In Norway a national committee set up under research council auspices has recommended against any specific new legislation; this committee, together with officers of the Central Public Health Laboratory, will continue to supervise any work, almost certainly using the revised NIH Guidelines. In Sweden, as a result of a report by a lawyer appointed by the Ministry of Education, no new legislation is contemplated since legislation on safety at the work-place already exists. A new committee with five scientists and five lay members is to be appointed by the Government to supercede the present committee which was established by the research councils, and again the revised NIH Guidelines will probably form the basis of decisions. In Denmark the present Registration Board[10], set up by the Medical Research Council, has proposed that responsibility for supervision of recombinant DNA work should rest with the Department of Health of the Ministry of the Interior, but, until the fate of a draft directive being discussed in the EEC is known, changes to the present voluntary system are unlikely.

Italy, Ireland, and Switzerland see no need for new legislation, the voluntary systems of registration presently used or to be introduced are considered adequate. In Switzerland the initial US Guidelines have been followed on a voluntary basis. The Committee for Experimental Genetics of the Swiss Academy of Medical Sciences maintains a registry of current recombinant DNA research and it has always insisted that the responsibility for the consequences of any research with micro-organisms rests with the principal investigator who must make the final decision about appropriate safeguards. The existing Law on Epidemic Diseases is considered an adequate legal background. Italian national guidelines have recently been adopted, although I have not yet seen them, they are apparently based on the revised US Guidelines, but emphasize flexibility, and compliance is voluntary for research not funded by the Government or research councils. The national committee, under the auspices of the Ministry of Health, will in 2 years time review the guidelines and decide whether or not any are still necessary. In Ireland new legislation is not considered necessary.

Outside Western Europe and the USA national guidelines or national committees to consider guidelines have been established in Australia, Canada, Israel, Japan, New Zealand, the Soviet Union and elsewhere. All of these have been influenced by the sequence of events in the USA, and to a lesser extent by events in Europe, and the guidelines reflect this. In general, the earlier the guidelines were drafted, the more stringent they tend to be. The South African Guidelines are in some respects even more stringent than the initial NIH Guidelines. The Canadian MRC Guidelines[11] are unique in that they also apply to work with cells in culture and animal virus research; also they specify six levels of physical containment.

In summary in most of Western Europe new legislation is not being contemplated; existing "voluntary" systems are considered perfectly adequate. The revision of the NIH Guidelines has had great influence and for the smaller countries the NIH model is easier to follow than that of the UK GMAG. The French Guidelines closely resemble the revised US Guidelines. The Federal German national committee is in the process of aligning its guidelines with those of the USA and France. In Norway and Sweden the likelihood is that the NIH Guidelines will form a basis for national decisions. In Switzerland the US Guidelines have so far been followed and it is hard to believe that the revised NIH Guidelines are without influence there.

The belief that the conjectural hazards associated with this research were initially over-exaggerated is gaining ground in virtually all countries. This is being reflected either in new guidelines or individual decisions about the appropriate containment for particular experiments. Indeed in some countries it is being increasingly recognized that the recombinant DNA technique can offer a comparatively safe way of studying the genomes of dangerous pathogens. But bureaucracy remains rampant and will probably increase as it finds less and less to do. The Commission of the EEC is still pursuing a directive which, if ever passed by the Council of Ministers, would enshrine bureaucracy by legally compelling member states to establish national committees. Any attempts to promote international agreement on guidelines or legislation should in my opinion be strenuously resisted. They would be doomed to failure and would simply waste time, money, and intellectual resources, of none of which is there a superabundance.

In conclusion the risks of accidentally creating a biohazard by recombinant DNA research remain entirely conjectural. In some countries, however, seemingly commonsensical and responsible proposals to divert scarce financial and scarcer intellectual resources to risk assessment experiments are being made. I believe, the only worthwhile way to achieve an assessment of the risk of this technique is to encourage its use in unfettered scientific inquiry.

The rate at which guidelines can be relaxed will necessarily vary from country to country because of political factors. In most countries, however, it should be possible to exempt from recombinant DNA guidelines all self cloning experiments which only mimic natural genetic events, and for the same reason all cloning experiments involving the DNAs of pairs of species that exchange genes in nature. In the USA these classes of experiments are already exempt - why not everywhere else? There is also a case for excluding from guidelines all experiments in which E. coli K-12 is used as a host with non-transmissible vectors because there is ample evidence that E. coli K-12 cannot become an epidemic pathogen.

If significant risks exist they will surely manifest themselves and become worth studying. If they are not significant they will not become manifest and aren't worth studying. As Dr. Johnson wrote in Rasselas: "Nothing 'replied the artist', will ever be attempted if all possible objections must first be overcome."

References

1) Report of the Working Party on the experimental manipulation of the genetic composition of micro-organisms (1975). H.M.S.O. London Cmnd 5880.

2) Berg, P., Baltimore, D., Boyer H., Cohen, S.N., Davies, R.W., Hogness, D.S., Nathans, D., Roblin, R., Watson, J.D., Weissman, S., and Zinder, N.D. (1974). Science 185, 303.

3) Berg, P., Baltimore, D., Brenner, S., Robin, R., and Singer, M.F. (1975). Science 188, 991-994.

4) Guidelines for Research Involving Recombinant DNA Molecules (1978). Federal Register 43, 60108-60131.

5) Recombinant DNA Research-Guidelines (1976). Federal Register 41, 27902-27943.

6) French Guidelines on Recombinant DNA Experiments (1977), Délégation Générale à la Recherche Scientifique et Technique, Paris.

7) Richtlinien zum Schutz vor Gefahren durch in vitro neukombinierte Nukleinsäuren (1978). Der Bundesminister für Forschung und Technologie, Bonn, F.R. of Germany.

8) Report of the Committee in Charge of the Control on Genetic Manipulation (1977). Koninklijke Nederlandse Akademie van Wetenschappen, Amsterdam, The Netherlands.

9) Report of the Working Party on the Practice of Genetic Manipulation (1976). H.M.S.O. London Cmnd 6600.

10) Rapport on Genetic Engineering (1977). Forskningssekretariatet, Copenhagen.

11) Guidelines for the Handling of Recombinant DNA Molecules and Animal Viruses and Cells (1977). Minister of Supply and Services Canada, Cat.No.: MR21-1/1977.

SUMMARY OF REPORT TO COGENE OF THE WORKING GROUP ON RECOMBINANT DNA GUIDELINES

S. N. Cohen

Departments of Genetics and Medicine, Stanford University School of Medicine, Stanford, California 94305, U.S.A.

The Working Group on Recombinant DNA Guidelines was established by COGENE in May 1977 and charged with the responsibility of 1) obtaining information about the status and content of Recombinant DNA Guidelines in different nations and 2) analyzing, comparing and evaluating the provisions of the various national guidelines and the premises on which these provisions are based.

Shortly after its formation the Working Group prepared and circulated a short questionnaire to the scientific liaison for each nation represented in the International Council of Scientific Unions. A second questionnaire was distributed to the appropriate scientific representative of each nation which had indicated that recombinant DNA experimentation and/or the preparation of guidelines governing this research was underway.

By the time of the meeting of the Working Group on April 7-8, 1978, completed questionnaires and/or copies of the national guidelines had been received from 39 countries having scientific societies affiliated with ICSU. The accumulated questionnaire data and the results of the Working Group's review and analysis of the guidelines documents themselves form the basis for this report. Other information available to the committee includes reports and compilations furnished by the European Molecular Biology Organization (EMBO), the National Institutes of Health, and other sources.

At the end of March 1978, approximately 365 recombinant DNA projects were estimated to be underway in 180 laboratories in 20 countries. Five of these were at a high level of containment, 75 were at a moderate level, and the remainder at a level of containment designated by the respondent as "low". Twenty nations had drawn up guidelines for recombinant DNA experimentation, and these were in force in 13 nations. Five of the nations had prepared their own guidelines, whereas the remainder had adopted or modified guidelines of the United States or the United Kingdom. Of the responding nations that had not drafted guidelines by the time of the response, 12 intended to establish guidelines, while 6 indicated that they did not plan to establish formal guidelines. Approximately 15 recombinant DNA projects were reported as being underway in nations where no guidelines existed.

The guidelines of one nation were reported as being entirely voluntary. Ten nations have guidelines that are enforceable to varying extents through a research funding mechanism, and 2 nations have legally enforceable guidelines. The mechanism for authorizing research varies in different nations: in some nations authorization is provided to individual scientists, whereas in others it is granted to specified laboratories or institutions. Six require specific training for researchers carrying out recombinant DNA experiments and 5 offer training courses. Although the guidelines of 8 nations indicate that the precautions required for recombinant DNA experimentation can be modified on the basis of experience in this area of research, only 4 nations specify any criteria or mechanism for such modification.

A tabulation of information obtained from completed questionnaires and from examination of guidelines is contained in the body of this report, which is based largely on information available to the Working Group in mid-1978. In addition, levels of containment specified for certain experiments in the revised United States guidelines of December 22, 1978 are shown for comparison.

The Working Group's analysis of various national guidelines and of the responses to questionnaires has led to the following conclusions:

1. Guidelines for the conduct of recombinant DNA experiments have been prepared by a number of nations whose national scientific bodies are members of ICSU. Some nations in which recombinant DNA experimentation is carried out have not promulgated formal guidelines, and some do not intend to do so.

2. The guidelines of most nations have been derived directly or indirectly from those of the United States, which were the first to be drafted, and the assumptions implicit or explicit in the United States guidelines have been incorporated in guidelines of most other nations. The U.S. guidelines have in turn been based largely on premises developed originally during discussions at the Asilomar Conference. These have included the following assumptions:

   a). While no specific indication has been found that recombinant DNA experimentation poses a unique hazard to man or to the environment, the absence of certainty that the research is safe makes necessary the establishment of special procedures to govern work in this area.

   b). Based on experience obtained in other areas of research, it is possible to avert potential hazards of microorganisms containing recombinant DNA molecules by using tested containment procedures to prevent their dissemination.

   c). The extent of potential hazards is determined by the biological origin of the DNA used. The potential risk presented by DNA of eukaryotic organisms is perceived to reflect the closeness of the evolutionary relationship between the donor species and the species at risk, which is taken by most national guidelines to be man. For this reason, DNA from plants and invertebrates is regarded as constituting a relatively low hazard, whereas DNA from mammals is considered to be the most hazardous. Viruses of eukaryotes are commonly classified according to the degree of evolutionary relatedness between their natural host organism and man, as well as the pathogenicity of the virus itself. DNA from prokaryotes unrelated to the commonly used recipient organism, *E. coli* K-12, requires a relatively high level of containment by most guidelines, despite the evolutionary remoteness of these organisms from man.

d). Classification of potential hazard is influenced by the purity and homogeneity of the DNA employed, and is reduced for pure and homogeneous preparations so long as such preparations are not known to encode harmful products.

e). The use of recipient organisms having reduced survival outside of special laboratory conditions aids in the containment of recombinant DNA.

f). The likelihood of success in containment decreases when increased numbers of organisms are used.

3. Although fine gradations of risk and containment are specified in most national guidelines, no scientific justification for such distinctions, or for the view that phylogenetic considerations should determine the extent of containment, is provided in any of the guidelines we have analyzed. Rather, the basis for such determinations appears from our analysis to be entirely conjectural.

4. Despite the patterning of guidelines of other nations after those of the United States, there are major differences in the various national guidelines. Some have employed an encyclopedic approach which attempts to anticipate and specify conditions for all possible forms of recombinant DNA experimentation, while others have adopted an approach that utilizes case-by-case analysis in determining containment. Some have the force of regulations, while others are treated simply as recommended practices analogous to those routinely used for work with certain non-recombinant microorganisms. Some guidelines relate only to organisms that contain recombinant DNA molecules, while others include work with the DNA itself.

5. The level of precaution required for identical experiments in different nations is strikingly different. In some nations, the guidelines specify the maximal level of precaution required, and a national committee may reduce containment from this level under certain circumstances. In other nations, the guidelines specify the minimal level of precaution required and this containment may be increased by the actions of local or subnational bodies.

6. The level of containment required for work with recombinant DNA that includes genetic material from known pathogens is often greater than the containment recommended for work with the pathogen itself, although recombinant DNA methods would appear to provide a means of studying medically important disease-producing organisms at reduced risk. Our analysis of guidelines has revealed no scientific justification for this practice, which has resulted in the effective prohibition in some nations of work with segments of DNA derived from medically important microorganisms.

7. In general, those national guidelines that have been prepared more recently are less restrictive than are the earlier guidelines. This situation appears to reflect a changing assessment of the risk of recombinant DNA experimentation, which, in turn, has been attributed by guidelines' authors to the accumulation of more knowledge and greater experience in these areas of research. While revision of the original guidelines promulgated

by several nations has occurred or is currently in progress, the guidelines of most nations do not have a time limitation clause.

The status of guidelines for recombinant DNA experiments is undergoing rapid change at this time. This appears to be the result of a changing assessment of the conjectural risks that led to the initial formulation of guidelines and regulations for research in this area, and the realization of some of the scientific benefits that had been anticipated earlier from the use of this research tool. The information contained in this report represents the state of affairs at the time the data were gathered. Because substantial changes in the guidelines of several nations occurred recently, some of the information contained in this report is no longer current; for this reason, the Working Group has issued a draft report, rather than a formal document. The information contained in that report may be of value to individual scientists as well as to recombinant DNA national advisory committees. A subsequent report by the Working Group will reflect recent changes in the various national guidelines.

The report distributed to participants at the Conference and this summary statement have been prepared by the Working Group on Recombinant DNA Guidelines, Committee on Genetic Experimentation, International Council of Scientific Unions:

S. N. Cohen (U.S.A.), Convener

G. L. Ada (Australia) L. Bogorad (U.S.A.)

A. A. Bayev (U.S.S.R.) P. Starlinger (F.R.G.)

G. Bernardi (France) J. Tooze (EMBO)

March, 1979

DISCUSSION

O.W. PROZESKY: I want to correct a possible wrong impression you might have got from naming the South African guidelines the most restrictive. There was only one category of P4 EK3 and the only reason why it was used was because we did not prohibit any experiments. Not wanting to prohibit any experiments we made it impossible to perform them!

S.N. COHEN: That practice has to be used in certain other nations as well.

O.W. PROZESKY: That is a public relations trick, I think. The next is, I am also a heavy cigarette smoker and that is why, perhaps, they are so restrictive.

D. WEISS: There are two questions I would like to ask that might possibly improve the breadth of your report, Dr. Cohen. One is the question of volume. This has not been addressed at this meeting. The amplification nature of DNA product production is a question and, as you know, the question has now arisen in the USA regarding the possible breaking of the ten litre limit. I wonder if some notation could be made in your report about the volume restrictions in the various guidelines.

The second item which would be enormously helpful would be an indication of the degree of separation of administration of guidelines from the question of enforcement of guidelines. This is a very serious political issue and a serious scientific issue. For example, in the USA we have tried very hard to encourage the NIH not to get into the enforcement business. It is a scientific agency and not an enforcement agency. I would hope that some comparative data could be accumulated for your report to indicate exactly how this is handled in different countries.

S.N. COHEN: I have comments about both of those points. The questionnaire we distributed included a question about specific guideline provisions that relate to industry or agriculture and a number of the respondents have dealt with the volume question. Their replies are indicated in the report. There was also a question that related to enforcement. Certain guidelines are legally enforceable; some have enforcement associated with revocation of a licence or certificate; certain countries have fines. No country has indicated that incarceration is a consequence of violation of guidelines. Enforcement is carried out at different levels in various nations.

Let me make a point that is a sort of an aside regarding the ten litre limit. It was particularly interesting to me to see the ten litre limit included in virtually all of the national guidelines, knowing as I do of the origin of that ten litre limit. I believe that the ten litre limit was originally proposed by the Plasmid Working Group in its report at Asilomar. The proposal arose during a meeting of the Working Group in November 1974 in New York, when we recognized the obvious fact that large numbers of organisms are more difficult to contain than small numbers of organisms. The question was where to put the cut-off. We concluded that one can spin down about $2\frac{1}{2}$ litres of culture fluid in a large centrifuge rotor, and four centrifuge runs seemed to be a convenient number to expect to do in the course of a given day's experiment. Larger volumes would probably require continuous flow centrifugation. So ten litres went in there and it is interesting to know that such an arbitrary judgement has now been incorporated as a major component of virtually all of the guidelines in the world. Presumably if the group had concluded at the time that six centrifuge runs could conveniently be done in a day rather than four, we would have a 15 litre limit.

J. SUBAK-SHARPE: To put the record right, the Ashby Working Party had considered this point and ten litres seemed to us an acceptable limit - undoubtedly also on somewhat arbitrary grounds.

W. SZYBALSKI: A claim for priority? I am surprised, since I would be embarrassed to claim priority for the ten litre restriction.

S.N. COHEN: I think we met before the Ashby report was issued. I would be willing to concede to Dr. Subak-Sharpe that priority claim.

W.F. BODMER: I would like to make two general points which have come up before and which I am sure will come up again. I think, Stan, in your opening you said something about absence of certainty. I believe in science, very rarely do we have certainty. The only certainty, perhaps, is that we shall die. There's no such thing as requesting absence of certainty and I do think in considering risks, and I am sure we'll hear much about that later, we must take that into account. The other thing which relates to John Tooze's comments is that I do believe that in judging the issues at the moment we must seriously try to separate the scientific from the political questions. If we really believe there are no risks in this as a scientific assessment, we should say so and the consequences of that for guidelines and other procedures follow after that, but we should not adjust the scientific judgement by the political.

S.N. COHEN: Walter, I'd like to make the point that the item you noted was not the view of the Working Group. It was simply one of the assumptions that the Group found to be implicit or explicit in the guidelines of various nations.

W. SZYBALSKI: The discussion is drawing to an end, so let me summarise my impressions. First, I have been very impressed with the candour of the top NIH officials who were directly or indirectly involved in promulgation of the 1976 and 1978 NIH Guidelines. They were very forthright and critical about both the guidelines and the several other aspects of the regulations. I have the highest respect for Drs.Fredrickson's paper, read by Bill Gartland, the Director of ORDA, Stetten's and Setlow's presentations, which clearly allude to the future dismantling of the bureaucratic edifice spawned by the present mandatory guidelines. I am only very sorry that Don Frederickson was not able to join us because of unexpected emergencies in the United States.

My impression is that the different kinds of recombinant DNA guidelines can be divided into three general classes:

First, encyclopaedic comprehensive set of rules, as represented by the NIH Guidelines; second, case-by-case approach, represented by the British system as administered by the GMAG and summarized in Sir Gordon Wolstenholme's presentation and third, general microbiological experience of the individual investigators and/or the heads of the laboratories.

The regulations based on these guidelines follow three general patterns: one, a principle of "prior restraint", represented by the USA, where research is prohibited until a formal approval is secured by the investigator from a proper regulatory agency; two, a less restrictive approach characterized by the absence of the "prior restraint" or "censorship" by specific regulatory agencies, as represented by Switzerland, Hungary and Poland. Under this system, the individual laboratory heads decide how to follow the officially accepted guidelines and assume all the responsibilities, and three, in the absence of officially accepted guidelines, it is up to the principal investigators and individual scientists to apply the most appropriate microbiological procedures for each type of experiment. This is a traditional procedure followed since the dawn of the microbiological sciences.

These comments stress the two aspects of the so-called guidelines: (a) their role as a code of good microbiological procedures, and (b) their regulatory and bureaucratic role. Their third, a surprising and unintentional aspect (c) is their political role as a "proof" of the "serious dangers" of the recombinant DNA technique. I believe that I sensed the following sentiments of most, if not all, of the participants. They do not object to the first (a) role, but they are against the regulatory edifice (b), compared by Hans Stetten to the "bastions" of the Tower of London. Moreover, because of the third (c) role, which is very misleading to the public, the hopes were expressed that either the guidelines would be soon abolished or all the recombinant DNA experiments which pose no significant risks, e.g. all DNA cloning in the *E. coli* K-12 host-vector systems, will be excluded from the mandatory guidelines.

Very thorough reviews of the guidelines and regulations in several countries were presented by Drs. John Tooze and Stanley Cohen. It appears that many countries have blindly followed the example of the USA regulations and NIH Guidelines, without ever assessing their necessity or the reality of the imaginary risks. From the historical point of view, this might be quite embarrassing to many countries. Special legislation was proposed in a few countries, but at present such legislative efforts have been abandoned in most countries, including the USA, because there are no scientifically valid reasons for legislation. The FRG has the dubious distinction of being practically the only country actively trying to introduce a special recombinant DNA legislation which is based on the false premise of epidemiological dangers and which probably will violate the FRG Constitution.

At the introduction of this session I posed several questions asking to specify the officially-adopted compelling reasons for regulating the recombinant DNA research. Maybe this question was impossible to answer since such reasons are non-existent. The lack of reason, however, has little effect on the existing regulations.

RE-EXAMINATION OF BASIC ASSUMPTIONS

CHAIRMAN'S INTRODUCTION

Maxine Singer

National Institutes of Health, Building 37, Room 4A-01, Bethesda, Maryland 20014, U.S.A.

The title of this session is "Re-examination of Basic Assumptions". If you look at the organisation of the session you will see that the organisers intended that we re-examine two different kinds of assumptions: first, scientific assumptions, and second, political assumptions. In keeping with Walter Bodmer's remarks a few minutes ago these have been separated in the plan of the session.

Several speakers during the last two days alluded to various aspects of the history of this whole situation. It is important, if we are going to re-examine, that we keep clearly in mind what some of those initial assumptions were. The first scheduled speaker will speak to part of the history. I was asked to talk briefly about the initial events, namely, the 1973 Gordon Conference on "Nucleic Acids" of which I was co-chairman. When I was elected to be co-chairman, I took that as an honour and looked forward to the Conference. In retrospect it turned out to be one of the worst things that ever happened to me and my life changed from that day.

Many of you will recall that the first successful recombinant DNA experiments were discussed at that Gordon Conference in the summer of 1973. They were presented in the course of the programme and immediately upon their presentation concerns were expressed within the group of people attending the meeting. Several of those people came to me and Dieter Soll, who was the other co-chairman, and said that they had questions about the construction of microorganisms containing genes not normally found in those same microorganisms. They thought that the Conference ought to discuss the question. Dieter and I spoke with some other members of the Conference. We finally concluded that while we were unable to discuss the substance of the question because we had a full programme scheduled, and because no one was fully prepared, we might at least raise the question of what could be done about this concern. A brief discussion was held on the last morning of the meeting. That discussion resulted in two decisions, both of which were taken by vote. One was to send a letter to the United States National Academy of Sciences explaining the kinds of experiments which had been described, indicating that questions were raised and asking the Academy to consider the substance of those questions in more depth. The second decision was to publicise that letter in *Science* magazine. That is a brief history of what happened at the 1973 Gordon Conference.

The next speaker will address himself to the manner in which the Academy responded to that letter. He will speak, in fact, of the next major incident, namely the publication in 1974 of the letter which is widely termed the "Berg" letter. But before he begins I would like to tell you that what in the United States would be called the "public image" of nucleic acids is, in fact, not all bad. You may recall that a few years ago there was a fad in the United States involving the use of shampoos that contained protein: advertisements stated that a protein in the shampoo invigorates one's hair. Recently one American manufacturer decided to go one better and bring out a "Nucleic Acid Hair Constructor" in the form of a "Nucleic Acid Reconstructive Shampoo". In case you doubt the seriousness of this, I can tell you that drawn on the shampoo box cover is a ribosome, and messenger RNA. There are squiggles representing transfer RNA molecules, an A site and a P site marked on the ribosome with a growing strand of portein, and the amino acids coming into the transfer RNA. In fact, all of this is described very accurately in the legend on the box, until you come to the last sentence which tells you that you don't need to use a protein shampoo because if you use this one you will make your own proteins on your head. More careful inspection, however, shows that only one enzyme is included in this nucleic acid shampoo, and that is DNA polymerase. That is very puzzling because DNA polymerase is not required for protein synthesis, and furthermore you can see that you never have to buy the shampoo again. Now, I left a great deal of money in a very fancy shop in Washington DC in order to buy this shampoo. Having done it I felt obliged to use the shampoo and I would like to recommend to all of you that you never buy this shampoo!

The next speaker is Jim Watson. He needs no introduction of any kind, either in the scientific world or in the world at large.

WHY THE "BERG" LETTER WAS WRITTEN

J. D. Watson

Cold Spring Harbor Laboratory, Cold Spring Harbor, New York 11724, U.S.A.

When John Tooze asked me to speak on the history of the "Berg" letter, I was pleased that it was I who was so chosen. Initially I thought I would ask everyone who was at MIT that April 1974 morning for their recollections so I could today present a collective opinion. But then I realized I would only re-fester old wounds, and it would be best to give you only my own memories.

My thesis today is that the "Berg" letter was largely a reflection of the anxiety created within a number of molecular biologists about working on tumour viruses. This anxiety was felt most by people who had grown up in molecular biology as distinct from medicine or microbiology. Many of us had started working in science with *E. coli*, and subsequently went into tumour virology both because it might be the best way to study cancer and because it seemed a relatively clean system to study eukaryotic genes. At that time viruses provided the only really pure DNA from higher cell systems. However, at the same time there might be biohazards created by such work. The facts on this latter point were scanty. No real evidence of any risk existed, and leaders in the tumour-virus field like Michael Stoker, Renato Dulbecco and Howard Temin were not perceptibly concerned. They worked with viruses on the open lab bench, and if you visited their labs when you were new to the game, you got slightly nervous because there were seemingly no precautions.

When I became the Director of Cold Spring Harbor in 1969 and wanted to start up a tumour-virus group, I, myself, had to face up to the question whether there might be any hazard to the scientists here or to the community of families that lived very close to the labs.

We, of course, suspected there was not any risk. To play it safe, however, we decided to fix up our then primitive labs so that we would reduce our exposure. Before the mid-1960's most previous tumour virus research had focused on the development of cellular systems for the study of either lytic virus growth or for cell transformation. In such small-scaled work microbiological safety precautions could be routinely carried out. By 1968, however, experiments on the molecular level were starting and these necessarily were beginning to require growing large volumes of cells. You must remember this was before the days of restriction enzymes and micro DNA sequencing. The question then became, how much money should we spend to protect us against a danger which might, in fact, not exist at all. Clearly, you could not devote all your money toward this objective or no experiments at all could be done. Our first requests for biohazard protection were thus as modest as our grant request which found itself looked over by the Genetic Study Section. One of its members was Charles Thomas who formally asked NIH to form a committee to consider whether tumour virology at the molecular level was safe.

After we got our grant in 1969 we went ahead, as did many other labs being equipped at that time, bought biohazard hoods and stopped mouth pippetting. We did not buy the most expensive biohazard-type centrifuges because they were very expensive, in very short supply, and, moreover, it was not at all clear they were necessary. They had just been developed with the National Cancer Institute's help, not for work with SV40 or polyoma virus, but with the belief that within the next year or so human tumour viruses would be isolated. If so you would really need to work under conditions of maximum precaution, and the fancy Fort Detrick-like new building at NCI (Building 41) was just being completed with the expectation that, in fact, you would need such facilities.

Over the next several years we went ahead and grew more and more SV40, and here I can speak only for myself, I was occasionally anxious in an irrational way but would not openly mouth it. If you admitted it to anyone, then your scientific staff or your technicians and secretaries would get worried. Moreover, we always could point out that through the Salk vaccine many millions of people, some of whom were tiny infants, had accidentally received SV40 and there was no obvious increase in either juvenile or adult cancer. So, in practice, SV40 might not be the slightest danger, and if anyone asked me in the community whether we were worried, I could say that we were working with viruses to which most of us already had been thoroughly exposed.

But because of the possibility of long latent periods, maybe we had not waited long enough, and this unease led to the small meeting in California at the Asilomar Conference Center in February, 1973. It was devoted to the potential biohazards of working with tumour viruses. It was an essentially quiet meeting which had virtually no impact at all. There were not enough data to say whether, in fact, tumour-virus work was dangerous, and the meeting ended with the conclusion that there should be some epidemiological study of people who worked in tumour-virus laboratories. If given people were followed long enough you might see whether, in fact, they had been at risk. It arose after a talk by a Harvard epidemiologist named Phillip Cole, and we left thinking he would start such a study. But as far as I know, it only led to the collection of some blood samples which are now stored somewhere at NIH. I could be wrong here - I do not think there has been any follow-up and certainly from the viewpoint of our lab we are not sending names to any central registry and there has been no systematic collection of data.

Many people felt slightly - and I was among them - dissatisfied at the Asilomar I's outcome because no proposals came from it for formal rules to govern how we work. Some of us then were using moderate precautions - others, some working right within NIH itself, were working with essentially no precautions. They were doing the same sort of experiments as we were, so it was not as if we, as a private institution, wer less cautious than a public institution. Individual components of NIH have their own budgetary problems, and SV40 was never considered that dangerous by NCI. It was in the lowest category of potential risk.

Then a few months later the first paper on recombinant DNA by Boyer and Cohen came out and almost immediately some scientists, largely of the liberal-left variety, began to worry that it could be used for bad as well as good purposes. Interestingly, the Singer-Soll, Nucleic Acid Gordon Conference appeal, that went into *Science*, called not only for a consideration of what the new recombinant DNA technology could do, but, also, that the Director of NIH should examine and set up some guidelines for work with tumour viruses. This expression of concern did not come from the tumour virologists - there were only a few of them at that meeting -

but was a reflection of those people who thought they might someday have a lab down the hall from someone who was working with tumour viruses. If one talked to graduate students at that time, there would be those who were eager to move from *E. coli* experiments to those with tumour viruses. But there were others who admitted to being scared by tumour viruses. I think it was this latter group who wanted to add on the warning about unrestricted tumour virus research. They wanted government regulations to prevent them from being needlessly exposed.

I certainly sympathized with their fears and wanted precautions taken in the biological labs at Harvard, when I was still commuting between there and Cold Spring Harbor. There were two scientists on my floor who proposed to work with tumour viruses, and I argued successfully that neither should do such work without safety factors like negative air pressure and biosafety hoods. One was Jack Strominger who wanted to set up a group working on EB virus, but there were in his lab no biohazard hoods, much less negative air pressure. I thought this proposal irresponsible, and told him I would get a lawyer if he went ahead. He backed off and found space in the Jimmy Fund Building at the Children's Hospital where there had been a long tradition of unrestricted research with pathogenic viruses. The other was David Dressler who, using viruses grown at Johns Hopkins, had made the very interesting observation of the terminal redundancy of adenovirus genome. I said really he should not take up such work in Cambridge until Harvard had a P3 laboratory. In fact, this helped set the stage a year later for an application by the Biochemistry Department to NCI for a P3 laboratory. Afterwards it became very celebrated on account of the fact that it took longer than any other P3 laboratory to build, because as an afterthought they decided that they could use it for recombinant DNA research. In the end almost three-quarters of a million dollars were spent for super elaborate safety factors.

I remember first hearing of the Gordon Conference worries from Rich Roberts upon his return from New Hampton. Initially I was dead-set against any formal rules for recombinant DNA, in part because we had no precautions in any legal sense for SV40. Moreover, it was becoming clearer to me that you certainly did not want precautions which prevented good experiments if you did not know whether risk was appreciable. More and more I was accepting the fact that our biohazard money was conscience money - you did not know really whether you were spending little, enough or too much. So you set some figure which was not super big or very small - maybe 10% of your capital budget. If someone said half your renovation funds should go to safety we would react negatively. On the other hand, if you said that perhaps 1% should go for biohazard precautions you would be thought flippant. In any case there was certainly the occasional anxiety among postdocs coming from the world of phage, one of whom in California got a high antibody titer against SV40 and wondered what it meant.

The first informal response in the tumour virus world to the Gordon Conference Singer-Soll warning was that we should not take it seriously. It seemed just another irrational leftish outburst against genetic engineering to which those of us who had been in Boston had become allergic after Jon Beckwith and Jim Shapiro in 1969 called a press conference to denounce themselves for their isolation of *lac* operon DNA.

Nonetheless, the Singer-Soll letter required a response by the National Academy, and that task was given over to Paul Berg. Several years before, he had stopped his experiment of putting SV40 DNA into lambda. Bob Pollack, then on the Cold Spring Harbor staff, led the initial opposition saying how do you know that *E. coli* K-12 carrying this vector will not somehow get back to humans and give us more cancer. Paul could not prove it might not happen, and without much hesitation dropped the experiment. An act which most of us then thought was benevolently responsible.

Because he had taken such a stand, he was asked to head the NAS Committee. Herman Lewis tells me that in early 1974, at the meeting of the NSF Human Cell Committee, Paul asked around whether others in that room could subsequently meet together to come up with a report. Most of those present declined, but Norton Zinder said he would join Paul at a meeting that eventually occurred in David Baltimore's office in late April of 1974. Maxine Singer was also to come, but at the last moment she did not, because one of her children was sick. The scientists who finally joined Paul, besides myself, were Sherman Weissman, Dan Nathans, and my former Ph.D. student, Dick Roblin, who had worked in Dulbecco's lab at the Salk and subsequently had taken an interest in bioethics. He was the young conscience of that meeting. Dave Baltimore was also of this ilk and saw the recombinant DNA problem partly as a moral one which we had no right to avoid. Yet at the time he came to MIT from the Salk, there were no biohazard hoods or anything like negative air pressure to greet him, and there were those *E. coli* types who argued that before he moved into the Cancer Center he was taking lousy, if non-existent, measures to prevent outsiders being exposed to his lab's viruses. But he was essentially working the way he had been trained under Darnell and Dulbecco, and most certainly he was not deviating from commonly accepted safe practices for his profession.

So the "Berg" letter was the product of five tumour virologists and Norton Zinder who was and still is in *E. coli* science. Soon after we came together that sunny bright spring day, we posed the question of what would happen if some silly person should start growing hundreds of pounds of *E. coli* containing an inserted herpes genome or something like that. We trusted ourselves, but would we trust someone else - I will not use any names. In this way we convinced ourselves that maybe some rules might be a good thing.

Now, at that time, there was still considerable mysticism as to how DNA tumour viruses made cells cancerous, and almost any scenario seemed as good as another. We knew there were T antigens, but we were not sure how they acted. We also did not know to what extent the human exposure to SV40 was an accident or whether there were not infections by similar potentially oncogenic viruses that were happening all the time. Then several human SV40-like viruses (BK and JC) had just been found, but on that day we knew almost nothing about the actual frequency with which they infected humans. So it sounded socially responsible to say that we should not needlessly expose people to a further chance of getting cancer even though that probability might be very low indeed.

Unfortunately, we did not initially bring infectious disease people to ask whether they thought it possible that K-12 could bring about an epidemic of cancer. We then thought it would be best to bring these types together with the molecular biologists all at one time. If, however, we had arranged in crash fashion a series of small informal meetings before sending off the moratorium letter, we would probably have ended up with a much more cautious appeal that might have generated much less public excitement. Some practising medical microbiologists, of necessity, would have told us that we were acting like immature boys and you cannot survive in this world if you instinctively assume the worst possible outcome of any new technological or medical advance. But that morning we felt in a hurry, since if anything was clear, it was that recombinant DNA procedures needed only a child's mind to master.

We also knew much less about RNA tumour viruses than we do today. The *sarc* gene had not been isolated and the concept that many RNA tumour viruses were recombinants between endogenous harmless viruses and normal genes may have been believed by someone but it was not yet party line. So one could argue that exposure to DNA containing endogenous "RNA tumour viruses" could be dangerous. And DNA might have it, and in shot-gunning eukaryotic DNAs we maybe should be worried that we might frequently be cloning endogenous tumour-virus genomes.

We also discussed whether we could use recombinant DNA technologies to promote biological warfare, say by giving *E. coli* the capacity to make botulinus toxin. This was a very easy threat to get excited about. There still is almost no non-classified expertise in the world. But I should have said not to get upset about these scenarios, because I had worked in President Kennedy's Science Office during the early days of the Vietnam War. I actually knew what had gone on at Fort Detrick, and after a few months' immersion in classified documents and briefings, I stopped worrying about BW as a realistic way to fight a real war. But arguing on the basis of classified knowledge is tricky, and I did not want to dwell upon my knowledge of dirty information.

For the most part, however, we worried about the spread of cancer, and soon we agreed that we would write a letter saying we should have a temporary moratorium on inserting either cancer virus genes into bacteria or other pieces of DNA which might conceivably carry cancer genes. At the same time we were most conscious that we did not want to stop the development of eukaryotic cell biology and carefully tried to have a moratorium which was not too broad. Nonetheless we opened up the possibility of subsequently banning higher vertebrate shotguns by calling attention to endogenous cancer gene scenarios.

We also thought it would be nice if someone took risk assessment experiments seriously. There had been much talk at the first Asilomar about doing real experiments in which we inject SV40, polyoma and EB and their various mutants into appropriate animals to see if we could learn anything about which genes might lead to cancers. Nothing had come out of that original push for knowledge, so why not again speak for the obvious value of intelligent risk assessment.

Then, as judged by the final letter, we must have said some form of guidelines would be necessary. Here my memory is vague, but since the problem is so open ended, I know we could not have agreed then on how we might decide which ones to adopt. But it sounded good that we should have them, and some of us already were half congratulating ourselves for taking such a noble unselfish step. In all this, however, I remember a total lack of enthusiasm for the whole process by Dan Nathans and Sherman Weissmann. They struck me as silent, old-fashioned medical types who were not happy talking about imaginary dangers. In fact, later I learned that during the 1973 Nucleic Acid Gordon Conference they had counselled Maxine Singer against any public statement about recombinant DNA. But that day they did not want to seem like bad guys, and there was no serious discussion that the best course of action would have been to do nothing and let recombinant DNA research develop in an unfettered manner.

In the initial draft of the "Berg" letter, there was a statement that recombinant DNA technology might be used for bacteriological warfare, but this phrase was removed at a later draft because it would raise an issue which would involve unavailable classified data. Moreover, no one believed that if the CIA or the army wanted to do such work that our moratorium would stop them. So the whole document in large part rested on whether we were taking a step to reduce the risk of unnecessary cancer.

For a long time after I believed the outcome of Asilomar II would not have been nearly so bad if the NAS had not held a press conference to coincide with publication of the moratorium appeal. But upon re-reading it, I do not think that is true - the moratorium letter is very strong - it not only says we should temporarily ban certain experiments which might lead to cancer, but that guidelines should be formulated. Moreover, no one who signed it could have been accused of doing so for reflexive ideological reasons - that is, if you do not include a mild form of liberal guilt as to whether the science we do might not be all for the good.

Now in looking back we have to ask the question of why we took so seriously Rube Goldberg-like schemes for getting cancer genes into us, when, in fact, we were already doing experiments which were infinitely more likely to put us at risk, if, in fact, any risk at all exists. The answer, I believe, is that in running our labs we had already used up all our potential for worry. Any more accusations of potential moral indifference were more than we wanted to take psychologically. Almost from birth we had thought of ourselves as nice liberal types who work for the common good. To seem identical to a steel magnate polluting Gary, Indiana, or stonewalling as if we were Nixon and Kissinger leading us into Cambodia, was not what we wanted. We wanted to show you we were good guys, and in doing so we gave you the moratorium.

DISCUSSION

W. SZYBALSKI: How did the word "potential" danger come into being? Why weren't the more appropriate and less frightening adjectives, e.g. hypothetical, conjectural or imaginary risks, used? "Potential" is usually defined as something which is likely to happen.

J.D. WATSON: I don't think we ever thought about these verbal distinctions. But I don't think it would have made any difference. When you say there should be guidelines affecting your work, when you say that you should not do an experiment because it poses a possible risk, I think you can use any words you want to and you will get the same effect. Semantics were not our downfall. I think we went public too soon. We should have arranged a crash series of smaller meetings over a matter of three months and hammered it all out first. I think we all looked forward initially, or at least a few of us did, to getting to California the following year. It seemed a very nice place. How could you go wrong in a very nice place?

M.G.P. STOKER: I have been wondering why, as Jim said, we didn't worry more about tumour viruses right at the start and before all this. Incidentally we were very careful about one thing. It's very difficult to make tumours with viruses unless you inoculate newborn animals, and we absolutely forbade newborn babies in our lab and, so far as possible, indirect contact with them. But, I think the reason why we did not think it was dangerous - and I still don't think it was or is dangerous - is because most of the people involved then originally were trained as medical or veterinary microbiologists - Renato Dulbecco, for example, and Harry Rubin. Many of the NIH tumour virologists and Ludwig Cross too were originally medical microbiologists or trained in medical laboratories. If you come up that way you are impressed with the safety of such laboratories. There are, of course, dangerous organisms in medical microbiology labs which cause infections such as tuberculosis and hepatitis and clearly not enough attention is paid to these. Generally speaking, however, I believe that one of the safest places to be is in a medical microbiology lab.

N.D. ZINDER: I am not going to discuss the details of Jim's picture of the meeting at MIT. Two weeks from now is its five-year anniversary. I agree with him on some things and on others I disagree, but the point I do want to make is in regard to infectious disease people. This is a story that, if we only had spoken with infectious disease people we would have been informed and relieved of our ignorance with regard to microbial spread. During the early period, I, for one, did speak to a large number of infectious disease people. They were then of no help. They did not understand enough of the new science to integrate it into their own area. In addition, Sherman Weissman knows a great deal about infectious diseases. Despite the fact that he just sequenced SV40, he still makes rounds at the hospital ward every day in paediatrics where there's lots of infectious diseases. My conversations at that time with infectious disease people were furthered since there was one other thing that I was doing that year. Partially at the instigation of Jim Watson I was chairing a review of the viral cancer programme and I was running around from viral oncology lab to viral oncology lab in medical schools. Many virologists heard about the new technology and, talking to them, I confess, did frighten me. And the infectious disease people that I spoke with were all quite concerned, and their concern was very simple. They tend to think about the portal of infection. They were most concerned that we were creating, or would create by our manipulations, a new portal of entry by which a particular infectious, or even non-infectious agent, could be come infectious.

The Stetten dictum of the greatest hazard being the hazard of the most dangerous component thus would be violated. Today, of course, we all think differently but what then is science if not enlarging our understanding of Nature and natural events?

BIOLOGICAL ACTIVITY OF POLYOMA-PLASMID AND POLYOMA-PHAGE RECOMBINANTS IN MICE AND HAMSTERS

M. A. Martin, H. W. Chan, M. A. Israel and W. P. Rowe *

Recombinant DNA Research Unit,
National Institute of Allergy and Infectious Diseases,
Bethesda, Maryland 20205, U.S.A.
***Laboratory of Viral Diseases,**
National Institute of Allergy and Infectious Diseases,
Bethesda, Maryland 20205, U.S.A.

One of the early concerns involving recombinant DNA experimentation centered around the potential danger of cloning intact animal viral genomes in *E. coli* K-12. One scenario hypothesized the production of infectious virus particles in bacterial cells, raising the possibility that *E. coli* K-12 might be capable of delivering animal viruses to organisms or tissues not ordinarily accessible to virions during natural infectious processes. If this model were correct, high degrees of physical and biological containment would be called for to prevent the release of enormous quantities of virions into the environment. A second scenario postulated that viral DNA would be replicated in bacterial cells but no infectious virions would be produced because of the differences in the synthesis and metabolism of RNA and protein in prokaryotic and eukaryotic cells. That is, the possibility existed that a bacterial cell, harboring a phage or plasmid recombinant containing a complete copy of an animal viral genome, could transfer the DNA to a eukaryotic cell in which the molecule could be infectious. Such a cell would have to possess the enzymatic machinery to precisely excise (or copy) the viral DNA insert and convert it into a biologically active form. Apprehension about the cloning of animal viral DNAs in *E. coli* K-12 systems was further heightened by the concern that latent viral genomes might be inadvertently propagated during shotgun cloning of mammalian DNA.

In December 1975, we were asked by the NIH Recombinant DNA Advisory Committee to plan and conduct experiments to assess some of the conjectured risks attending the cloning of eukaryotic viral DNA in *E. coli* K-12 host vector systems. Polyoma virus (PY) DNA was chosen as the source of the viral insert in such studies for several reasons. Both the virus and its DNA are small and have been well characterized. The injection or feeding of PY to laboratory mice reproducibly results in infection but only rarely produces overt disease. Parenteral inoculation of mice with a single tissue culture infectious dose is sufficient to initiate an infection that results in antibody production (Rowe and co-workers, 1959). The testing of injected mice for serum hemagglutination-inhibiting (HI) antibodies, the mouse antibody production (MAP) test, provides a simple and unambiguous assay for PY infection (Rowe and co-workers, 1959). Since the antibody is directed against viral capsid proteins (and not viral DNA), the detection of such antibody indicates that the viral genome entered one or more cells and established productive infection. In a laboratory setting PY can induce tumors in hamsters, rats, and mice provided the virus is injected into animals one to two days of age. Older animals do not develop tumors, chiefly because of their ability to mount an immunologic response. PY fails to replicate in primate cells presumably because the synthesis of progeny viral DNA is blocked. The virus has been handled in many laboratories since 1953 with no evidence of human infection.

In the experiments to be described, PY DNA was cloned in derivatives of E. coli K-12 using plasmid and phage vector systems. Live bacteria containing PY recombinants, recombinant PY-phage particles, or PY-phage or PY-plasmid recombinant DNA preparations were then fed to or inoculated into weanling mice. Animals were bled 3 and 6 weeks following administration and their sera examined for the presence of HI antibody. The cloning of PY DNA in E. coli host-vector systems, the inoculation and feeding of mice with bacteria containing PY recombinants, the maintenance of test animals for 6 weeks, and the collection and analysis of serum samples were all done under P4 containment as specified by the NIH guidelines. Only certified EK2 host vector systems were used in the cloning of PY DNA.

PY-plasmid (pBR 322) (Bolivar and co-workers, 1977) and PY-λ (λgtWES.λB) (Leder and co-workers, 1977) recombinants containing a single copy of viral DNA inserted in either of two possible orientations (with respect to the vector) were prepared. In the first group of experiments, mice were inoculated either subcutaneously or intraperitoneally or were fed concentrated suspensions of log-phase bacteria containing the PY-plasmid recombinant (Table 1). Other animals were injected intraperitoneally with 10^8 E. coli DP50 SupF (Blattner and co-workers, 1977) containing PY-λ recombinants or PY-λ phage particles (Table 1). None of the animals receiving live E. coli harboring PY-plasmid or -λ recombinants developed a viral infection.

In further experiments, purified preparations of PY-plasmid or PY-λ DNA were inoculated into weanling mice and their sera assayed for the presence of HI antibody to PY. In no case did the injection of intact recombinant DNA molecules containing a single copy of the PY genome result in a viral infection (Table 2). However, when these recombinant DNA preparations were cleaved, prior to inoculation into animals, with the same restriction enzyme as was used in their construction, nearly all of the mice developed PY infection (Table 2). This latter result clearly indicates that PY DNA is faithfully replicated in E. coli and is infectious when it is no longer a component of a recombinant molecule. Our failure to detect PY HI antibody in animals injected with live bacteria containing PY recombinant DNA or with intact monomeric PY-plasmid or PY-λ DNA preparations indicates that: a) no PY virions or free PY DNA was present in the E. coli used in these experiments and b) mouse cells were unable to excise the PY DNA insert out of the PY recombinants. It should be noted that similar results were obtained with infectivity tests done in tissue culture (Israel and co-workers, 1979a).

In contrast to recombinant molecules containing a single copy of a viral genome, those containing a head to tail dimer would, in theory, be capable of generating infectious PY genomes by recombinational mechanisms. Despite repeated attempts, we were unable to identify or isolate PY-plasmid recombinants containing a dimeric insert of PY DNA. This was probably related to the presence of an active recA+ gene in the X1776 (Curtiss and co-workers, 1977) used in these experiments. During the propagation of λ-PY recombinant phage however, we noted that one preparation derived from a single plaque (λ-PY30) gave rise to progeny phage particles that banded at two different positions after isopycnic centrifugation in CsCl. The recombinant DNA purified from the denser phage band (λ-PY30-5B) contained a head to tail dimer insert of PY DNA as judged by restriction endonuclease digestion and analysis of heteroduplex molecules by electron microscopy (Chan and co-workers, 1979). The less dense phage band (λ-PY30-5T) contained a monomeric PY DNA insert. None of the mice fed or inoculated with live E. coli containing the dimeric phage recombinant developed a viral infection

(Table 3). On the other hand, PY HI antibody was detected in a number of animals inoculated with a concentrated lysate of *E. coli* infected with the dimeric phage recombinant, with purified DNA from the dimeric recombinant phage, or with DNA prepared from *E. coli* DP 50 SupF harvested 20 minutes following infection with the phage (Table 3).

In these experiments we have monitored the infectivity of PY DNA following its administration to mice in a variety of forms. Compared to PY virions, recombinant plasmids and phage containing a single copy of the viral genome, are at least 5 x 10^8 to 5 x 10^{10} times less infectious, whether administered to test animals in live *E. coli*, or in the form of phage particles, or purified DNA (Table 4). Although parenterally injected phage particles containing a dimeric insert of PY DNA did cause virus infections in mice, the preparations analyzed possessed 10^6 to 10^7-fold less biological activity than PY virus particles (Table 4). Further, injection or feeding of mice with *E. coli* infected with such phage preparations did not result in PY infection.

These experiments provide answers to several questions that have been raised regarding the cloning of animal viral genomes in *E. coli* host vector systems. First, they clearly show that infectious virus particles as well as free viral DNA are not formed in *E. coli* harboring recombinant DNA molecules consisting of a prokaryotic vector and a complete copy of a eukaryotic virus. Second, they indicate that potentially infectious constituents of intact recombinant DNA molecules are not excised following their inoculation into sensitive animals. Third, the administration of recombinant molecules containing a head to tail dimer of PY DNA provides a means for evaluating the extent to which DNA molecules can be transferred from *E. coli* to mammalian cells, since recombinational mechanisms rather than precise excision or replication could generate infectious PY genomes from a linear DNA template. The inefficiency of such a process, particularly when the PY recombinant is contained within live *E. coli*, is quite remarkable.

We have also used another feature of the PY system (viz. the ability of PY virions and PY DNA to induce tumors) to further evaluate the possible transfer of recombinant DNA molecules from bacterial hosts into mammalian cells. Experiments recently completed in our laboratory have indicated that subgenomic portions of PY DNA can induce tumors in inoculated newborn hamsters. The polyoma-hamster system thus provides a biological model in which the transfer of only a segment of viral DNA rather than the entire viral genome can be evaluated. Live bacteria, recombinant phage particles, and purified recombinant DNA preparations were inoculated subcutaneously into 1-day old hamsters and the animals were monitored for the development of tumors over a 5 month period. Inoculation of very large doses of *E. coli* containing recombinant plasmids did not result in tumor production in any of the 32 animals injected (Table 5). Intact recombinant DNA preparations as well as purified phage particles induced tumors in 5-19% of inoculated animals, a value similar to that observed following the injection of supercoiled PY DNA (19%) (data not shown). When recombinant plasmid DNA molecules were digested with the same restriction enzyme used in their construction, they exhibited tumorigenic activity comparable to DNA derived from PY virions.

Although these tumorigenicity experiments should be extended to other host vector combinations, the results we have obtained complement the findings of mouse infectivity studies. Potentially infectious or tumorigenic recombinant DNA molecules are not transferred out of EK2 hosts into mammalian cells in either of the animal model systems we used. Even in those cases in which animals inoculated with recombinant phage or purified recombinant DNA did develop a viral infection or tumors in worst case experiments conducted in a laboratory setting, the biological activity of recombinant preparations was at least a million-fold less active than PY virus particles. Taken together these results indicate that it is far safer working with *E. coli* containing recombinant viral DNA than intact virions and are highly reassuring with respect to the safety of cloning in such systems.

Table 1. Inoculation of Mice with Live *E. coli* Containing PY-Plasmid or PY-λ Recombinants

	No. infected/No. inoculated		
Inoculum	Intraperitoneal	Subcutaneous	Oral
X1776 (pPB1)	0/5		0/4
X1776 (pPB3)	0/5		0/4
X1776 (pPB4)	0/5		0/5
X1776 (pPB5)	0/5	0/5	0/5
X1776 (pPB6)	0/5	0/5	0/5
X1776 (pPR18)	0/5		0/5
X1776 (pPR21)	0/5		0/5
λ-PY3 infected *E. coli*	0/5		
λ-PY63 infected *E. coli*	0/5		
λ-PY3 phage	0/5		
λ-PY63 phage	0/5		
PY DNA I (0.1 ug)	9/10		
Saline	0/9		

Weanling mice from a PY-free colony (Charles River Laboratory) were inoculated parenterally with approximately 1 x 10^9 to 3 x 10^9 *E. coli* X1776 containing PY-pBR 322 recombinant plasmids. Plasmids pPB1 and pPB3 through pPB6 contain the PY genome inserted at the Bam H1 site while plasmids pPR18 and pPR21 contain PY DNA inserted at the Eco RI site (Israel and co-workers, 1979a). Other animals received 1 x 10^9 to 3 x 10^9 bacteria suspended in milk and absorbed on bread following an overnight fast. Feedings were repeated two to five times at approximately weekly intervals. *E. coli* DP50 SupF (in the log phase of growth) were infected with λ-PY recombinants. Following a 15 minute adsorption, the bacteria were grown at 37°C for 20 minutes, then chilled in ice and injected into test animals. λ-PY3 and λ-PY63 are recombinant phages containing a single copy of PY DNA inserted in opposite orientations at the Eco RI site (Chan and co-workers, 1979). Recombinant phage was prepared from bacterial lysates by precipitation with polyethylene glycol and isopycnic centrifugation in CsCl. At 3 and 6 weeks following inoculation, sera were collected and assayed for PY HI antibody. Titers greater than 1:20 were considered indicative of a PY infection.

Table 2. Inoculation of Mice with Intact and Cleaved PY Recombinant DNA

Inoculum	Dose (ug/mouse)	No. infected/No. inoculated
pPB5	5	0/5
pPB6	5	0/5
pPR18	5	0/5
pPR21	5	0/5
pPB5 + Bam HI	0.02	5/5
pPB6 + Bam HI	0.02	5/5
pPR18 + Eco RI	0.02	5/5
pPR21 + Eco RI	0.02	5/5
λ-PY3	1.4	0/4
λ-PY63	1.4	0/5
λ-PY10	1.4	0/4
λ-PY25	1.4	0/5
λ-PY3 + Eco RI	1.4	5/5
λ-PY63 + Eco RI	1.4	5/5
λ-PY10 + Eco RI	1.4	5/5
λ-PY25 + Eco RI	1.4	5/5
PY DNA I + Eco RI	0.1	5/5
Saline		0/5

Mice were injected with the indicated recombinant PY-plasmid or PY-λ DNA preparations. pPB5 and pPB6 are recombinant plasmids containing PY DNA inserted at the Bam HI site in opposite orientations whereas pPR18 and pPR21 contain the PY genome in opposite orientations at the Eco RI site. λ-PY3 and λ-PY10 are recombinant phages containing PY DNA inserted in one orientation at the Eco RI site while λ-PY63 and λ-PY25 contain viral DNA inserts in the other orientation. Sera were collected and assayed as described in Table 1. An uninoculated control mouse was included in each cage; all failed to develop PY HI antibody.

Table 3. Inoculation of Mice with Phage Recombinants Containing One or Two Copies of PY DNA

Inoculum	Dose	No. infected/No. tested	
		Subcutaneous	Oral
E. coli infected with λ-PY30-5T			
DNA	16.4 ug	0/5	
λ-PY30-5T Phage			
DNA	3.6	0/5	
E. coli infected with λ -PY30-5B			
Live bacteria	2×10^6PFU/5×10^7 *E. coli*	0/28	
	4×10^6PFU/1×10^8 *E. coli*		0/13
DNA	23 ug	2/5	
λ-PY30-5B Phage			
Concentrated lysate	2×10^8PFU	6/11	
DNA	2.8 ug	2/5	

E. coli DP50 SupF were infected with λ-PY30-5B at a multiplicity of 0.04 PFU/cell and allowed to adsorb for 15 minutes at room temperature. The indicated numbers of bacteria were inoculated into or fed to test animals. DNA was prepared from *E. coli* DP50 SupF 20 minutes after infection with λ-PY30-5T or λ-PY30-5B (MOI = 20). The λ-PY30-5B-phage lysate was prepared as described in Table 1 except the CsCl centrifugation step was omitted. Serum samples were assayed at 3 and 6 weeks following inoculation as described in Table 1.

Table 4. Infectivity for Mice of Polyoma DNA in Various Forms

Form of DNA	Infectivity: Parenteral	Infectivity: Oral	No. of PY DNA molecules in MID*: Parenteral	No. of PY DNA molecules in MID*: Oral
Virions	+	+	10^2 to 10^3	10^8 to 10^9
Extracted from virions				
Supercoiled circles	+	-	7×10^7	$> 7 \times 10^{12}$
Relaxed circular	+		10^8	
Eco RI linears	+		4×10^8	
Bam HI linears	+		5×10^8	
Cloned in plasmid pBR322 (monomer)				
Live X1776	-	-	$>5 \times 10^{12}$	$>10^{12}$
Plasmid DNA	-		$>5 \times 10^{12}$	
Plasmid DNA, cleaved	+		$\leq 4 \times 10^8$	
Cloned in λ, monomeric insert				
Live coli in latent period	-		$>5 \times 10^{10}$	
Total (coli) cellular DNA	-		$>3 \times 10^{11}$	
Purified phage	-		$>2 \times 10^{10}$	
DNA from purified phage	-		$>4 \times 10^{11}$	
DNA from purified phage, cleaved	+		$\leq 5 \times 10^9$	
Cloned in λ, dimer insert				
Live coli in latent period	-	-	$>3 \times 10^{10}$	$> 8 \times 10^{10}$
Total (coli) cellular DNA	+		3×10^{11}	
Purified phage	+		10^9	
DNA from purified phage	+		10^{11}	

*Mean infectious dose.

These estimates were calculated by the Poisson distribution from the data in this report and from experiments conducted in our laboratory (Israel and co-workers, 1979b). The following assumptions or data were also used. (i) Polyoma virions have a ratio of particle to infectivity of about 3×10^2. (ii) One microgram of PY DNA is 10^{11} molecules. (iii) The PY-plasmids are carried at ten copies per bacterium. (iv) The PY-plasmid is 58 percent PY DNA, the monomer PY-λ DNA is 14 percent, and the dimeric PY-λ DNA is 25 percent PY DNA. (v) At 20 minutes, the time when E. coli lytically infected with PY-λ were harvested for the "total E. coli DNA" tests, the cells contained ten PY DNA equivalents per cell in the case of the monomeric inserts, and 20 in the case of the dimeric inserts. (vi) These estimates are also assumed for the PY-λ infected E. coli administered to mice by injection or feeding; this very low estimate is used because of the inefficiency of the lytic cycle under anaerobic conditions.

Table 5. Inoculation of Hamsters with PY-Plasmid or PY-λ Recombinants

Inoculum	Dose	No. with tumors / No. tested
Plasmid System		
Live Bacteria		
X1776 (pPB5) + X1776 (pPB6)	1.8×10^8 cells	0/23
X1776 (pPR18) + X1776 (pPR21)	1.7×10^8 cells	0/9
Mixture of pPB5 + pPB6 DNAs		
Uncleaved		1/16
Cleaved with Bam HI		3/8
Mixture of pPR18 and pPR21 DNAs		
Uncleaved		1/11
Cleaved with Eco RI		6/9
Lamdaphage System		
Phage particles		
λ-PY30-5T	1.0×10^{10} PFU	0/8
λ-PY30-5B	0.7×10^{10} PFU	2.12
DNA		
λ-PY30-5T	0.4 ug	0/20
λ-PY30-5B	0.2 ug	3/16

One day old Syrian hamsters were inoculated subcutaneously with approximately 2×10^8 *E. coli* X1776 containing PY-pBR322 recombinants. The orientation of the PY DNA insert in the recombinant plasmids is described in Table 2. λ-PY30-5T and λ-PY30-5B are recombinant phage that contain monomer and dimer inserts of PY DNA respectively. Injected hamsters were observed for tumor development over a five month period.

REFERENCES

Blattner, F.R., B.G. Williams, A.E. Blechl, K. Denniston-Thompson, H.E. Faber, L. Furlong, D.J. Grunwald, D.O. Kiefer, D.D. Moore, J.W. Schumm, E.L. Sheldon, and O. Smithies (1977). Charon phages: safer derivatives of bacteriophage lambda for DNA cloning. Science, 196, 161-169.

Bolivar, F., R.L. Rodriguez, P.J. Greene, M.C. Betlach, H.L. Heyneker, and H.W. Boyer (1977). Construction and characterization of new cloning vehicles. II. A multipurpose cloning system. Gene, 2, 95-113.

Chan, H.W., M.A. Israel, C.F. Garon, W.P. Rowe, and M.A. Martin (1979). Molecular cloning of polyoma virus DNA in Escherichia coli: lambda phage vector system. Science, 203, 887-892.

Curtiss III, R., D.A. Pereira, J.C. Hsu, S.C. Hull, J.E. Clark, L.F. Maturin, R. Goldschmidt, R. Moody, M. Inoue, and L. Alexander (1977). Biological containment: The subordination of Escherichia coli K-12. In R.F. Beers, Jr. and E.G. Bassett (Eds.), Recombinant Molecules: Impact on Science and Society. Raven Press, New York. pp. 45-56.

Israel, M.A., H.W. Chan, W.P. Rowe, and M.A. Martin (1979a). Molecular cloning of polyoma virus DNA in Escherichia coli: plasmid vector system. Science, 203, 883-887.

Israel, M.A., H.W. Chan, S.L. Hourihan, W.P. Rowe, and M.A. Martin (1979b). Biological activity of polyoma viral DNA in mice and hamsters. J. Virol., 29, 990-996.

Leder, P., D. Tiemeier, and L. Enquist (1977). EK2 derivatives of bacteriophage lambda useful in the cloning of DNA from higher organisms: the λgtWES system. Science, 196, 175-177.

Rowe, W.P., J.W. Hartley, J.D. Estes, and R.J. Huebner (1959). Studies of mouse polyoma virus infection. I. Procedures for quantitation and detection of virus. J. Exp. Med., 109, 379-391.

INFECTIVITY OF POLYOMA VIRUS-PLASMID pBR322 AND POLYOMA VIRUS-LAMBDA RECOMBINANTS IN MOUSE FIBROBLASTS

M. Fried, W. Boll*, C. Weissmann*, B. Klein**, K. Murray**, P. Greenaway*** and J. Tooze****

Imperial Cancer Research Fund Laboratories, Lincoln's Inn Fields, London WC2, U.K.
***Institut für Molekularbiologie, Zürich, Switzerland**
****University of Edinburgh, Scotland, U.K.**
*****MRE Porton Down, Salisbury, U.K.**
******EMBO; Heidelberg, F.R.G.**

There has been a great deal of speculation on the possible hazards inherent to hybrid DNA technology. An important question in this context is whether a viral DNA, covalently linked to a vector, can be transferred from within a bacterium into an animal cell, either *in vitro* or *in vivo*, and initiate viral infection. As a first step, it is necessary to determine whether a hybrid containing eukaryotic viral DNA can, on penetration into a permissive cell, give rise to virus formation. As part of a risk-testing programme sponsored by EMBO, this question was investigated with mouse polyoma viral DNA. Polyoma virus will readily grow in mice; infection can be initiated by a single infectious particle as well as by naked viral DNA, and can be easily detected by monitoring the production of anti-viral antibodies. Virus production, and not tumour formation, was chosen as a criterion for biological activity because, although polyoma virus will transform cells *in vitro*, tumours are produced *in vivo* only when large quantities of virus are inoculated into immunologically immature or immuno-suppressed animals.

This report describes the infectivity of various plasmid pBR322 and phage λ derivatives containing polyoma DNA in mouse fibroblast cells. The only recombinant molecules so far obtained which are infective as intact molecules were plasmids with dimeric polyoma DNA inserts.

Hybrid plasmid molecules were made with DNA from a readily identifiable mutant of polyoma by digestion with R.*Eco*RI or R.*Bam*HI, each of which cleaves at a single site, and ligation to appropriately cleaved pBR322, or a number of bacteriophage λ derivatives some of which could lysogenise their host, some which had temperature sensitive repressors, and some which carried functional genes for phage-mediated recombination (*red*).

Restriction enzyme analysis of the plasmid-PY recombinants showed that the polyoma virus sequences were inserted in both orientations. Two of the recombinants joined via the *Bam*HI site were shown to contain oligomers of polyoma. One, D3, contained a dimer of PY in a head-to-tail array whereas the other, D5, was similar to D3 except that one of the PY molecules contained a deletion of about 10% of the PY sequences. PY molecules were inserted into λ in both orientations. In the initial experiments with λ no stable recombinants containing oligomers of PY were isolated, although more recent results suggest that such molecules can be obtained.

The infectivity of the polyoma-pBR322 hybrid DNAs before and after cleavage with the restriction enzyme used to generate the joined fragments was compared to that of polyoma virus DNA, using the DEAE method for transfection of mouse embryo cells. Table 1 shows the specific infectivities of some of the polyoma-pBR322 hybrids joined via the *Bam*HI site. Hybrids containing one unit length of polyoma virus DNA gave less than 1 PFU/μg of hybrid DNA (less than 1 PFU/10^{11} DNA molecules), but cleavage of these hybrid DNAs with R.*Bam*HI resulted in a specific infectivity (in terms of unit lengths of polyoma DNA) similar to that of the viral DNA cleaved with R.*Bam*HI. Similar results were obtained with 5 out of 6 hybrids that contained monomers of polyoma virus DNA joined to pBR322 via the *Eco*RI cleavage site (results not shown). In no case was the uncleaved DNA infectious, but 5 of the hybrid DNAs were infectious after cleavage with R.*Eco*RI, which liberated a unit length linear viral genome. The reason for the lack of infectivity of the remaining hybrid after restriction remains obscure; a non-infectious molecule may have been cloned, or infectivity of the polyoma DNA moiety may have been lost upon passage of the hybrid in bacteria.

In contrast to the intact hybrids containing one unit of polyoma DNA, those containing 2 or 1.9 tandem head-to-tail genomes of polyoma virus DNA *were* infectious. The hybrid containing a dimer of polyoma virus DNA was one tenth as infectious as intact form I viral DNA (in terms of PFUs per polyoma DNA molecule), whereas the hybrid containing 1.9 viral genomes was about 1% as infectious as viral DNA. The specific infectivity after cleavage with R.*Bam*HI of the two hybrids containing more than one unit of polyoma virus DNA was similar to that of similarly treated viral DNA in terms of PFUs per unit length polyoma DNA molecules.

The results for the infectivity of the polyoma-λ hybrids before and after restriction were essentially the same as those for the monomer polyoma-pBR322 hybrids. Less than 1 PFU/μg of hybrid DNA (less than 1 PFU/10^{10} DNA molecules) could be detected after infection of mouse cells with any of the uncleaved polyoma-λ hybrids that contained a single copy of the polyoma virus genome, but cleavage with the appropriate restriction enzyme led to an infectivity similar to that of a molar equivalent of restricted polyoma DNA (Table 1). Of the polyoma-λ hybrids constructed via *Eco*RI sites, 7 of the 8 different types of DNA molecule were infectious after digestion with R.*Eco*RI, but DNA from one hybrid (λ-PY 48), which was shown to have a deletion at one of the sites of attachment, was not infectious after such treatment.

The fact that cloned polyoma DNA, when excised at the site of joining, has an infectivity very similar to that of form III DNA derived from polyoma virions testifies to the remarkable fidelity of the cloning procedure and the stability of a eukaryotic sequence replicated in a prokaryotic host. The finding that hybrid DNA containing only one unit length of polyoma DNA is not infectious to cells permissive for polyoma shows that intracellular excision of full-length, circular or linear polyoma DNA from the hybrid must be a very rare event. In contrast, excision of polyoma DNA from a hybrid containing a head-to-tail dimer is relatively efficient in mouse fibroblasts and probably results from legitimate recombination. A hybrid containing a head-to-tail dimer of polyoma DNA with a 800 base pair deletions in which, due to the location of the deletion only 1.3 units of polyoma DNA are available for excision of a unit length viral DNA had 10% of the infectivity of the plasmid containing the full dimer. This suggests that the frequency of excision may be related to the target size for possible recombinational events. The production of SV40 virus from adenovirus-SV40 hybrids has also been observed to be dependent on the extent of duplicated SV40 sequences present.

Consistent with the *in vitro* results reported here it has been observed that only PY-λ hybrids containing a proportion of recombinant molecules with head-to-tail dimer polyoma sequences are biologically active *in vivo* (see Martin, Chan, Israel and Rowe, this volume).

Risk-testing experiments designed to test the transfer of recombinant DNA from *E. coli* to an animal cell or to an animal can only be meaningful if the recombinant DNA itself is potentially infectious. The demonstration that hybrid DNA containing head-to-tail dimeric polyoma DNA is highly infectious to mouse fibroblasts while that containing a monomeric DNA insert is not justifies an extension of the experiment to *in vivo* studies in which bacteria containing such hybrid plasmids and phage are administered to cultured mouse cells and mice which are then scored for polyoma infection. Such experiments await suitable animal quarters.

TABLE 1

DNA preparation		Specific infectivity * Uncleaved DNA		Restricted DNA **	
Prep. No.	Units of polyoma DNA per hybrid	PFU/µg total DNA	PFU/mole-cules	PFU per µg full-length polyoma DNA $\times 10^{-4}$	PFU per complete linear polyoma molecule $\times 10^{7}$***
a) *Polyoma-pBR322 hybrids					
A2	1	< 1	$< 10^{-11}$	3.2	1.9
D6	1	< 1	$< 10^{-11}$	2.8	1.6
E8	1	< 1	$< 10^{-11}$	2.6	1.5
G3	1	< 1	$< 10^{-11}$	3.7	2.2
D3	2	7.5×10^{4}	1.2×10^{-6}	2.3	1.4
D5	1.9	8.7×10^{3}	1.4×10^{-7}	2.2	1.3
PY	-	2.2×10^{6}	1.3×10^{-5}	7.9	4.6
b) *Polyoma-λ hybrids					
12	1	< 1	$< 10^{-10}$	1.6	0.94
35	1	< 1	$< 10^{-10}$	2.6	1.5
37	1	< 1	$< 10^{-10}$	1.8	1.1
48	< 1	< 1	$< 10^{-10}$	$< 10^{-4}$	$< 6 \times 10^{-5}$
57	1	< 1	$< 10^{-10}$	2.3	1.4
PY	1	9×10^{-5}	5.2×10^{-6}	6.4	3.8

* The DNA was cleaved to completion with *Bam*HI endonuclease (experiment a) or *Eco*RI (experiment b) and the enzyme was inactivated by heating at 60° for 15 min.

* The DNA was cleaved to completion with *Bam*HI endonuclease (expt a) or *Eco*RI (expt b) and the enzyme was inactivated by heating at 60° for 15 min.

** Appropriate dilutions of unrestricted and restricted DNA, respectively, were assayed on mouse embryo cells for plaque formation using the DEAE method. PY, polyoma DNA; PFU, plaque-forming units.

*** The molecular weights used for the calculations are polyoma DNA, 3.51×10^{6}; pBR322, 2.89×10^{6}; λ DNA, 32×10^{6}.

DISCUSSION

J.K. SETLOW: I would like to ask Dr. Martin how in the world we are going to convince the non-scientific public that this experiment does not have to be done over again on all possible animal viruses.

M.A. MARTIN: Actually, Jane, before I answer that, I had a question to ask you in return! We feel that the results of this experiment can be extrapolated to many other groups of viruses, if not to all DNA animal viruses. In fact, one could make a good argument that, with the exception of two categories of RNA viruses (the retroviruses and positive-strand viruses) our experiments apply to the cloning of all animal viruses in K-12 systems. We have recently cloned Harvey sarcoma virus as a collaborative effort between Wally Rowe, Ed Scolnick and myself at the NIH, but are unable to inject this material into animals. Under the revised U.S. guidelines cloned retroviral DNA could be considered as a potential vector in a mammalian cell. The present U.S. guidelines don't specifically address this question and we have been awaiting the decision of the Director of NIH on this matter. Do you have any suggestions how we can move ahead on this experiment?

D.O. HAINES: I would like to ask a question which is the opposite to the one which was just asked. Are these experiments sufficiently extensive to postulate a system of testing the biological crippling of the host systems that are being proposed?

M.A. MARTIN: We did not set out to evaluate the leakiness or the enfeeblement of the *E. coli* strains used to propogate polyoma DNA recombinants. Our task was to evaluate whether recombinants containing a viral DNA segment could transfer out of *E. coli*. The answer to that question is an emphatic "No!".

We have recently completed a group of studies that focused on *E. coli* K-12 from a yeast shotgun experiment. In a nutshell we were trying to determine whether the presence of a piece of foreign DNA in *E. coli* K-12 could alter its rather low pathogenic properties. This study, which was done in collaboration with Daniel Botstein, MIT, involved the injection of newborn or weanling mice with live K-12 from a yeast shotgun experiment. The results indicated that after five serial passages, the LD_{50} of the *E. coli* was indistinguishable in K-12 containing no recombinant DNA. Approximately 10^7 - 10^8 organisms had to be injected intra-peritoneally and intracerebrally to effect the death of 50% of the test animals. Interestingly enough what did change was the heterogeneity of the yeast recombinant DNA present in the *E. coli*. The recombinant DNA evolved from a very heterogeneous pattern to one which was very simple and yet the virulence of organisms carrying yeast DNA did not change.

S.D. EHRLICH: I think these experiments illustrate very nicely a point that was raised by Walter Bodmer and which I think has a general significance. Insertion of foreign genetic information into an organised genome is not likely to increase its evolutionary value.

A. CAMPBELL: The types of experiment you have done, potentially, can add to the information on recombinant DNA in two respects. One is the one you have addressed principally, which is the likelihood of causing any kind of infection from a cloned virus. The other side of the coin, if you will, is that they provide a method for testing the question some of us have worred about, "Does DNA ever get out of bacterial cells and find its way into animal cells?". By extrapolation from your results I would say that probably it does - if you had a bacterium in the gut which was lysogenic for P1, say. So it was occasionally sending particles with pieces of the bacterial genome out, since the DNA from the phage particle seems to be taken up intact by the animal cell, although rarely. Is there a way of testing that more directly in a more sensitive manner?

M.A. MARTIN: As I mentioned in my talk, one of the hypothetical hazards following animal viral DNA in *E. coli* K-12 systems involved the transfer of recombinant DNA molecules out of *E. coli* into mammalian cells. We interpret our findings as indicating that *E. coli* does not deliver recombinant DNA to animal cells in a biologically active form, although the inoculation of purified recombinant DNA containing a head-to-tail dimer of polyoma did result in mouse infection, the inefficiency of recombination processes in the mouse which was required to render such preparations biologically active is quite striking. Further, we could demonstrate that the injection of 10^8 PFU of polyoma-λ recombinant particles resulted in mouse infection. However, inoculation of animals with live bacteria infected with phage recombinants consisting of a polyoma dimer insert again resulted in no evidence of polyoma infection in the mice. This was somewhat unexpected, since this latter experiment was designed to deliver the same amount of phage, not as purified particles but as phage present in live bacteria, their "natural" hosts. This negative result may indicate that phage replication is severely inhibited when the bacteria are present in a subcutoneous environment or that the ability of phage particules to gain access to mouse tissues is inhibited by bacterial cell components. The bottom line is clear. A potentially infectious DNA molecule did not move out of *E. coli* and cause an infection in a sensitive animal host.

SECOND REPORT OF THE COGENE WORKING GROUP ON RISK ASSESSMENT

A. M. Skalka

Department of Cell Biology, Roche Institute of Molecular Biology, Nutley, N.J. 07110, U.S.A.

COGENE's Working Group on Risk Assessment was established in May 1977 in response to concerns regarding the safety of recombinant DNA technology. Its members include: Drs. Giorgio Bernardi (France), Vittorio Sgaramella (Italy), and more recently, Dr. Natalie Teich (U.K.) with myself as convener. At its formation, the Group outlined two objectives. One was to gather information to be considered in the assessment of risks. The second was to identify possible areas from which additional relevant information might be obtained, and, if studies in such areas were not already underway, to consider their sponsorship.

In July of 1978, COGENE made public the first report from this Working Group[1]. This two part, 319 page report contains a variety of information including responses to a questionnaire sent to various international organizations and evaluation of their significance, reports of various meetings and hearings as well as comments on relevant information from the literature and opinions of individual scientists.

Analysis of this information indicated that the concerns regarding recombinant DNA research could be grouped into three major categories:

1. Organisms carrying recombinant DNA may spread in the natural environment and disrupt existing ecological equilibria.

2. These organisms might produce some toxic or noxious substance, or otherwise cause disease.

3. By exploiting this technology, scientists may be crossing some hypothetical barrier to DNA exchange between eukaryotes and prokaryotes and thus affect pathogenicity or dispersion of pathological agents.

Reports from the Falmouth Meeting (June, 1977) and COGENE-sponsored analyses of *E. coli* K-12 systems, both discussed in the report, showed that many relevant experiments had already been conducted by scientists in such closely related fields as epidemiology and infectious diseases. Their observations that *E. coli*

---

[1]Copies of the report are available from COGENE (c/o Dr. W. J. Whelan) P.O. Box 016129, Miami, Fla. 33101, U.S.A.

K-12 has a limited potential of survival and their consensus that this bacterium cannot be converted into an epidemic pathogen by laboratory manipulations with DNA inserts, has done much to satisfy the first concern mentioned above. As recently reported in the literature, (Petrocheilou and Richmond, 1977) results from long-term monitoring of laboratory workers who routinely handled K-12 organisms carrying transmission-proficient plasmids showed no bowel colonization despite the fact that the work was carried out without any special precautions. If the organisms which carry recombinant DNA cannot spread in the natural environment, then clearly the other concerns must also be diminished. The conclusion of a NIH/EMBO Animal Virus Workshop (in Ascot, Jan. 1978) was that recombinant organisms carrying viral inserts cannot be any more hazardous than the viruses themselves, and in some instances may provide an opportunity to work more safely with virulent agents.

As indicated in our report and in the proceedings of this meeting, important information on the nature and possible consequences of the manipulations used in recombinant DNA technology comes from experiments in a variety of fields. For example, attempts to understand the biological role of restriction enzymes have revealed that prokaryotic and eukaryotic DNAs can recombine *in vivo*. Other studies, including those with *Agrobacterium tumefaciens* and plant cells, also show that in some cases there is no barrier to exchange and expression of genes across the hypothetical prokaryotic-eukaryotic barrier. On the other hand, studies of animal virus genomes and eukaryotic genes for differentiated functions have revealed previously unsuspected possible barriers to their expression in prokaryotic backgrounds. It is clear that further studies of the molecular genetics, pathogenicity, ecology and other properties of organisms will continue to provide information relevant to the assessment of risk and that this source is no less important than specifically targeted experiments.

Replies to our Questionnaire indicated that there were several projects underway directed specifically towards risk assessment. These included the two separate polyoma virus experiments sponsored by EMBO and by NIH and several NIH contracts aimed at testing and verification of EK2 and EK3 systems. Preliminary results, as well as conclusions from the NIH-EMBO Virology Workshop indicated that these studies were unlikely to reveal unknown hazards. However, these and other experiments aimed at elucidating various aspects of the ecology and natural history of microorganisms should provide useful information and will be important in the development of host-vector systems other than those based on *E. coli* K-12.

In summary, as of July 1978 our analyses had revealed no scientific findings to justify any of the three concerns listed above; no risk unique to recombinant DNA research was identified. Available evidence indicated that recombinations of the type made possible by this new technology could occur in Nature. Evaluation of *E. coli* K-12 showed that this bacterium is essentially harmless and that insertions of segments of foreign DNA into its genome could not alter this property. With few exceptions, it seemed likely that the same would prove true of other host bacteria.

Some of the results discussed in our report were also documented in material which accompanied publication of the December 1978 version of the U.S. NIH Guidelines. This material was included as part of the justification for relaxation in containment requirements for permissible experiments.

With respect to the second objective of our Working Group, at the time of our first report it appeared that additional information on results from feeding experiments and monitoring for acquistion of laboratory *E. coli* strains would be useful. In these efforts we were fortunate in being able to enlist the aid of three distinguished investigators, Drs. S. Falkow (Seattle), Dr. M. Richmond

(Bristol) and Dr. E. S. Anderson (London) who had been conducting routine monitoring of their personnel involved in bacterial and plasmid research. These three scientists agreed to collate the results of their surveys for COGENE. Their conclusions, which were summarized by Dr. M. Richmond during this meeting, are included as Appendix II of this report. In our view these findings are of special importance because they record experience from real working situations in which no more than accepted microbiological practice has been employed. The results showed that under these conditions no acquisition of bacteria or even transmissable plasmids could be detected.

Additional data which bear on the assessment of risks has become available since our first report. These include results from the "polyoma" experiments now documented in the literature (Israel and co-workers, 1979; Chan and co-workers, 1979) and in preceding contributions in this volume. These studies, meant to be "worst case" tests using an extremely sensitive assay, show that polyoma-containing recombinant hybrids or hybrid-containing bacteria are either non-infectious or are less infectious in mouse than the virus itself. Thus, they constitute strong support for the conclusions from the Ascot Animal Virus Workshop mentioned above.

Finally, as mentioned above, specifically targeted projects designed to provide information on the survival potential of recombinants or recombinant-containing _E. coli_ in aerosols, sewage, and the intestines of mouse and man have been the subject of several contracts supported by the U.S. NIH. Although these studies are still not complete, progress reviewed at a March 19, 1979 meeting at NIH also tendsto confirm original estimates on the relative enfeeblement of those systems. Appendix I of this report is a summary of that meeting prepared by Dr. J. Setlow, Chairman of the NIH Recombinant Advisory Committee, who presided.

How do these findings effect the future goals of our Working Group and what advice can COGENE offer its' sponsoring scientific unions and national organizations at this time? We conclude, as in our first report that no special risk inherent to recombinant DNA research has been defined. In the one case most thoroughly studied, polyoma virus, the recombinants clearly present less risk than the virus itself. With respect to experiments involving _E. coli_ K-12 systems, it appears that little if any grounds for concern remain. Regardless of the nature of the cloned fragment or even its ability to be expressed, accepted microbiological practice should be sufficient to prevent escape and survival. These conclusions are especially reassuring because they serve to restore confidence in the ability to predict the outcome of recombinant DNA activities based on available knowledge of the biology of the organisms involved. Similar facts relating to other potential cloning systems should prove equally predictive and fewer verification experiments should be required.

However, despite our confidence in the relative safety of recombinant DNA techniques as they are currently used in basic research, our Working Group does not consider it justified to extrapolate from experience with _E. coli_ to all possible experiments in the future. As technology related to genetic experimentation evolves, there are sure to be operations and applications that we can not even envision at this time. Thus, it is important that a Committee of scientists international in scope, continue to follow developments which relate to genetic experimentation, that it continue to collect information to be made broadly available and that it stand ready to sponsor specific projects which may be necessary for further evaluation.

REFERENCES

Chan, H. W., M. A. Israel, C. F. Garon, W. P. Rowe and M. A. Martin (1979). Molecular cloning of polyoma virus DNA in *Escherichia coli*: Lambda phage vector system. *Science* *203*, 887-892.

Israel, M., H. W. Chan, W. P. Rowe, M. A. Martin (1979). Molecular cloning of polyoma virus DNA in *Escherichia coli*: Plasmid vector system". *Science* *203*, 883-887.

Petrocheilou, V. and M. H. Richmond (1977). Absence of plasmid or *Escherichia coli* K-12 infection among laboratory personnel engaged in r-plasmid research. *Gene* *2*, 323-327.

APPENDIX I

REPORT ON THE PHAGE AND PLASMID SUBCOMMITTEE ON MEETING WITH RECOMBINANT DNA CONTRACTORS

March 19th, 1979 in Bethesda

Jane K. Setlow, Chairman

SUMMARY OF REPORTS FROM CONTRACTORS

Testing of Host-Vector Systems in Sewage. (Bernard Sagik, James Walker and Charles Sorber, University of Texas at San Antonio)

The testing was based on a model system set up in a laboratory, capable of taking 60 liters/day of waste. San Antonio waste was brought into the laboratory to be put into the system. Data were presented to show that the various parts of the model system behaved quantitatively like those of real sewage processing systems (with respect to such parameters as detention time, pH, temperature, removal of solids, etc). The organisms to be tested (χ1776 and DP50*supF* and Charon 4A) were added continuously for 48 hours. Samples were taken at intervals up to 120 hours from nine parts of the treatment plant. For assays of DP50*supF*, the antibiotics nalidixic acid and trimethoprim were used, which kept the indigenous bacterial population from entering appreciably into the counts. These plus cycloserine were also used for the assay of χ1776. The Charon 4A phage was distinguished from the indigenous phage by plating it on XG agar, such that the plaques of these phage had blue-green halos. Considerable effort was put into verifying that the colonies obtained from the putative χ1776 that had passed through the treatments were really χ1776. Seventy-five colonies were tested for other markers such as phage susceptibility, efficiency of plating in the presence of bile salts, and various nutritional markers.

DP50*supF* to some extent multiplied during passage through the first parts of the treatment plant, although it would not be able to compete with wild type *E. coli*. There was little or no multiplication of χ1776. The titer of DP50*supF* dropped from a high of around 10^7/ml down to around 10^2 in the last effluent during the time of the experiment, and was somewhat lower for the treated sludge. Comparable values for χ1776 were about an order of magnitude lower. Charon 4A phage titers dropped about three orders of magnitude.

Testing of Host-Vector Systems in Mammals (Rolf Freter, University of Michigan)

Cells (1-5 x 10^9) were placed in the stomachs of mice, and at various intervals the stomach, small and large intestines were assayed for the presence of colony-forming units of the test microorganisms. Feces were also assayed. A non-absorbable radioactive tracer and spores of thermophilic *Bacillus subtilis* were added at the same time as the cells to monitor the passage of material through the digestive tract. Five types of *E. coli* were tested: DP50*supF*, χ1776, a wild strain derived from mouse intestine, a wild-type human strain C25, and another wild type, χ1666. The results showed that whereas the wild strains could multiply in the gut, especially when passage time was slowed by morphine, DP50*supF* and χ1776 were killed in the gut, particularly when passage time was increased. The amount of cells recovered in the feces was less than one per cent of the original inoculum of χ1776, and up to 10% of the DP50*supF*. Germ-free mice were used in some experiments, which showed that even in such hosts χ1776 could not become a resident in the intestine unless it was a revertant no longer sensitive to bile salts, but even then it did not readily establish itself. DP50*supF* similarly only became implanted when it had lost its bile sensitivity.

Testing of Host-Vector Systems in Mice and Humans (Stuart Levy and Bonnie Marshall, Tufts University)

Two strains of *E. coli* were tested χ1776 and χ2236 (the latter χ1776 plus the plasmid pBR322). The mouse results were similar to those of the Freter study, in that no colonization was observed, even χ2236 in germ-free mice. No survival of χ2236 was detected after 24 hours in either germ-free or conventional mice.

The human study involved groups of four individuals who were kept free of other human contacts for the ten days of the experiment, in a room previously prepared for bone marrow transplants. In addition to assay of stool samples, hand, throat and nose swabs were tested for the presence of the microorganism previously ingested in milk (around 10^9 cells). No recovery of the test organisms was detected from skin, and there was recovery from throat and nose swabs only the first day after ingestion. In the case where three of the individuals were given χ1776 but not the fourth, no transfer was observed to the fourth. After 4 days χ2236 survived only to the extent of 5×10^{-6} in its passage through the digestive tract, and χ1776 somewhat less. No recoveries were detected after 4-5 days. It was not clear whether the plasmid pBR322 really slightly increased the survival of χ1776 in the human gut (nor was it certain that the plasmid-containing χ2236 had no properties different from those of χ1776 in addition to the presence of the plasmid). Two hundred isolated colonies obtained from stools 4 days after ingestion were tested by the contractor and also colonies obtained from stools 4 days after ingestion were tested by the contractor and also colonies were sent to Roy Clowes and John Donch for further checking that the cells indeed had the characteristics of the test organisms. The presumed χ2236 isolates contained pBR322 and no other plasmid by the criterion of restriction analysis, and in no case was there detectable mobilizability of the plasmid. In all other respects (antibiotic resistances, nutritional requirements) the isolates had the expected properties.

Testing of Host-Vector Systems in Situations Simulating Accidental Spills (Melvin Hatch and Mark Chatigny, Naval Biosciences Laboratory)

Survival of χ1776, DP50 *supF*, wild type χ1666, an *E. coli* strain 980 isolated from an infant with diarrhea and the EK2 Charon 4A phage was tested at various relative humidities and temperatures on various types of laboratory surfaces, and also in aerosols. The latter measurements were made by obtaining the ratio of survival of the test organisms to that of *B. subtilis* spores resistant to aerosolization. As expected, χ1776 survives better under conditions not permitting any attempts at growth in the absence of diaminopimelic (dap) acid and thymine or thymidine (which could result in dap-less or thymineless death). The aerosol survival of χ1776 is only slightly lower than that of strain 980 at 50% relative humidity and 22°C, but much lower than that of χ1666. The two EK2 hosts, χ1776 and DP50*supF*, survived considerably less well on laboratory surfaces than did the wild type χ1666. Charon 4A phages were even more readily killed under comparable conditions.

CONCLUSIONS OF THE PHAGE AND PLASMID SUBCOMMITTEE CONCERNING THE CONTRACTORS' REPORTS

1) The data show no gross discrepancy between natural environmental decay of the EK2 microorganisms and the *in vitro* testing performed earlier (in which it was required that "no more than one in 10^8 host cells should be able to perpetuate a cloned DNA fragment under specified nonpermissive laboratory conditions").

The natural environment is actually a succession of individual environments, and when the loss of survival each of these is taken into account and multiplied together, the 10^{-8} figure is clearly obtained.

2) The original emphasis for the testing contracts was concerned with the possibility of converting EK2 systems to EK3. Since the revised Guidelines do not specify EK3 containment in any experiments, it was concluded that what was needed was information on survival of cloned fragments in EK1 experiments (such as in χ1666). Such experiments would, of course, be for risk assessment rather than in preparation for certification of vectors.

APPENDIX II

SUMMARY OF CONCLUSIONS FROM FEEDING AND MONITORING EXPERIMENTS[2]

M.H. Richmond, Bristol, UK

1. *E. coli* K-12, even prototrophs, fed to man and livestock in high doses will generally survive about 4 days, although in a few individuals the organism can be detected by selective methods for nearly 2 weeks. The organism may multiply to a limited extent within 24-28 hours after administration but generally does not undergo more than a few doublings.

2. If the *E. coli* K-12 that are fed carry a self-transferring plasmid that is de-repressed, the plasmid will be transmitted to the recipient flora. Transconjugants are found at a frequency of about 1×10^{-10} of all *E. coli* shed per day. No evidence for the mobilization of non-transferable plasmids from *E. coli* K-12 has been found for naturally occurring F group or I group R plasmids. We think it unlikely that such transfer would occur with approved cloning vectors. This conclusion is supported by *in vitro* mobilization experiments. *In vivo* survival of EK2 strains and their potential transfer of either conjugative or non-conjugative plasmid must be considered to be below the current limits of experimental detection.

3. *E. coli* K-12 strains which survive passage through the human or animal gut rarely show evidence of having acquired plasmids from the indigenous flora. In calf-feeding experiments, only 1 out of well over 10,000 K-12 strains which were screened had acquired colicinogeny.

4. Over a period of 26 months in Bristol, 12 months in London and 12 months in Seattle involving 64 subjects, there was no evidence that laboratory workers, or members of their families, acquired either bacterial strains or plasmids that were employed in the laboratory. This was true even in a few individuals working in the laboratory who received short courses of antibiotic therapy. Monitoring of the oropharynx failed to show any evidence of colonization by laboratory strains. Individuals with excessive facial hair likewise showed no evidence of carriage. In a few instances, individuals who had done routine experimental procedures (vortex mixing, or plating, etc.) showed transient numbers of *E. coli* K-12 on their laboratory garments, facial skin or hands. These organisms were few in number ($4/cu^{mm^2}$) and disappeared within three hours. Simple washing with soap and water resulted in instantaneous loss of these organisms. No evidence whatsoever was found for even transient survival of *E. coli* EK2 derivatives.

---

[2]The following Report was presented by M.H. Richmond to the Meeting of the Risk Assessment Subcommittee of COGENE which met at The Royal Society, London, on Friday, March 30th, 1979. The Report was compiled by Dr. Stanley Falkow on the basis of information provided by himself, by E.S. Anderson and by M.H. Richmond. The information contained in the Report represents the situation as it was at the end of October 1978, save that the data from Bristol have been updated to include information up to the end of 1978.

5. Monitoring of the laboratory work bench, water baths, and centrifuges, etc., occasionally revealed detectable numbers of EK1 *E. coli* K-12. Such organisms could be found immediately after performing laboratory procedures such as flaming bacteriological loops, replica plating or transferring cultures (even with manual pipetting devices). Such contamination was instantaneously sterilized by simple disinfectants, including 70% ethanol. Deliberate spills of EK1 *E. coli* K-12 had become sterile within six hours after drying. Reliable numbers for the survival of EK2 hosts in the laboratory are not available to us but survival, if any, was much reduced relative to ordinary *E. coli*.

6. Individuals working with *E. coli* K-12 infected mice and calves were not found to more likely become colonized with *E. coli* K-12 than workers who did only *in vitro* experiments.

DISCUSSION

S.D. EHRLICH: I wonder whether there is a figure available on the cost of the risk assessment experiments up until now?

A.M. SKALKA: The cost of the risk assessment experiments?

S.D. EHRLICH: I'd like to have a global estimate on the amount spent on risk assessment experiments since the recombinant DNA question was raised.

A.M. SKALKA: I don't have that figure. What information do you hope that would provide?

W.J. GARTLAND: I wanted to mention that when the NIH Guidelines were issued on December 22nd, there was a directive to NIH to conduct a risk assessment programme. The directive says that further actions under the Guidelines must be based on data gathered from a risk assessment programme. NIH has had an internal working group which has proposed a risk assessment plan which is about to be published in the *Federal Register*. The proposal lists some types of things that could be done. One of the central features of the proposal is the recruitment of an eminent scientist who would come to the NIH and who, in addition to directing the programme, would do a lot of data-gathering to try to pull together a lot of information which is already in the literature and perhaps prod people to do some simple risk assessment studies. The proposed plan will be published for comment very shortly.

D. HABER: This is something I'd like to ask someone from the NIH about because it's just a rumour I've heard, so it may or may not have any foundation. I have heard that there was a meeting of an NIH subcommittee on vectors on March 19th at which some evidence was provided - and this evidence is supposed to have been put forward by NIH Private contractors - that EK2 microorganisms survive in sewage. I wonder if someone from the NIH could elaborate on that - confirm, deny, or explain it.

ANNA MARIE SKALKA: Yes, the contractors were Drs. Sapik, Walter and Sorber. The report is available for you to look at and it is included as an appendix in my report.

D. HABER: I would like to direct myself again to the NIH and to the risk assessment activities of COGENE, in particular of EK2 strains K-12. I wondered what note is being taken, or advice being given, or if, in fact, there has been any consideration of evidence that the properties of the EK2 organisms seem to change when grown in large quantities - or could possibly change? Some evidence has been published to that effect and I wonder whether that's been considered.

W.J. GARTLAND: As far as I know that question has not arisen but it might be a logical thing for someone to comment on in the proposed risk assessment plan, if this is viewed as something serious.

M. SINGER: I suppose one might make the following very general comment: in maintaining the identity of any strain, when you change conditions, one requires additional and more careful testing, to be sure, for example, in a large fermentor, that the strain maintains its identity. It would be my impression from my own experience with such things, outside of recombinant DNA research, that those kinds of precautions are relatively routine.

J.K. SETLOW: A whole lot of those mutations in χ1776 have been published and are well-known, so these are matters that can be calculated.

WHAT LESSONS DOES THE RECOMBINANT DNA DEBATE TEACH US A ROUND TABLE DISCUSSION

Maxine Singer (Chairman), W. Szybalski, M. H. Richmond, R. H. Pritchard, W. J. Peacock and J. D. Coombes

W. SZYBALSKI
McArdle Laboratory, University of Wisconsin, USA

It is human nature to find some redeeming features even in the worst debacle. In this spirit I will try to answer the question "What lessons did we learn while being actively or passively involved in the so-called recombinant DNA debate?". Debate it was, but unfortunately not a very rational debate, more like a medieval dispute involving more of the irrational and imaginary elements than rational facts.

The lessons we learned were in a variety of fields, including history, sociology, politics, legislation, regulations, law, literature and even science.

In the field of history, we learned that history repeats itself and that irrational or imaginary dangers were seriously considered not only in the middle ages, but that some ghosts are being regarded as real even now, in the seventies (see Stetten, 1978. *Gene* 3, 265). Also, we became more familiar with the trials and tribulations of Copernicus and Galileo Galilei, and I have a feeling that if they would have lived now and have discovered the recombinant DNA technique they still would have been called to testify at some committees, even if it could not have beeen the Holy Inquisition. Actually the latter statement is misleading in respect to Copernicus, who wisely decided that the time of publishing his revolutionary discoveries should coincide with his death, so as to avoid a necessity to testify at some control commission.

In the field of sociology, we learned several lessons. One was that we should not ask too many irrelevant questions and ask the public to answer these. We also learned that we should be prudent when dealing with the lay press, and be careful as not to mislead the reporters by presenting hypotheses in such a manner that they could be mistaken as facts. The scientists have a serious responsibility to educate the lay society with the help of the press, and not to confuse the public. We also learned that crying wolf is dangerous not only in fables but also in science, especially if there is no wolf. Another lesson is that some segments of society consider it as a sign of untrustworthiness when a person in general or scientist in particular changes his mind. It was a surprise to me, since I was trained to regard an open-minded approach as a virtue, because it is the basis of a scientific method that the hypothesis has to be either radically changed or adjusted as one more closely examines the logical structure based on either the existing facts or on the new experiments.

The latter lesson is closely related to politics and legislation, since in this area any major change in the attitudes of an elected official appears to be a source of distrust and embarrassment. On the positive side, I learned how the legislative bodies in the USA work. It was very inspiring and interesting to walk the halls of Congress, to visit the individual Senators, Representatives, and their ever-changing staffers who were either disinterested or passionately partisan about the issues involved. We learned that the staffers wield often a tremendous power in influencing the decisions of the Congress, but unfortunately their zeal is frequently not matched by their scientific knowledge and wisdom. It was also my privilege to meet several very wise and dedicated Congressmen, of whom honourable Ray Thornton, who is present here, was a shining example.

A very important lesson which we should all have learned is that scientists are expected to provide facts and carefully considered, logically derived opinions based on the hard facts. Neither facts nor opinions should be coloured by their political beliefs and personal anxieties, or by their often naive or perhaps arrogant assessment of "what the public wants or is ready to accept". It is up to the elected official and policy makers to represent the public and make decisions on the basis of the facts provided by the scientists. In the recombinant DNA controversy, the public had often the impression that scientists have been disagreein on the facts, whereas the truth was that they disagreed only on the degree of their personal anxieties, their political theories and hastily conceived opinions not based on hard facts. Unfortunately, some of them performed a disservice by confusing or misleading both the public and the administrative and legislative bodies

We learned that some scientists have tried to play politicians or outguess the politicians, and did that very poorly. A good example of such poor judgement was the philosophy behind the first draft of the NIH Guidelines by the subcommittees of which, I have to admit, I was a member. We were instructed to draft a very restrictive set of guidelines, as to show the social responsibility of the biologists, their willingness to impose self-regulation and their lack of self-interest. It was said, that in this manner biologists would avoid any future restrictive regulations or legislation imposed by some outside agencies. We soon learned, however, that the restrictive nature of the guidelines was taken as "proof" of the real dangers and consequently resulted in a multitude of legislative proposals, as described in Don Fredrickson's lecture.

As far as the regulations are concerned, we learned that some segments of our society cherish the regulations and consider them as a panacea for all real or imaginary social ills. The "true" regulators never ask whether regulations make any sense but their main preoccupation is how to enforce the regulations and uncover the violators. During my three years on the NIH Recombinant Advisory Committee, I heard rarely, if ever, a detailed and productive discussion (1) on the risk assessment of recombinant DNA, (2) on the costs of the regulation, and (3) on the new dangers created by the regulations. When proposing the regulations, it is quite important to examine both sides of this coin, since regulations could create new dangers instead of helping to alleviate the conjectural risks. Actually, a recent Appeals Court ruling requires a scientific cost-benefit assessment, before any regulation could be proposed (see *Science*, 1979. 203, 1324).

I heard about several kinds of dangers created by the NIH Guidelines (*TIBS*, 1978. 3, N243) in at least one of the laboratories on our campus. One involved my only daughter who was working as a dishwasher and media maker in a recombinant DNA laboratory since she was 14 years old. When the NIH Guidelines were introduced in 1976, they required a great increase in the amount of autoclaving and the result was that a gasket blew out in the overworked autoclave burning my daughter's arm. So she was a victim of regulations based on the NIH Guidelines, and not of the recombinant DNA technique. Other examples of novel laboratory dangers created by regulations are first, inhalation of harsh chemicals spread on large surfaces, the purpose of which was to inactivate recombinant DNA and second, the possibility of electrocution and other injury by the laboratory machinery in the cramped confinement of the special containment laboratories. It appears that we need now a new set of safety guidelines designed to protect us from the novel laboratory hazards created by the NIH Guidelines for recombinant DNA research.

In the field of law we learned that science is very vulnerable even if the freedom of basic research is under Constitutional protection. We learned that the traditional academic freedoms have to be carefully protected, since they could easily be lost under the pretext of imaginary dangers. We learned that in general the judges and lawyers are less interested in the scientific reasons for proposing or abolishing the regulations than in interpreting or enforcing the regulations and seeing that the due administrative process was followed.

There are two different worlds: one consists of the research scientists who are very sceptical about any scientific laws and who constantly question and modify the accepted dogmas. The second is the world of the jurists whose role is to guard the accepted legal doctrines and protect society from those who do not follow the law.

As far as the scientific facts are concerned, some of us knew and many learned that pathogenicity is a complex aglomeration of many finely tuned mechanisms, and not only is it practically impossible to enhance the natural virulence by at random additions of DNA fragments, but it is even difficult to preserve the natural pathogenicity under the laboratory conditions. Also, plasmids, especially when carrying foreign DNA, are easily lost from the host cells, unless being cultivated under highly selective conditions in special laboratory media. We also learned that, as theoretically predicted and then experimentally proven, the cloning of viral DNA in the phage or plasmid renders it either non-infectious or much less infectious than the virus or its DNA. No instance of infectivity of viral DNA was seen when recombinant DNA carried by *E. coli* was intestinally implanted. Thus, the recombinant DNA technique should always be applied in order to make the virus research much safer (see Stetten, 1979. *Science* 203, 1292). We even heard the opinions that the recombinant DNA regulations might have contributed to the Birmingham smallpox tragedy, since if the use of the recombinant DNA technology would have been encouraged, instead of instituting discouraging regulations and prohibitions, the research on comparing the DNA sequences and proteins of DNA viruses could have been done in a very safe manner and more revealing one, by cloning the fragments of smallpox DNA in *E. coli* K-12 host-vector combinations. It is interesting to see how the pendulum of opinions swings, and that the proponents of regulations are now cast as the villains. This development, however, could have been expected since the regulations had an irrational basis.

To allude to the literature, three stories come to my mind. I already alluded to the "crying wolf" story, and now the Hans Christian Andersen fable about "The Emperor's New Clothes" comes to my mind. The important difference is that in the fairy tale it was one little boy who said "The Emperor is naked", whereas at present quite a few not-so-little boys are saying that the "dangers" are only the imaginary ones. However, the "bastions" of the recombinant DNA regulations did not crumble as yet. This appears strange, because even in the frequently cited Mary Shelley's horror story, the dangers of Frankenstein's monster were not imaginary but well proven, and Victor Frankenstein became concerned about it only after his own brother was killed by the monster.

Yes, we learned very much, both new and obvious, during the recombinant DNA debate. But was it worthwhile? I doubt, since the price we paid for these lessons was very high.

R.H. PRITCHARD
Department of Genetics, University of Leicester, UK

The greeness of the grass depends on which side of the fence one sits. I have two fences in mind: that between Mark Richmond, who is a member of GMAG, and myself who most certainly is not, and that between Britain on the one hand and the United States on the other.

It is obvious that is the American scientific community which must take responsibility for the excesses of the recombinant DNA controversy. Equally, as someone working in Britain, I have been immensely impressed at this conference by the confidence and openness and flexibility with which American scientists, and even more especially their administrators, are tackling the re-entry problem. One cannot fail to be impressed by Fredrickson's unequivocal statement that his objective now is the ending of the guidelines.

Dr. Stetten praised the literary style of the Ashby Report and the fact that it did not lay down rigid guidelines, unlike Asilomar. But from my side of the fence our guidelines appear to have been interpreted with the flexibility of a crow-bar. Sir Gordon Wolstenholme speaking of the Williams' Report recalled its intention that there should be a dialogue between GMAG and the scientific community. While I have been impressed by the evidence of such a dialogue in the United States, there has been no such dialogue in Britain. There is no mechanism for such a dialogue. I do not know how GMAG reaches its decisions because they are arrived at with all the openness of a papal enclave. I would like to hear from members of GMAG at this conference how they have arrived at their latest guideline revisions which retain an interest, for example, in self-cloning. Why is self-cloning more hazardous here than in the USA? In view of the extent and openness of the American efforts to revise their guidelines safely, why can we not simply accept their conclusions? If GMAG has serious grounds for disagreeing with these conclusions, has it not a duty to inform us of them now? If it does not, then let us for the moment adopt the United States position without further ado.

So the first lesson I have learned is that what is on paper is less important than the way it is interpreted and that in this respect British is not best.[1]

I have been puzzled by the comments I have heard about re-entry, or relaxation of the guidelines. Why, if all are agreed that the Stetten formulation would cover every conceivable concern, do we have to go on vomiting for another 18 months? (The time-consuming validation procedures could be justified as long as a significant body of informed opinion remained convinced of the existence of hazards. If this has ceased to be the case, how can you demand of your RAC members that they continue to engage in an activity which is then reduced to window dressing?)[2] In this connection I am interested in the position of the 11 people who signed the "Berg" letter. The credentials of the bodies who control recombinant DNA research stem from that letter. This may not be too apparent in the USA, but in Britain the following recent statement in a report by GMAG will illustrate the point.

---

[1](It has been suggested to me that criticisms of GMAG by people on my side of the fence may be due at least in part to the shoestring budget under which it has to operate compared with the RAC.)

[2]Inserted in published version to clarify previous sentence.

"The question of hazards was raised by responsible scientists and we feel it essential that, however strong an individual worker's estimate, we have a right to expect, until we know more of the reality of the situation, that he should do the work either with scrupulous care or not at all". Mere workers apparently, who disagree with distinguished and responsible scientists, can only have intuitive reasons for doing so.

I have quoted this to illustrate our dilemma. If the hazards hypothesis is challenged, the response will always be the same. Why should we believe you who have a vested interest in dismantling controls rather than the eminent scientists who signed the "Berg" letter? The "Berg" letter is a petition. Its force rests not on the arguments contained in it but on the eminence and authority of those who signed it. Consequently, for those who are impressed by it, its message can be countermanded only by the authors themselves. I believe it would be helpful if the authors of such an authoritative letter were to consider again the possibility of writing collectively a second letter which would state their current position on the scientific issue and I should like to hear the views of those who are at this conference.

Let me make clear that in suggesting a second letter I do not for one moment presume to assert that they were wrong to publish the first one. If it reflected their considered views it would be wrong not to have done so. Jim Watson has, of course, been quite unequivocal in stating his position. But for the reasons already alluded to by Walter Bodmer, I think there is some responsibility on the authors to state their current position collectively if it has changed, and if it would help to accelerate the relaxation of guidelines.

It has been said publicly by members of GMAG and by others at this meeting that to argue for a significant relaxation of substantial modification of guidelines now would be counterproductive. The British phrase is that it would "undermine public confidence" in the scientific community. This seems to me an arrogant position to occupy. It is true that public confidence in the infallibility of scientific experts might be undermined but that is surely a gain. I don't see how we can have genuine public involvement unless the public can be confident that the expert is not always right. Also by stating the truth as we see it honestly and frankly and unequivocally we may demonstrate that we have made asses of ourselves but public confidence in our integrity is surely more important.

That is why I support the position taken by Bodmer and why I am disturbed by the implications of what people have been saying about smallpox and power stations. It is implied that others cannot see distinctions that we can. Of course, they will not see the distinction if we do not make it and if the press persists in obscuring it. But it is surely arrogant of us to assume that if the public does not see the distinction between real hazards and science fiction, it is because it is not clever enough to do so rather than because we are not careful enough to make the distinction clear.

The second lesson, then, that I have learned is that there is only one thing the public deserves to hear from us and that is the truth as we see it.

M.H. RICHMOND
Department of Bacteriology, University of Bristol, UK

I think I should begin by saying - as many of you know - that I am a member of the British Genetic Manipulation Advisory Group. But, I am in no sense here as a spokesman for that group. What I am saying is my own opinion and should not be attributed to them at all.

What I am going to say, I am sure, is very substantially coloured by a UK point of view. One can't have served on GMAG for two and a bit years without bearing some marks of this. So when you listen to what I have to say you must put it into context and perhaps set it against the situation in other parts of the world where some of what I say may apply and some may not.

If we look at the UK, we've actually passed through a very rapidly changing situation over the last few months. The publication of the new NIH Guidelines has altered the situation from one where, in many respects, the regulation of genetic manipulation work in the UK was laxer than in the USA; but now rather suddenly we find ourselves in a situation where the governing rules here are in many respects more severe. It's worth mentioning, in passing, I think, that just by looking at the Guidelines the full difference doesn't always become apparent. One of the great misconceptions that we have noticed on GMAG is the way in which people equate the American P3 with the British category 3. In a category 3 lab you work in a glove box all the time, and I think you will see, at least those of you who have ever tried to work in glove boxes, that there is a very considerable difference between that and trying to work in a laminar flow hood. Therefore, the British situation, particularly relating to category 3 experiments is, I think, a very much more severe condition under which to work than is P3.

In this country we also have a law which governs this work. In addition to that, we now have an Advisory Committee, and this is coupled to a Governmental Agency whose job it is to see that the work that goes on in this country is carried out safely.

Now, all that's fine and I wouldn't want to quarrel with that arrangement at all myself. The point that has become very acute in this country, and may become acute in others as well, is that groups of this kind which have a very important say, like GMAG, in the way in which scientific experiments are carried out, often have a membership which comprises representatives of the public interest. In the UK, GMAG also has representatives of other bodies such as the trade unions; and it also has a few people who are scientific experts. The letters of appointment speak of these people as being *scientific experts*; and it's a moot point whether these "scientific experts" *represent* the practising scientific population or not. That's one point which, from time to time, I find difficult.

The upshot of this is that you have a fairly large committee on which the scientists have voices, and I think, perhaps I flatter myself, significant voices on matters of scientific importance, but they don't by any means rule the situation as far as decisions about the ways in which the work should go forward are concerned. The public interest groups in this country have a very big say as to what is done and the conditions under which it is done - and that is by means of membership of a government committee. The result of this is that if scientists wish to change things, and we'll assume they do - if only to have a harmonious set of rules across the world - the question is how do you make the change?

The big difficulty - and this is one lesson that I think is to be learned - is that the only way in which things can be changed is by the scientific community putting up information which will allow a group of people, and in the British Isles, of course, it's a very wide representative group of people, to change their opinion as to the way in which work should be done.

The difficulty is that the people who are not scientific experts very frequently ask the following and, I think, very difficult question: they ask what has happened to make the Americans suddenly change the level under which experiments should be done from a very severe position to a much, much looser one? Has the actual amount of concrete evidence that has come forward been enough really to justify that? One of the things that worries me is that I feel the main basis on which it has been done is only a change in the climate of opinion. I don't think you can really say that the amount of risk assessment work that's been done, and the amount of scientific realisation that has come to us, is solely responsible for the change in position.

I happen to feel personally that the level at which things are operating under the new NIH Guidelines are much more correct than they were post-Asilomar. But, I think it is primarily an emotional change of view and that this has to be explained to the non-scientists. I am as guilty as any: I've changed my emotional position considerably. Another thing that is often said is that you will never get an objective opinion from the scientists because there is too much self-interest, too much at issue for them. There's the issue of the kudos to be derived from scientific research and there's the question of a large number of scientists involved in this field setting up small private companies in an attempt, presumably, to make a lot of money. How, the public interest people say, can such people possibly be objective? How can they really have a say in the way in which the work out of which they are going to benefit should be done?

I think the other thing that, as scientists, we cannot escape is that from time to time things do go very wrong in the world. We may be able to look at these events and say that that's the fault of the technologists, not the fault of the scientists. But in practical terms it is very hard for people who don't distinguish technology from basic science to make those distinctions.

So I think the way forward is, as Walter Bodmer says, to try to distinguish science from politics. But the lesson I have learned is that in publishing the letter and by having the Asilomar Conference, and also to a certain extent, I put it to you, by having this meeting today, and talking in the way we are, that we are not separating science from politics. The actual publication of the letter, I think, was a political act. It has to be seen as a political act but I absolutely believe there was not a political intention. But having seen a political act at one stage, the subject has, as it were, lost its innocence; and I think it's going to be very hard for the scientists who are involved in that work to regain the situation where their views can be regarded as objective on the matter.

So this is the problem I can see before me. In a situation where one has to influence a group of people, only a proportion of whom are scientists, how are you going to do it? How are you going to assure them that you are not arguing from self-interest but that you are now arguing objectively - particularly, when looked at from the outside, you seem to make enormous quantum jumps between what you require for conditions at one time and what you require for the same experiments a couple of years later?

W.J. PEACOCK
CSIRO Division of Plant Industry, Canberra, Australia

I would like to make two suggestions, stemming from the recombinant DNA debate. The observations come from a somewhat parochial viewpoint. I would like to tell you a little bit about what has happened in Australia.

The Australian Academy of Science some years ago formed a committee of scientists, some directly in the field and some not, to shepherd this brand of science into our country. We adopted guidelines which were a hybrid of the NIH Guidelines and the Williams' Report. These were voluntary in Australia but have been followed by all participating scientists and their institutions. We have experienced a painless introduction of this area of science to our scientific and non-scientific community. Every now and then there is an expression of contra-feeling and just recently there was a call in the national press from a Melbourne University group for a total moratorium on this field of work. At times like this the members of the Academy Committee and other scientists are asked to comment and are then quoted defending the area of work and trying to straighten out statements which are often based on misinterpretation or are incorrect. The point is they are always defending and they are defending a situation which has been criticised with screaming headlines the day before.

I suggest that the original open publication of the letter was perhaps a mistake - this is a retrospective view, of course - and that in the future for these sorts of situations, and they will turn up in other parts of our science, that a small group of scientists should not take what could be considered a rather flamboyant and emotion-inducing step. I urge that a body like the National Academy, The Royal Society, or a body representing an international group of societies, should respond to the concern of scientists, and should set up an evaluation by people with the appropriate knowledge needed to look into the situation. At the same time they should issue a public statement saying that concern has been expressed and describe what they are doing about it. Thus, if the National Academy had issued a statement about the discussion of the Berg Committee, the outcome may have been different.

The other point is that the scientific bodies should continue to issue statements. The newspapers should be given statements reporting progress in the field. The swing should be presented to the public not only in a defensive way after heavy criticism. It is the duty of national scientific bodies to make findings known publicly and to reflect changing opinions based on increased factual information. The voice of the individual scientist is often weak. The scientist's voice rarely carries the same media power as a church leader, a union leader or even an activist for some particular cause.

I think one lesson we should take then is that the scientific bodies that we set up because we respect their aims and their members should have a direct responsible role in relating our science to our society.

J.D. COOMBES
Hoechst Pharmaceutical Research Laboratories, Milton Keynes, UK

Genetic engineering is but one of the many fears that have been created in the mind of the public about the activities of scientists. There is now a strong public belief that scientific - or technological - progress can only be achieved at the cost of risk of harm to people or the environment. These fears find their rational expression in attempts to gain public control and public accountability of science; and politicians just as inevitably respond to this collective public desire by introducing legislative controls over science. What has changed perhaps is that these controls are now being directed, not to the applications of science, but to the practice of scientific research itself and laying down conditions on what research may be done and under what circumstances. The FDA Good Laboratory Practice Regulations and the NIH Recombinant DNA Guidelines directly affect the conduct of scientific work in the laboratory.

To my mind the lessons of the DNA debate are that we need to adopt a more rational attitude to risk and we need to define more clearly the roles of the scientific expert and the public in the decision-making process relating to scientific issues.

The problem with the recombinant DNA debate was - as other speakers have indicated - that the risks of genetic engineering were not really analysed or defined before public concern was stimulated by the "Berg" letter, and because they were not defined they were not subjected to the normal scientific peer review process. Nor was a distinction made between the risks of the technique itself and the difficult moral and social issues that might arise from some of the potential applications of the technique in human biology. Surely much of the concern about recombinant DNA related to the re-emergence of suggestions based on eugenics.

But, now I should like to turn to the role of the scientific expert in technological decisions. Any major scientific decision affecting the public involves the integration of facts and value judgements. As scientists we are trained to analyse facts and draw conclusions, but we must recognise that so far as making value judgements are concerned, our voice - properly - carries no more weight than any other member of the community. It is not up to the scientific community to make societal preferences on behalf of others.

If this distinction is accepted, then it is the responsibility of the scientific community to lay the facts before the public and it is the duty and responsibility of the public - through their elected representatives, the politicians - to express their preferences and enact decisions. At this point I would suggest that the professional societies representing scientists might take a more positive and active role in publicising the consensus view of the professional experts on the facts relevant to the issue.

I would like to finish by making a few comments about GMAG, the body which effectively controls genetic manipulation in the UK. As we have heard at this meeting, most countries are adopting the NIH Guidelines as the basis of their control systems, whereas in the UK GMAG is pursuing an independent line. Surely any system of risk control must be international and reasonably uniform if it is to be credible and not lead to the creation of "risk havens" where work can be undertaken with the economic advantage of minimal interference or control. As an Englishman I regret very much the direction GMAG is taking, for once again we are in danger of becoming the "odd man out" and excluding ourselves from playing a part in the process of international harmonisation of regulations.

The economic penalty we will pay for this will not be small.

Finally, I should like to make a few comments about GMAG as a concept. The fact that GMAG includes amongst its members representatives of the trade unions, industry, and the general public, had led to the suggestion that it should take into account not only safety considerations, but the scientific merit and social desirability of the proposed research. The suggestion has even been made that bodies of similar composition to GMAG should consider such difficult scientific issues as new drug introductions. Suggestions of this kind originate from groups seeking to influence scientific policy and the direction of scientific research towards what they would perceive as more socially desirable objectives.

In conclusion, let me say first, we must expect public fears about science to be expressed in a wish for greater public control of scientific research and public accountability. Second, as scientists we must distinguish between facts and value judgements. Third, professional scientific societies should be active in presenting the facts to the public on major issues. Finally, on many scientific and regulatory issues an international perspective is desirable, and bodies such as COGENE can fill a key role in this respect.

DISCUSSION

D. WEISS: I would like to make two observations, quite the converse of many of the observations you have heard this afternoon. I hear in this discussion a great deal of self-berating and self-bemoaning. Despite the awkwardness of the recombinant DNA debate, and awkwardness that has been imposed upon our laboratory experimentation, it has had two major positive effects. First, regarding the "Berg" letter and the Asilomar Conference, I believe that it represented one of the finest and most unique displays of scientific responsibility on the part of the signatories because they saw at that time, in their considered judgement, a clear and present danger, and they wished to inform the scientific and lay public of their views. The fact that they changed their minds on the basis of reconsideration and new information does not in any way say that the original decision was a bad one. The second great advantage of the recombinant DNA debate has been an absolutely unique opportunity to educate the public, the lay public, about science and the way science works, and it has been through that education in the United States, I am convinced, that the possibility arose for the relaxation of guidelines and the decline of interest on the part of our legislators to enact restrictive legislation.

E.L. WOLLMAN: I have hesitated before intervening in this discussion, because my command of English is inadequate compared to that of the fifteen preceding speakers on today's programme who all come from English-speaking countries. However, since this is supposed to be an international meeting, I have assumed that it would not be completely inappropriate to express the point of view of a non-sophisticated layman coming from a developing country.

Whereas most of the preceding speakers have, to some extent, been actors in the game of *in vitro* DNA hybridization, an expression which I prefer to that of DNA recombination which applies to all naturally occurring *intra species* processes, I have only been, since the onset, a spectator interested, not only in the scientific developments of the research, but also in its sociological and psychological implications.

When I first read what is generally referred to as the "Berg" letter, I must admit that I was completely misled. Naively I had thought that those biologists who had been most directly involved with these important new developments meant, with a somewhat new sense of the social responsibility, to warn the scientific community and the public in general of the possible misuses that could come out of this new technology. I thought that they were seeing themselves in the position of atomic scientists who, if it had not been in the pre-war situation, would have met to consider the possible dangerous applications of nuclear fission and dissociate themselves from such harmful uses.

It was in that spirit that I went to Asilomar, and was extremely disappointed when I heard the introduction by David Baltimore saying that the conference would not touch on possible harmful applications of the recombinant DNA technology for industrial or military purposes, but would only deal with the hypothetical risks that could come out of ordinary work in a research laboratory. To take again the comparison with atomic scientists getting together in 1938, it is as though they had been told "Let us not worry about possible atomic weapons or power plants, but let us restrict ourselves to household problems of not getting ourselves contaminated and keeping our labs clean". These ancillary problems are important, since laboratory work exists, but they do not seem to call for worldwide publicity and a moratorium on research. In my opinion, and in the words of the French proverb "La montagne avait accouché d'une souris".

The way in which the problem of *in vitro* DNA hybridization research consequences has been handled seems to me to have carried two main drawbacks. First, not having stresssed that a harmful consequence of a scientific achievement cannot be a chance consequence of the research done by the academic scientist, but a harmful consequence of a hazardous application of the research, which is the decision not of the scientist, but of either a political or economic power. Second, it has revived the old beliefs of the scientist, sorcerer apprentice, from whose hands can come new harmful living beings, made *de novo* by chance assembly of harmless parts, a kind of new avatar of the everlasting belief in spontaneous generation.

There is one last point I would like to make: that of the special responsibility with which, willingly or not, our colleagues from the United States are invested. It is a fact that many of the things that come from the other side of the Atlantic have a special impact and are likely to spread all over the world. We have thus witnessed various sorts of world epidemics ranging from rock and roll to the wearing of blue jeans. Due to its strength and dynamism, biological research carried in the United States deserves its high reputation. Therefore, statements coming from the United States have more consequences than if the same statements were made by scientists elsewhere: they have, besides the credit given to renowned scientists, the special momentum due to their American origin. There is no doubt in my mind that, if the statements contained in the "Berg" letter had originated in another country they would not have had the same publicity. One may wonder why the recombinant DNA affair did spread so easily in Europe. There is first, of course, the role of the press: it is always tempting to revive irrational fears of the unknown in the public's mind. There is also, one must admit, the fact that scientists everywhere prefer to conform rather than to stand up and disagree.

The history of the DNA hybridization controversy will be an exemplary one in the history of biology. It teaches us a lesson on how biologists must learn the real meaning of a sense of social responsibility.

I.S. JOHNSON: It seems an appropriate time to make a comment that I was really going to make tomorrow. We have heard from the panel comments about the public being able to make a judgement, and I think that's correct. I took comfort from the dialogue in Cambridge, Mass., because I think I saw lay people come to a reasonable decision that this work should go on. The public's judgement, though, is coloured by the type of information they get. This has been particularly brought home to me in the last couple of days, because I have seen circulated among some members of this meeting copies of an article in *Nature*. I think it is probably one that Dr. Weiss referred to earlier this afternoon, in which, quote: "There are some companies in the United States who are going to break the ten litre limit and do something bad." Our company was discussed in this article at great length, and one of my supervisors, an executive vice-president of the corporation, was quoted. The article was written as if it were an interview with Dr. Pettinga and no such interview took place. We had a telephone enquiry from this reporter about what kind of problems we saw in industrial application of recombinant DNA. We said we saw some which were largely based on the proposed regulations of the Commissioner of the Food and Drug Administration which were published in the *Federal Register*. We gave him a copy of our comments. The proposed regulations were published by the Commissioner for public comment. Portions of our comments to the Commissioner on his intent to issue regulations were excerpted and taken out of context in this article as if they were a quote from Dr. Pettinga. These were not general comments on guidelines but specific comments on proposed regulations

I have little quarrel with the points raised in the article. We found many things in these proposed regulations to be objectionable. There is still a major problem, from a parochial viewpoint, an industrial one in the United States, and it has largely to do with the protection of proprietary information. There are no provisions in the guidelines for protection of such information. We are in a highly competitive free enterprise system and patent protection and trade secrets are important to us. In his intent to issue regulations, the Commissioner suggested that if you could not prove you were in complete compliance with the letter of the guidelines from the inception of a project, to what you were doing now, to what you intended to do in the future, that he might withhold approval of an IND and NDA.[1] That sort of retroactive regulation seemed to me, and to several others, to be unusual. I am aware of a law in Israel which is retroactive as applied to Nazis, but I do not remember many examples of retroactive law. We have complied voluntarily with levels of biological and physical containment. We suggested that whenever the regulations actually applied to us, and they have not been issued yet, that the clock should start at that time and not some date way back in the past.

For example, one of the ways we got interested in recombinant DNA was through a small programme we had in viral oncology where we were fractionating pieces of oncogenic DNA and seeing whether they would cause transformation or not. It occurred to us that these could be used as vectors and so on. Then, when restriction enzymes became available it was very easy for us to see some things that we ought to be doing. The accuracy of the information that the public gets is going to have a profound effect on their decisions. My guess would be that this article in *Nature* may well impell some action on the part of regulatory officials in the United States to meet with us concerning this issue of proprietary information. I would welcome that. We have been seeking such dialogue and conversations with both the NIH and FDA. My concern is whether what will come out, due to pressure based on an article which is inaccurate and uses information taken out of context, may not be what we like, and I would appeal that information provided to the public at least be as accurate as possible.

O.W. PROZESKY: I have a small practical suggestion. The point has been made that we have to present a unified front, that it is much easier if we all had the same regulations, and I think that is a point well made by Professor Pritchard, but I think it is time that we should stop having our own little national prides and committees and swallow them, and let's all go for the United States revised guidelines. I think these guidelines are good enough if we all try and follow them. Many people here can make these decisions or help influence them and if we get a unified guideline system - and the United States ones, I think, are the closest - I think this would work very well.

J.D. WATSON: I think there is a much simpler solution!

D. HABER: I am a member of GMAG, but I am also a Labour union representative, and I would like to speak as a representative of my union rather than as a representative of GMAG, especially as I haven't been authorised by GMAG to speak as their representative. I have been listening all day today and all day yesterday to your point of view, to scientists' points of view, and I would ask you for a moment to listen to

[1]IND = Investigational New Drug; NDA= New Drug Application

our point of view, and to try to understand our point of view in this situation. Firstly, I might say that our union was one of the first to state publicly that we would support and encourage recombinant DNA work and that we hoped it would go ahead given safe conditions under which it should be done. The last couple of days, especially yesterday, I think, have provided very interesting information on this field, most of it obviously already published, but the work seems to me anyway, as a non-scientist, to be going ahead and to be very exciting and a very exciting field for scientists to be in at the moment. It seems to me to be very successful work and to be being done very well, and being done in countries which are using guidelines. As a trade union, and I am speaking in a British context now, we have had experience over the years of members of our's contracting infections from laboratories. We have had experience of members contracting hepatitis, TB, typhoid, and lately we have had a member who died of smallpox from a laboratory in Birmingham. Now that does affect, and it can't help affect, our point of view towards laboratory safety. I don't agree with the comment that was made earlier, with the experience we have had, medical microbiology laboratories are the safest place to be in. They certainly are not the safest place to be around since the laboratory in Birmingham was a medical microbiology laboratory and, frankly, trust is not at a very high level. I think that this situation wasn't true when the "Berg" letter was published. I think that that commanded a lot of respect from non-scientists such as myself, and a lot of the people that I represent felt that scientists were coming out and telling us of their fears about work that they were undertaking. I think that that commanded a lot of respect.

Now there is, as Mark Richmond said before, a new climate of opinion. I think that with the "Berg" letter a lot of people convinced us that there were risks in this research. I think you now, in a sense, have a responsibility to convince us that there aren't, if that's what you are trying to say, and it seems to me that that point of view has been building up over the past few days at this conference. I realise, and you are going to get up and say that it is not possible to prove that there are absolutely no risks in this research.

J.D. WATSON: There is no logical way to prove that what we are doing is safe. All I can do today is say I was a jackass to sign the "Berg" letter. It was not a well thought document, but a silly emotional response which we as scientists should all be ashamed of. I would willingly go to jail, say for thirty days, to atone for the harm I've caused to others. But you cannot ask me for a logical impossibility, no matter how much wanted by the leaders of your trade union.

D. HABER: If that is your opinion, then I think that's what you have to say to the public. Rather than coming along and saying that this research is safe, which a number of people have said. I think you have to come along and say "It's impossible to prove that it's safe. We do have evidence to the effect that it doesn't seem as dangerous as it used to be" or whatever arguments one wants to put forward. I think you have to do this to the public, just as you gave the "Berg" letter to the public, and I think that that would be an advance in this field.

When I came to this meeting I wasn't aware of sufficient evidence to wind up guidelines in GMAG and RAC and all the advisory bodies tomorrow, and I haven't seen any further evidence come out of this meeting that has given me anything new that wasn't known before. There has been talk about evolution. I don't really think that is relevant to the safety of the laboratory worker, although I was advised that in the States the safety of the laboratory worker hadn't entered into the debate. It certainly entered into the debate in Britain. I don't know about the fact that this recombination now appears to occur naturally in very low frequencies. Not being a scientist I would like to hear more about that argument. I don't believe that there's necessarily a case to say that just because something occurs in law frequencies naturally doesn't mean it's not more dangerous to do it in higher frequencies in a laboratory. I don't think it's good enough to laugh at people who have opposed genetic engineering research or who have at least insisted on control of genetic engineering research. Some scientists have been telling the public that they have over-reacted, that there has been an over-reaction in terms of hazard, an over-reaction in terms of the "Berg" letter and the possible risks. Well I think scientists have over-reacted. I think scientists have over-reacted to these "jail-bars", as somebody described them, that are being constructed, as it's obvious that the work really hasn't been stopped and it seems to me that it's gone on and gone on very well. I think that a meeting like this is not a way to reassure the public. I think if you want to wind up guidelines, if you want to wind up GMAG, you have to reassure the public that this can be done safely and a meeting like this, way off in the corner of the country, with a list of speakers who are known beforehand to be opposed to guidelines and regulations is not really a way of reassuring the public.

I don't think you can separate politics from science any more and I think that is a lesson that should be learned out of this debate. A meeting like this should be held in the middle of London. Everybody should be invited, it should be reported extensively, and then I think you might perhaps win your aim.

W. SZYBALSKI: I have a very short proposal that came to my mind when I was listening to this debate. Since there are good reasons to believe that the present regulations result in serious harm to science and society, how about proposing a moratorium on all guidelines and regulations until the dangers of regulations and benefits versus risks of the recombinant DNA technique are carefully assessed?

M.H. RICHMOND: I think there is one positive thing that could be done and that is to involve the professional scientific societies in the matter. In the UK context, if the Society of General Microbiology and Biochemical Society were to consider the matter in detail and come forward with a balanced view, I think that would have a lot of impact.

R.H. PRITCHARD: I should appreciate an answer to the questions I directed specifically to members of GMAG who are attending this conference.

M. SINGER: Is there anyone who would like to respond to the questions raised by Professor Pritchard? Someone in the GMAG?

J. SUBAK-SHARPE: I am a scientific member of GMAG, appointed presumably because of my past interest in virology. Before addressing myself to one of the points that has been raised by Professor Pritchard, I cannot help but make one comment about Miss Haber's discussion which went earlier. It seems to me that Miss Haber's interest has been very much to create a power base for her union within GMAG, and I feel that some of her remarks must be seen in this context.

Bob Pritchard's remark specifically asked why GMAG did not immediately accept the United States guidelines. The situation which has developed is really the following, and I am trying to be as chronologically accurate as possible. I am Chairman of the Subcommittee on the Validation of Safe Vectors, one member of which is Sydney Brenner. At its meeting on 12th Janaury, 1978, Sydney Brenner verbally outlined new ideas on hazard assessment. They had not been fully evolved, but appeared to incorporate a consistent and workable logic that would allow rational assessment. From then on the Subcommittee spent much of its time trying to identify whether this logic might be workable. Sydney produced a series of three papers, the first on 8th February which was very general, just giving a basic outline of the approach; the second of 15th February was much more definitive and this, plus the third on 18th May, presented his views in a way which persuaded the Subcommittee to devote a lot of its time, instead of looking at the validation of safe vectors, to this particular project, giving it very serious consideration. The Subcommittee reported back to GMAG and was encouraged to continue this investigation. We ended up with a teach-in at the Subcommittee meeting in Glasgow on 30th June, 1978, given by Peter Rigby who had done a very careful evaluation looking at nine pre-determined examples. In each he evaluated the likelihood, for given activities of the DNA, of successfully escaping, progressing across all nodes of a generation pathway and then achieving the particular undesirable outcome. The base-line for comparison was seen as a cononical experiment, one providing a direct route from the recombination experiment to the undesirable outcome. Rigby's examples included human shotgun experiments and virus experiments. This exercise persuaded the Subcommittee that, though there was a difficulty in precise quantitation, the logic itself probably presented a sound way of analysing such problems. At that point the matter was taken out of my Subcommittee's sphere of work and all subsequent development taken on by a small *ad hoc* committee chaired by John Maddox which held six meetings and presented recommendations to GMAG on 1st October, 1978.

Because of the impending retirement of the GMAG Chairman, Sir Gordon Wolstenholme, and our knowledge that the American guidelines were likely to be soon revised, the Maddox Committee was under pressure to complete its task quickly and in my view its final sessions did not have quite sufficient time to do this elegantly. Their recommendations were discussed by GMAG and the finally published version in *Nature* was perhaps not produced in such a way that it might have been acceptable to people like Bob. But I think this résumé puts the record straight and demonstrates that GMAG, after considerable and responsible enquiry, was committed for some time to the initiative and course of action it has taken.

W.J. PEACOCK: There is confusion in Miss Haber's remarks with regard to demonstrated dangers and potential hazards. Also, I don't see why we should take up a polarised situation at this meeting. All of us here are here to consider a field of research, concerning both scientists and their technicians. Our conclusions will affect all of us.

J.D. COOMBES: I'd like to make two points. First, I think we must try to develop uniform international risk control measures rather than try to defend national differences, the very existence of which casts doubt on our scientific knowledge and understanding of the risk question. As Bob Pritchard has said, if the NIH Guidelines are generally acceptable, let's adopt them rather than going along a new path all by ourselves because we cannot persuade others to join us. Secondly, we should distinguish where the scientist speaks with authority and where his views carry no more weight than that of any other concerned citizen. As scientists we should not presume to make value judgements for other people.

M. SINGER: The 20th century has taught us that scientific and technological advances, like any other human activity, have both positive and negative aspects. Dr. Coombes mentioned various examples of materials that were originally presented as desirable but which also turned out to have untoward effects. Such examples urge us to consider the possible negative and positive consequences of developments in science and technology before they come about. Therefore, the scientific community raised the question of recombinant DNA.

It is essential that we recall that the letters of 1973 and 1974 asked a question. No one ever stated that a hazard existed. Yet it was widely perceived that an imminent danger was present. In part the response reflected the timing of public statements. The technology was not yet fully developed, and both hazards and benefits were conjectural. In the absence of certain benefits, the question of possible hazard was raised in the absence of independent advocates. There was no one, except the scientists themselves, who could say "Yes, this may involve potential dangers, but it also provides many substantial benefits". The scientific community was left to describe both sides of the coin on its own. Thus, the recombinant DNA situation was very different from the situations regarding most of the examples listed by Dr. Coombes. In those instances there was an establishment, including consumers, which saw the benefits, and could argue for them, thereby providing balance to the debate.

We have then learned that to raise a question in an open and unprepared community is very difficult. We have learned that rational discussion in a neutral environment cannot be expected. We have learned that most people assume that when you ask a question you must really be stating an answer. Thus, we face a serious dilemma.

Many of us would like to be responsible members of democratic and open societies and yet preserve the independence of science as well as the atmosphere we know to be essential to scientific and scholarly research. The customs of our scientific community arose within an elitist tradition and at a time when a scientist needed only to satisfy himself and the few individuals who personally supported his work. Now our support comes from complex societies with multiple and conflicting interests. In recent years those societies have demanded an increasing role in determining what we do and how we do it. Yet, widespread ignorance about science and scientific methods, as well as anxieties and fears, have tended to make research the victim of ill-informed public decisions. This has been exacerbated by the activities of groups that represent special interests, groups which often manipulate, for their own irrelevant ends, the political naïveté of scientists. Furthermore, the press has frequently been accomplice to those activities and has thus thwarted the well-meant intention of scientists to deal honestly and publicly with our dilemma.

Our own community's efforts to find a reasonable path through this dilemma must always be made in the context of the current political and social atmosphere. When a problem like recombinant DNA comes up again it cannot necessarily be handled or evaluated in a way that may now look appropriate on the basis of our experience over the past five years. New issues will arise in a new and different world and in different contexts. This is also one reason why our experience has been so varied in different countries.

In the United States the recombinant DNA issue was raised at a time when there were increasing demands for public participation in the day-to-day decisions of government. We raised the subject at a time when there was tremendous distrust, and for good reason, of government, of science, and of technology. We raised it at a time when there was a lack of knowledge about science and therefore irrational fear even among highly educated people. Freeman Dyson is a physicist at the Institute for

Advanced Studies at Princeton who long ago openly criticised the molecular biology community for the public actions regarding recombinant DNA. Nevertheless, in a book he is now preparing he acknowledges that the public sees the face of Dr. Moreau in those of molecular biologists. In the United States there are deep-seated antagonisms to biology and that should be recognized by people in other countries who try to evaluate what happened in the United States. Dr. Szybalski mentioned the Scopes trial, which took place in 1925. But even at the present time, the content of biology textbooks used in secondary schools in the United States is still influenced by religious groups who do not wish evolution to be taught. A law suit was brought, this year, in Federal Court, to enjoin the Smithsonian Institution from mounting a major exhibit on evolution. The suit was unsuccessful, but I mention this to show you the problems that are still current.

Certainly we have learned some lessons, although most of them are rather negative. We have learned, perhaps, what not to do. But given the dilemma, it is not at all certain that we would act in a wiser fashion in the future.

It does seem clear to me that most people are in fact anxious to reap the substantial rewards that science has to offer. There is a lot of evidence for that statement, although we tend to overlook the fact. The public and its representatives have a very great interest in supporting research and ensuring that good scientists are trained and able to do science. The public and its representatives also must depend on scientists for early and straightforward discussion of the benefits to be expected as well as the possible negative corollaries of particular research. It is, therefore, essential that the same public work with us to engender atmospheres that are hospitable and encouraging to open discussion. We need to attend carefully to instances where scientific interest seems to diverge from the public interest and we need that attention not only from scientists but from people in situations of public responsibility.

WHAT OPTIONS ARE NOW OPEN?

CHAIRMAN'S INTRODUCTION

M. G. P. Stoker

Imperial Cancer Research Fund Laboratories, Lincoln's Inn Fields, London WC2, U.K.

The Conference really changes character as from now. As I said at the beginning, ICSU, and therefore COGENE, represent the world's scientists and at a non-governmental level. We have been urged here several times - for example, by Walter Bodmer - that we should give our views on the facts as scientists and not be influenced by whatever we think the reaction of other people is going to be - non-scientists and the public generally.

Dr. Coombes said yesterday that the job of the scientist was to identify hazards. At its simplest, the situation is as follows: five years ago a group of scientists said "Here's a new technology. There might be a hazard - let us stop and look carefully at all the information before we go on". Now, five years later, after acquiring some new information, which is not negligible, and re-assessment of old information - looking up papers in sewage magazines and so on - the view of many scientists is that the hazard to human health of recombinant DNA techniques is vanishingly small. But in the meantime a lot has happened. Public opinion has been fashioned, various organisations, especially governments, have taken action and so on. It is not up to us now to say what are the options that are open - which is the title of this morning's session. We have to ask what others think - representatitives of other branches of society, including other branches of science, such as public health.

We heard yesterday what the view of a union might be; today we are going to hear the views of lawyers and public health experts, historians and experts concerned with risk perception.

Now, there is a slight change in the programme since, unfortunately, Dr. Thomas, who was going to sum up the whole Conference has not been able to come - he is sick, I am afraid - and the change in order therefore is as follows: Congressman Thornton is going to speak last and he is going to not only give his view as a legislator but also sum up the Conference or at least give his impressions. We are very grateful to him for agreeing to do this.

We are, therefore, going to start with Dr. Weiss from Washington. Dr. Lennette is not able to get here but Norton Zinder is going to read his communication. We have asked Roger Lewin, who is Chairman of the Association of British Science Writers, if he would give some comments on the views of the pressas he sees them. Then, we have the historian, Dr. Weiner, and Sir Frederick Warner who is concerned with risk assessment, and finally Congressman Thornton.

INTERNATIONAL LEGAL ASPECTS OF RECOMBINANT DNA RESEARCH

Edith Brown Weiss

Georgetown University Law Center, 600 New Jersey Avenue, N.W. Washington, D.C. 20001, U.S.A.

New technological developments which promise exciting benefits may at the same time pose serious risks or carry with them profound economic, political, social and ethical implications. Increasing international exploration of such technologies as nuclear power, weather modification, pesticides, drugs, and other chemicals give rise to international responsibility to create a legal regime that will at the same time maximize and disseminate benefits from the research, assess and minimize risks associated with it, and handle disputes about liability for damages resulting from it. The development of genetic engineering, including recombinant DNA techniques, presents us with another instance of a technology that raises these international legal and political issues.

Scientists deserve the thanks of the world for warning of potentially serious risks which may be associated with recombinant DNA research, even before the extent or even the existence of these risks was definitely established. In the initial landing on the Moon, there were elaborate precautions to protect our Earth against the possibility of contamination by materials from the Moon, even though subsequent discoveries revealed these precautions were no longer necessary. Recent findings in recombinant DNA also minimize the conjectured hazards, and some scientists have wondered whether they should have warned of risks at all. But, if potential hazards turn out to be well-founded, the scientists who warn of them are heroes. If they foresee possible risks, but do not warn, they get the chance to be wrong only once. It is in the scientific community's interest to build a long-term base of public confidence in the safety of their research, because ultimately basic and applied research requires political support at the national and international level.

Recombinant DNA research raises at least five issues which are international in scope and merit attention:

1. What is and should be the role of proprietary rights in fostering the development and application of the technology

worldwide, and what special measures are needed to facilitate its availability to developing countries?

2. What measures are available to minimize any occupational and environmental hazards associated with recombinant DNA research?

3. What measures are available to minimize any risks that novel organisms, accidentally or intentionally introduced outside the laboratory, will infect other populations, either human, animal, or plant? And what will be the impacts from widespread intentional introduction of such organisms?

4. What international institutional arrangements are appropriate for coordinating or managing recombinant DNA research?

5. What measures are available to deter the development and application of recombinant DNA research for biological warfare?

1. Proprietary Rights

Companies faced with substantial investments for producing microorganisms commercially are concerned about proprietary rights to the results of their research and development efforts. In some instances this takes the form of concern about protecting "trade secrets" if they are subject to national research guide lines. In many cases the concern is with obtaining a patent for a microorganism *per se* developed through recombinant DNA techniques. The question is whether we should patent new forms of life. In the United States and in other countries this question is under review. The new European Patent Convention, which provides for one European patent to be effective in member states, allows patent protection for microbiological processes and the products thereof, as do the patent laws of the United States, West Germany, the Netherlands, Scandinavian countries and Japan. The United Kingdom and West Germany may be the only countries to date in which the microorganism itself may be patentable, if the inventor shows a reproducible method for obtaining it. This issue is now before the United States courts. (See *In re Bergey* and *In re Chakrabarty*, decided March 29, 1979, in which the U.S. Court of Customs and Patent Appeals extended patent protection to microorganisms).

Legal commentators have sometimes proposed that patent protection for recombinant DNA products could fall within already established provisions giving patent protection for plants. The problem is that few countries provide for plant protection; many specifically exclude it. The European Patent Convention excepts from patent protection "plant or animal varieties or essentially biological processes for the production of plants or animals." (Article 53(b).) (This exception does not apply to microbiological processes or the products thereof.) Thus, we return to the question now before the U.S. courts, whether the microorganisms *per se* are patentable.

It could be argued that the fruits of recombinant DNA research should be unpatentable and should belong to all mankind. One criterion should be whether the incentive of patent rights is needed to encourage investment in commercialization of innovations which will emerge from research in this field.

One concern is that patenting new forms of life could be a precedent for extending patents to things that society views as undesirable. Most national patent laws in Western Europe contain an explicit prohibition against patents that would be against "ordre public" or against morality. This provision could be used to reassure a concerned public that providing patent protection to recombinant DNA products need not lead to patent protection for undesirable new forms of life. Alternatively, as a matter of social policy governments could hold the patent rights for certain kinds of biological advances, as they do for the fruits of atomic energy research, where the developments were thought to be too powerful to be left to private control.

We should also examine how national patent laws and existing treaties affect the access of developing countries to the results of recombinant DNA research, and more generally to the results of genetic engineering.

We need to encourage a common policy among countries as to whether or not novel organisms are patentable. Otherwise companies have an incentive to shop around for countries offering favorable patent protection, which in turn could make it more difficult to enforce guidelines on the research. The Budapest Treaty on the International Recognition of the Deposit of Microorganisms for the Purpose of Patent Procedure outlines procedures for the deposit and international recognition of microorganisms. This is a useful step in facilitating national patent protection and could serve as the instrument for developing a uniform patent policy.

2. National Occupational and Environmental Hazards

We can identify two different kinds of risks that have been associated with recombinant DNA research:

A) Risks to workers in the laboratory.

B) Risks connected with release of novel organisms outside the laboratory (in greenhouses or otherwise), or with the effects of intentional large-scale introduction of novel organisms into the environment.

Most attention has focused on measures to keep the organisms in the research laboratory. The solution has been national guidelines which feature biological and physical containment measures for varying levels of risk. Internationally the issue is that not all countries have adopted guidelines, the guidelines vary in content and application between countries, and they do not apply to private industry in all countries.

A major international issue is whether and how to harmonize the provisions of the various national guidelines. Harmonization may be essential, both for the protection of society and for scientists, who may fear that whether they will be the first to make a scientific breakthrough will in part depend upon where their laboratory is located. Variations in guidelines between countries also pose problems for industry, particularly since the variations most frequently relate to low-level risk categories requiring P1 or P2 facilities. The lack of harmonization could induce some companies to shop for the country with the fewest and least expensive restrictions, at the expense of protection to the international community.

COGENE should continue its initiative to harmonize the guidelines, particularly for the lower risk categories. Alternatively, or in addition, the World Health Organization could issue uniform minimum safety measures in the form of an international health regulation. One possibility would be for the International Labor Organization to draft safety standards for workers in recombinant DNA laboratories.

The basic problem in drafting guidelines governing any laboratory research is how to take account of scientific uncertainty regarding the risks involved. All recombinant DNA guidelines need provisions for updating to take account of reassessments of the risks for given experiments. COGENE could make an important contribution by regularly reassessing the risks, disseminating this information to the appropriate national bodies, and assisting in the redrafting of guidelines in light of these new assessments. This function needs to be carried out at the international level. COGENE might also recommend or even help design experiments to resolve any uncertainties about the risks of recombinant DNA research.

3. International Hazards

The major potential hazard that has been associated with recombinant DNA research is that novel organisms might escape from the laboratory and infect human, animal, or plant populations. If national guidelines on laboratory safety exist, they should prevent the organisms from getting out of control. But even with adequate guidelines in some countries, there will still be problems internationally because of countries where laboratory guidelines do not exist, are inadequate, or are not followed or enforced. Field research raises a separate class of risks. If the experiments cannot be satisfactorily confined within a greenhouse or other ecological island, but require releasing the organism into a larger ecosystem, the concern will be greater. What safeguards should apply to the intentional release of such organisms in the field for experimental purposes? And how can they be enforced?

Countries are responsible for ensuring that activities within their borders do not cause injury beyond their borders. This is liable regardless of fault, may apply. This principle has begun to appear in some international conventions: conventions governing nuclear hazards (such as the Brussels Convention on the Liability of Operators of Nuclear Ships), outer space hazards, and certain forms of ocean pollution. Are certain classes of genetic engineering so inherently dangerous that the party engaging in it must bear all the risks? Negligence would apply to research and development which involves risks at the lower level (P1 or P2 facilities), but strict liability might possibly be imposed upon those experiments conducted in P4 facilities. The latter principle could seriously hamper research by creating the possibility of large damage claims. To forestall this possibility, if international experiments in recombinant DNA are conducted in P4 facilities, states might consider an international provision which limits the amount of damage for which the states could be liable. The Warsaw Convention, for example, limits airline liability in crashes. United States law provides a ceiling on the total liability for damage arising out of a single accident at a nuclear power plant.

Companies which intentionally introduce new organisms into the environment which cause ecological damage beyond a country's border, and states which sanction such action, may be liable under national laws for such damage, or even under international law. The extent to which environmental damage is included within those items which are compensable varies among countries.

Impacts from DNA products

When recombinant DNA products are marketed, novel organisms will in some cases be intentionally introduced into the environment, e.g. to clean up oil after a spill. We need to determine whether there are any unacceptable ecological side effects to such introduction and whether adequate quality controls exist in the production of these organisms. A comparative analysis of national measures applicable to products to encourage ecological safety would be useful.

Conventions adopted under the Food and Agricultural Organization which place limitations on the importation of plants may be applied to international traffic in some recombinant DNA products directed to the agricultural sector. We might also explore the extent to which an institutional arrangement similar to CODEX, which is used for pesticides, could be appropriate for some recombinant DNA products. It will be important to develop and to exchange scientific information about the ecological safety of introducing certain novel organisms into our environment.

4. The Institutional Framework

Several international organizations have taken initiatives to coordinate recombinant DNA research programs: ICSU, WHO, means that if they engage in activities which might cause such injury they need to consult with each other and to cooperate in developing appropriate processes for managing these effects. These may include procedures for settling disputes about paying for damage.

The basic rule in international law is that a state (i.e. a national government) is responsible for damages or for injuries resulting from the violation of an international obligation. In classic international law at least two principles would apply: a principle of good neighborliness, which imposes upon states the duty to protect other states from injuries caused by acts within the state or by persons under its jurisdiction, and the related Abuse of Rights doctrine which makes a state liable when it exploits a right "in such a way as to inflict upon another State an injury which cannot be justified by a legitimate consideration of its own advantage." (Oppenheim) Hence it is reasonable to conclude that a state might be held liable for damages caused to other states by organisms which had escaped from its laboratories. Indeed this could apply whether or not these organisms were produced by DNA research or by any other research - e.g. the experiments which resulted in the release of African honeybees in South America.

Recent legal fora have tended to extend these principles to environmental problems. Principle 21 of the Declaration on the Human Environment, adopted at the 1972 U.N. Conference on the Human Environment in Stockholm, declares that "States have, in accordance with the Charter of the United Nations and the principles of international law, the responsibility to ensure that activities within their jurisdiction or control do not cause damage to the environment of other States or areas beyond the limits of national jurisdiction." Principle 22 says that "States shall cooperate to develop further the international law regarding liability and compensation" for such damage. Several international cases, such as the Trail Smelter Arbitration between Canada and the United States, point to a similar obligation, with liability for damages for the violation of that obligation.

Leaving aside the question of how states may enforce international law - if microorganisms cause damage beyond a state's borders in violation of an international obligation the traditional rule in international law makes a state liable if there was negligence, i.e. if the state failed to live up to expected standards. Violation of internationally accepted minimum guidelines, whether expressed as a formal international agreement or as informal consensus, may suffice to show negligence, and compliance with them to preclude a finding of negligence.

It may also be possible for a state to be liable for damage caused to another state by microorganisms, regardless of whether there is negligence, if the activity is regarded as ultra-hazardous. In such cases strict liability, which means that the party

UNESCO, EEC, European Molecular Biology Oranization, and the European Science Foundation. In addition, the Council for International Organizations of Medical Sciences (CIOMS) proposed in 1972 an international ethics body to consider moral and ethical issues raised by new and forthcoming developments in biology and medicine. This could be expanded to include recombinant DNA research. The Committee on Biotechnology and Applied Microbiology of the UNEP/UNESCO/ICRO Microbiology Program, which is intended to help advance microbiology primarily in developing countries by providing for effective utilization and proper management of genetic stock of microorganisms, might be appropriate for disseminating some of the benefits of recombinant DNA research. We need to examine the appropriate mix of international institutions and functions for coordinating national efforts in recombinant DNA research.

It may be appropriate for the International Council of Scientific Unions, in concert with the WHO and the FAO, to designate an interdisciplinary panel of experts on biological research, with an initial focus on recombinant DNA research. COGENE could serve in this capacity. The panel should consist of people in the natural and social sciences who would be available to countries to perform some or all of the following functions upon the request of individual states, the WHO, FAO, or other intergovernmental bodies. It should also be linked with the existing scientific networks, such as the microbiological research network. Members could coordinate and disseminate data assessing the risks associated with all levels of recombinant DNA experiments, coordinate national guidelines, and assist national governments in bringing guidelines into conformity with the latest risk assessments; offer technical advice to countries as to the utility of the research and the likely impacts from the intentional introduction of new organisms; and perhaps assist as a fact-finding body if disputes arose regarding international liability for damage from novel organisms.

ICSU might sponsor an annual workshop in which scientific researchers, representatives of industry, and government representatives could meet to discuss means by which they could facilitate research and development in this field, disseminate the benefits, minimize the risks associated with it, and enforce measures designed to minimize risk. Such a workshop could be an appropriate vehicle for encouraging both cooperation between States and coodination of national measures.

5. Military Applications

Recombinant DNA techniques may be developed and applied for hostile purposes, either by a country's armed forces or by terrorist groups. Will these new biological weapons offer significant advantages over those that can be produced with conventional technology? One concern is that we may develop by recombinant DNA, organisms that attack plants and animals and spread so rapidly that it could be difficult to develop defenses against them. We need more knowledge to assess carefully the prospects that such organisms could become significant weapons. According to Callahan and Tsipsis, a speculative concern is that the techniques might be used to develop genetic weapons which target specific racial types. But even if this were ever to become possible, the weapons would be subject to the same problems associated with conventional biological weapons, e.g., it would be difficult to limit them geographically.

Do the Biological Weapons Convention of 1972 and the Geneva Protocol of 1925 apply to the development and use of recombinant DNA techniques for hostile purposes? If not, even if recombinant DNA techniques offer no advantages over conventional technology for developing biological weapons, states may apply them for hostile purposes, since this would not be banned. The Geneva Protocol bans the first use in warfare of bacteriological weapons. The Biological Weapons Convention of 1972 prohibits the development, production, stock piling or other acquisition of microbial or other biological agents or toxins for other than peaceful purposes. The United States position is that a biological weapon produced by recombinant DNA techniques "clearly falls within the scope of the Convention's prohibitions." Parties to the Biological Weapons Convention should confirm this understanding. It is in the interest of scientists, pursuing research for peaceful purposes, that this be done.

There is one class of weapons which does not clearly fall within the Biological Weapons Convention: toxic chemicals produced by recombinant DNA techniques. Callahan and Tsipis have suggested adding a provision to any treaty negotiated on chemical warfare to cover "toxic chemicals synthesized using reagents, or processes or biological origin." If this is desired, it may be a somewhat easier and more appropriate route to propose a clarification or an amendment to the Biological Weapons Convention (under Article X) providing that toxic chemicals produced by recombinant DNA techniques are covered by the Convention.

BIOLOGY IN INTERNATIONAL LAW

International lawyers are only beginning to come to grips with the special properties of natural systems, such as weather, whose effects spill across national boundaries. Biology gives rise to many such effects, of which recombinant DNA research is only the most recent and most arresting.

But recombinant DNA research is only one of the many advances in biological research. We should also address the social, legal, and other implications of the more traditional means of genetic manipulation and of the introduction of new organisms into the ecosystem. The problems are similar, and in many cases the risks more serious. For example, what responsibility do countries have for the consequences of introducing new species into their ecosystems? Is Brazil responsible in some way to other American states for the negligence of its citizens in allowing the African honeybee to escape from the laboratory? Do countries have a responsibility to do what they can to control pests that may spread to other countries, such as locusts or disease vectors? Might there be circumstances such that they have an obligation to join international efforts to control such pests? Might these obligations translate into legal liability?

In the end, international law is enforceable only to the extent that sovereign states wish to be bound by its provisions. The task of the international lawyer is to call attention to new problems and to cooperate with scientists and other experts and with concerned citizens to devise legal regimes that resolve them in a manner that is on balance in the interest of all concerned.

REFERENCES

Callahan, J.B., and Tsipsis, K., (1978). Biological Warfare and Recombinant DNA. The Bulletin of Atomic Scientists, Vol. 34, No. 9, pp. 11, 50.

Oppenheim (1955). International Law, 8th ed., Lauterpacht.

Cases

Baker's Yeast, Decision of the Bundesgerichtshof, 6 11C 207 (1975).

In re Bergey, 563 F.2d 1031, 195 USPQ 344 (1977), petition for cert. filed, 46 U.S.L.W. 2574 (U.S. April 20, 1978) (No. 77-1503), vacated, 385 P.T.C.J. A-4 (U.S. June 26, 1978) (No. 77-1503), Decision issued March 29, 1979).

In re Chakrabarty, 571 F.2d 42, 197 USPQ 72 (1978), petition for cert. filed, 47 U.S.L.W. 2086 (U.S. August 8, 1978) (No. 78-145), vacated, Decision issued March 29, 1979).

Red Dove, Decision of the Bundesgerichtshof, 1 11C 136 (1970).

The Trail Smelter Arbitration (U.S. v. Canada), 3 U.N.R.I.A.A. (1905).

Conventions

Convention on the Grant of European Patents, 5 October 1973, entered into force 7 October 1977.

International Plant Protection Convention, December 6, 1951, 150 UNTS 67, TIAS 7465.

North American Plant Protection Agreement, October 13, 1976, TIAS 8680.

Convention on the Prohibition of the Development, Production and Stockpiling of Bacteriological (Biological) and Toxin Weapons and on Their Destruction, April 10, 1972.

Protocol for the Prohibition of the Use in War of Asphyxiating, Poisonous or Other Gases, and of the Bacteriological Methods of Warfare, June 17, 1925.

World Health Organization International Health Regulations, July 25, 1969, TIAS 7026, 21 UST 3003.

World Intellectual Property Organization, Budapest Treaty on the International Recognition of the Deposit of Microorganisms for the Purpose of Patent Procedures, opened for signature, April 28, 1977.

DISCUSSION

M. SINGER: Yesterday, I talked about the general problem of raising questions which other people then turn into automatic answers. Although, I think it was unintentional on your part, one of the ways in which you expressed something is often done both by scientists and non-scientists and is something we might all avoid. You talked about the special liabilities that might be connected with experiments in a P4 laboratory and you actually used, as many of us do, P4 as an adjective describing an experiment. In fact the danger in an experiment which is classified into P4 containment is no more a danger than is any experiment that's classified as requiring P1 containment. Therefore, to talk about the liability increasing because the experiment has been so classified, in the absence of any known hazard, seems to me to put the wrong emphasis - if not indeed a totally wrong impression - on things.

The real problem with the faulty grammatical construction is that the press and the public then believe that an experiment that is done in a P4 laboratory is a terrible hazard when, as we all understand, the reverse is true. So I would make a plea to everyone that we stop using those classifications as adjectives to describe experiments - that's not what they are - and that furthermore we be very careful to avoid this erroneous implication.

E.B. WEISS: I think it's a fair point I would note that there was a Bill introduced in the U.S. Congress that would have applied strict liability to all recombinant DNA experiments.

A. CAMPBELL: One of the situations we've become conscious of with the recombinant DNA business is that it's still an area in which both the perception of risk and the technology itself have been evolving very rapidly. To make any kind of guidelines to cover them in a sensible way requires that they also evolve very rapidly, and the principal limitation has been the slowness with which it has been possible to change them. At this point I don't think that international control of recombinant DNA is necessary at all, but in terms of the broader issue you raised of some technology in the future which might really be dangerous, what disturbs me about the possibility of international agreements is that they would make the situation even more inflexible and harder for anyone to make changes that keep things current and realistic. You mentioned annual workshops and things - but I didn't really understand a mechanism whereby a legal agreement between states can be rapidly altered as knowledge changes. Would you comment on that.

E.B. WEISS: Yes, the proposal for a workshop was intended as a means of disseminating information and achieving international coordination of national measures. This is something short of an international agreement. Under international law, states have certain responsibilities, an obligation to consult, an obligation to, if necessary, engage in an appropriate arrangement for managing something, but this does not necessarily mean an international treaty. Indeed, if I may suggest, that to the extent that you can agree on processes for fulfilling your international responsibility, rather than looking to a formal treaty that lays down very precise substantive standards and requirements, you are much better off. That is why I prefer meetings in which you agree upon adopting certain ways of proceeding, which will minimise your international legal liability, and stop short of an international treaty. You are quite correct that once you negotiate an international treaty, it is difficult to modify it, although there should be amendment procedures. I for one have always argued for having some provision for flexibility to keep up with new technological advances or new developments, but this is difficult to achieve.

J.D. WATSON: I am not aware of any way in which recombinant DNA procedures will make possible selective racial genocide. Unless one has real facts to back this idea, I find it totally irresponsible for a respected person to raise it as a potential consequence of our research. So I would like you to please tell us who is the authority you are quoting.

E.B. WEISS: It was an article by Callahan and Tsipsis on recombinant DNA and biological warfare that appeared in the *Bulletin of the Atomic Scientists*. The reason I raised that is not to give validity to their example, but to say that there is public concern and that I felt it was in the interest of all scientists engaged in the research to make sure that their research is in fact used for peaceful purposes and that no-one can raise the spectre that it might be directed towards military purposes.

J.D. WATSON: Who are Callahan and Tsipsis?

E.B. WEISS: Tsipsis is a physicist at MIT, and Callahan is an engineer there.

J.D. WATSON: What do these physicists know about this matter? They can be silly, too! You have to be very careful who you quote. There's a little book called "Who Should Play God", which in the United States has sold three quarters of a million copies. It raises bizarre possibilities like this without any reason at all, and thus adversely affects the public consciousness. After a talk like yours, some reporter in pure innocence could go out and say that this is going to happen. I think you're crying "fire".

E.B. WEISS: The point that I am trying to make - if I may delete this example, Dr. Watson - is that I think we ought to make it clear that the Biological Weapons Convention which bans the use of biological weapons in warfare does extend to any product that could be produced by recombinant DNA techniques. This is not clear now.

J.D. WATSON: There is no biological basis for talking about biological weapons that will bring about selective racial genocide and say let the British solve their "Irish problem". It's absurd and you should not go around talking absurdity.

E.B. WEISS: But there is good reason for talking about biological warfare. I gather than in the original "Berg" letter there was a statement about biological warfare that was deleted.

J.D. WATSON: There was absolutely nothing about selective racial genocide.

E.B. WEISS: No, I'm not talking about that, I'm talking about the Biological Weapons Convention.

M.G.P. STOKER: Well, we'll just have to see what any reporters make of it - but clearly that's not the intention.

P. STARLINGER: I was a little bit surprised that in your exposition you mentioned, if I understood correctly, genetic engineering including recombinant DNA research. What does that mean? To my knowledge, the scientists have brought up the question of recombinant DNA research and we have all heard that they no longer think it is a big risk - but now you say this is just a little field of something broader and I would like to know what you mean by that. What is genetic engineering outside of recombinant DNA research and which experiments do you foresee regulated? I have seen such suggestions sometimes in the press and I think there is a real danger that after a while people will come to every genetics laboratory breeding flies or *Neurospora* and ask for special guidelines because this is such a difficult job. I think we have to be very careful not to extend the meaning of what we are talking about beyond what is really thought that should be regulated.

E.B. WEISS: I was referring to the more conventional means of conducting genetic engineering.

UNKNOWN SPEAKER: It's called biology!

E.B. WEISS: Right, which is why I concluded the discussion with biology in international law. (Most of biological research certainly does not raise international issues. But there are some cases which could - for example, the escape of the African honeybee when it headed north from Brazil. International law is concerned with a state's responsibility in such cases.)[1]

---

[1]This is a clarification of the original answer.

R. CAPE: There is this perception by the public: where there's smoke there's fire. I'd like to just extend briefly what Maxine Singer said. The perception of the public is frequently non-scientific and, although we applaud such developments as the citizens of Cambridge, Massachusetts, ultimately deciding that the work should go on, they did do illogical things such as, thinking that a little more safety wouldn't hurt, stipulating that all experiments in P3 laboratories should be done with EK2 systems. There's absolutely no logic to that and I think that's the sort of thing Maxine was alluding to. It is the stimulation of these illogical public reactions that concerns us here and causes us to question this kind of presentation.

S.N. COHEN: It seems to me that Dr. Weiss is proposing a series of additional regulatory approaches, which, as Professor Starlinger has noted, would extend not only to recombinant DNA but to other forms of genetic research, and it's not clear to me what the basis is for such regulations. If there were some public need aside from the treatment of anxiety that was not being met by the current situation it would be reasonable to propose additional regulations, but in the absence of some demonstrated need that is not presently being adequately looked after, the status quo would seem to be more than adequate. I am wondering what specifically Dr. Weiss is addressing in proposing that international legal action should be undertaken in this area.[2]

E.B. WEISS: If you review the talk, you will see that I did not propose new regulations. What I did was to review international law regarding the responsibility of states and regarding potential international liability as applied to this area of research. There are proposals for an annual workshop, a proposal for an inter-disciplinary panel of experts, some proposals for national coordination. One of the problems is that lawyers tend to be equated with regulation and I think that's a very unscientific conclusion! (With regard to your question about international legal action, I was referring, for example, to the need for processes for updating national measures to take account of re-assessment of risks and benefits; for provisions which encourage immediate disclosure of new host-vector systems without foreclosing patent protection, for measures to protect against potential risks from intentionally releasing organisms in larger ecosystems, and for quality control procedures for large-scale production of new organisms.[3])

W. SZYBALSKI: I have two specific legal questions. The first is what are the legal obligations of the agency that proposes regulations? Do they have to provide some compelling and logical reasons for regulations? Or could they just cite some imaginary scenarios, hypothetical dangers or other capricious reasons, as was done in the case of the recombinant DNA regulations? What criteria should be satisfied? Is a statutory basis required for regulation? In the case of the compulsory 1976 and 1978 NIH Guidelines, no statutory basis was ever cited in the *Federal Register*; does this make the regulations illegal and unenforceable?

---

[1]These comments were edited by S.N. Cohen following the Conference.

[2]Added by E.B. Weiss in response to the edited remarks of S.N. Cohen.

My second question is whether the agency has to provide a proof that the regulations do not create novel dangers? In the case of the regulations specified by the NIH Guidelines, many novel and serious dangers were created, as I discussed last night. Is there any rapid legal remedy for suspending regulations which are not only unnecessary but which create serious dangers and are damaging to society? Who could apply for a court injunction?[4]

E.B. WEISS: The NIH Guidelines on recombinant DNA research were issued under the authority of Title III of the Public Health Services Act, sections 301 and 307 (42 U.S.C. 241 and 242L). These sections outline broad duties of the Secretary of Health, Education and Welfare regarding health research generally and international co-operation in biomedical and health services research. In making research grants to institutions and individuals the Secretary has the implicit authority to allocate the scarce funds in any manner consistent with his statutory mandate under section 241. It follows that he may impose reasonable rules on those who wish to compete for NIH grants. The Environmental Impact Statement prepared by the NIH for the guidelines on recombinant DNA research indicates that the objective of the guidelines is the protection of laboratory workers, the general public, and the environment from infection by possibly hazardous agents that may result from recombinant DNA research. This has been taken to be a reasonable objective.

The short answer to the second question is that a regulation can be challenged in court but the procedure is tricky and requires a detailed analysis of the specific situation. If you are planning to do it, I'm afraid I must advise you to retain a good lawyer.

V.R. OVIATT: Early in your presentation you mentioned that WHO was attempting to, or was involved with, coordination of recombinant DNA research. To set the record straight there's no attempt at this sort of thing and I see nothing in my experience that WHO should or would take this sort of action. The programme of which I am the coordinator - its genesis was with recombinant DNA work - after several consultations decided that recombinant DNA research was not the situation we should be looking at. Rather our concern is biological laboratory practice and safety in a general broad sense. We do have a mandate, however, to monitor the state of the research, to see where it's going and to advise the member states regarding both positive and negative public health aspects. So in this context we've been working with COGENE and its Risk Assessment Committee to carry out this activity. Our position is formulated in co-operation with COGENE, scientists active in the research, and public health authorities.

M.G.P. STOKER: We should really finish this discussion now. We thank Dr. Weiss very much. Originally we, or some of us, frightened the public. Now we ask some of the public, in this case the legal side, for a reaction and in return they frighten a good many of the audience.

---

[4]These questions were re-worded by W. Szybalski after the Conference. E.B. Weiss accordingly recast her answer to correspond to the revised questions.

RECOMBINANT DNA: A PUBLIC HEALTH VIEWPOINT

E. H. Lennette

Viral and Rickettsial Disease Laboratory, California Department of Health Services, Berkeley, California 94704, U.S.A.

During the past two days of this meeting, consideration has been given to delineation of recombinant DNA technology as a biological tool and (1), to placing it into appropriate perspective with other biological phenomena, naturally-induced or laboratory-contrived; (2), to the achievements of recombinant DNA research thus far; (3), to some of the practical benefits which may accrue through application of this technology; and (4), to guidelines and legislation, existing or proposed. As part of this session on "What Options are Now Open", I should like to comment, in a general way, on some of these aspects, merely touching upon some but emphasizing others to some degree since, in large part, there has been relatively little input from those concerned with medical microbiology, infectious diseases, epidemiology and public health.

Why the Asilomar (California) meeting was held, and why the "Berg" letter was written, has been discussed by Dr. James Watson. It remains only to state that these were received with some surprise (and even astonishment) by medical and public health microbiologists and by health professionals concerned with the clinical and epidemiologic aspects of infectious disease. The Asilomar conclusions reached, and the decisions made, were essentially those of investigators with a background in biochemistry or in genetics, and whose expertise lay in the application of these disciplines to gene transplantation but not in comprehension of the nature of pathogens; of how to work with disease incitants; and how to evaluate the relative risks and hazards involved.

The concerns expressed, some with reason and logic, at Asilomar ultimately engendered, as might have been anticipated, considerable fear and distrust in the public mind through the generation of scenarios ostensibly illustrating the "potential" hazards but actually dealing with hypothetical, speculative and conjectural "hazards". Fear was aroused that unbridled application of recombinant DNA techniques would give rise to "novel" organisms with pathogenic properties or new pathogenetic propensities and which, should they escape from the laboratory, would find an ecological niche conducive to their biological needs, replicate, and wreck devastation on other forms of life (and especially man). That an impressive array of microbiological forms endowed with new properties could be generated via recombinant techniques seem inescapable from the computations of the theoretical biologist. As Danielli (1972, 1974) points out, "a vast variety of possible organisms (species certainly, phyla probably) have never existed on the earth", and the organisms currently inhabiting this planet can in no sense be regarded as a representative set of those organisms...that are capable of existing". With respect to hazards, deliberate attempts to confer known pathogenic properties upon a microorganism are unacceptable and it is generally agreed that such experiments are interdicted. At the other end of the spectrum are those recombinant DNA experiments which pose no conceivable hazard and so require no unusual precautions. Between these two extremes lies that large unknown area that has given rise to an eschatology concerned with conjectural and hypothetical hazards. An important facet of the scenarios reflecting the fears of recombinant DNA technology is the presumption that on the basis of current knowledge, our ability to contain these engineered organisms, biologically or physically, is inadequate.

Let us devote a few moments to containment. With respect first to biological containment, the Workshop on Risk Assessment of Recombinant DNA Experimentation with *Escherichia coli* K-12, held at Falmouth, Massachusetts, in June 1977, presented impressive evidence that a host organism, such as *E. coli* K-12, crippled by a multiplicity of built-in genetic lesions would find it impossible to survive outside the laboratory environment specifically tailored to its requirements for replication and growth. With respect to the second, i.e., possible escape or inadvertent transport outside the laboratory, medical microbiology has had a long history of dealing with pathogenic and virulent microorganisms, and over a century of experience has accumulated a body of knowledge and expertise that has served to protect the worker and the public against infection. There are those who feel the record in this respect has been imperfect or poor, and offers no comfort against events in the future. They point to the former U.S. Army Biological Warfare facility at Fort Detrick, Maryland, as the "best microbiological containment facility ever built in the United States" (King, 1977), and "yet over a period of twenty years there were over 400 cases of lab workers getting a serious infection from the organisms with which they worked". However, so far as I am aware, there is no evidence that these laboratory-acquired infections resulted in outbreaks outside the installation. Others will point to the smallpox episode at Birmingham University as an answer to the question of how we might expect to contain theoretically hazardous recombinants when we were unable to contain a highly dangerous pathogen like variola virus. An investigation into this affair by a committee appointed by the Department of Health and Social Security, and headed by Professor Reginald Shooter, revealed that breaches in physical containment gave rise to an infection hazard outside the immediate confines of the laboratory. Professor Shooter's findings are well discussed in articles by McGinty (1979) and Hawkes (1979) who describe the technical and administrative factors which gave rise to this unhappy incident. The inadequacies and manifold infractions involved make evident the desirability of an overview mechanism for laboratories working with dangerous pathogens and, by extension, with the "gray area" of recombinant DNA research mentioned earlier. This will unquestionably call for establishment of extramural safety review committees to examine the methods, techniques and operations of the laboratory staff with respect to safety; to check for proper functioning of the physical modalities affecting containment (Hawkes, 1979); and with authority to enforce compliance.

As to laboratory staff, there is in the United States no definitive set of guidelines, except for recombinant DNA technology, concerning work with dangerous pathogens in clinical, public health or research laboratories but only the body of knowledge and wisdom acquired over the years and proved effective in practice. To illustrate such effectiveness, there occurred in 1971 a pandemic of a new disease, acute hemorrhagic conjunctivitis, caused by a new enterovirus, type 70 (Mirkovic and co-workers, 1973). Millions of cases were reported from Asia, Africa and parts of Europe. Although this has been regarded as perhaps the most highly contagious and rapidly transmitted virus of man, I am not aware that under normal containment precautions any laboratory-acquired infections occurred among staff of the laboratories in Berkeley, Houston, Lyon, Singapore, or Tokyo which worked with this newly-discovered agent. While it seems likely that formalized or regulatory guidelines may be established at some future time (Hawkes, 1979), equally relevant to the containment problem is a redirection of microbiology toward increased emphasis on the teaching and practice of pathogenic microbiology -- something more than a course in general microbiology is required of practitioners coming into this discipline from other fields or using it as a tool for the furtherance of their own related researches. As concerns physical barriers, there is no guarantee that the finest P3 or P4 laboratory built will not encounter erosion when its maintenance and upkeep lie not in decisions of the principal investigator, but in the hands of lay decision makers from backgrounds in office management, accounting, personnel, law, or others equally far removed from science. The lay administrative heirarchy should share with the scientist full responsibility for the proper physical maintenance of the laboratory and the function of its equipment. While some such control has been exercised in the United Kingdom under the eye of the Dangerous Pathogens Advisory Group, this Committee apparently has not lived up to its mandated charge and there is now talk of transferring its administrative and inspection functions to the Health and Safety Executive. Should inspection and approval in the United States of laboratories handling dangerous pathogens be undertaken, it would probably fall within the purview of the Occupational Safety and Health Administration; under the provisions of the Clinical Laboratory Improvement Act; or both.

In addition to the Fort Detrick and Birmingham University episodes, the question has further been raised how containment of presumably hazardous recombinant microorganisms can be expected when we are confronted with the important problem of hospital-acquired infections -- "bacteria are constantly getting into places where they shouldn't, and this is in an environment where people are supposedly trained to control such spread" (King, 1978a). Nosocomial infections, of which some two million are estimated to occur in the United States annually (Dixon, 1978), have a diverse etiology, come from endogenous, or a variety of exogenous, sources and arise under a diversity of conditions and circumstances, so that sweeping generalizations cannot be made nor simplistic conclusions reached (cf. Hewitt and Sanford, 1974). To give a simple, unqualified figure for the number of individuals with surgical wound infections (e.g., 90,000 (King, 1978b)) leaves the impression that the hospital environment is totally at fault and does not bring out that "the patient is a more frequent source of infection than all the exogenous sources taken together" (Polk, 1978); this is because the "human aerodigestive and genital tracts are regularly colonized by pathogenic bacteria and the urinary system frequently so" (Polk, 1978). It is virtually impossible, for example, to achieve microbiological sterility in genital and peri-anal areas.

The occurrence in premature infants of meningeal infections with *E. coli* K-1 has also been attributed to environmental inadequacies of the hospital (King, 1978a). Neonatal meningitis is not a common disease. Since this meeting is sited in England, data from the Registrar General's Statistics for England and Wales seem pertinent; neonatal meningitis accounts for approximately one death per ten thousand live born infants (Anonymous, 1976). The case fatality rate is 40% with a poor prognosis in the survivors. Coliform bacteria, especially *E. coli*, are the predominant etiologic offenders. However, Robbins and co-workers (1974) found that 84% of their *E. coli* isolates from the cerebrospinal fluid of newborn infants with meningitis possessed the capsular (K-1) polysaccharide and that this envelope antigen was immunochemically indistinguishable from that of the group B meningococcus; it was this property that endowed *E. coli* with the invasive properties of *Neisseria meningitidis*. The source of the organism and the pathogenesis of the infection remain to be elucidated.

In essence, containment problems of the laboratory and the hospital are not parallel nor comparable. Containment within the confines of a laboratory, with its physical barriers such as laminar flow hoods and negative air pressure; with a staff physically well, under continuous health surveillance and monitored for use of medications; strictly restricted and controlled access; etc., represents a highly controlled situation that cannot be attained in a hospital. It should be remembered that any hospital population represents in large part an aggregation of individuals with severe medical or surgical disabilities, e.g., burns, premature infant nursery, intensive cardiac care, intensive respiratory care, organ transplants, cancer, etc. (Hewitt and Sanford, 1974; Maki, 1978). These underlying conditions lead to some perturbation of the host's defensive mechanisms, which may be further compromised by supportive or therapeutic regimens.

In summation of safety, there are legitimate concerns apropos containment, and how microbiology is (or is not) being taught in our universities and professional schools. Protection against self infection requires on the part of the laboratorian not only appreciation and understanding of the theory and practice of physical containment, but also the acquisition, under the guidance of an able and competent preceptor, of the skill to carry out bench procedures essentially as a reflex, without need to think of each successive step in execution of the sequence. Such teaching and preceptorship in pathogenic microbiology is sadly lacking today; it seems to have been lost in recent years when cell biology became the major concern of departments of microbiology; indeed it would be difficult today, were it not for the identifying logo over the entry portal of such departments, to distinguish their activities and interests from those of departments of biochemistry and genetics (Neu, 1978). We might be well advised to consider restoration of the classical departments of microbiology, and include journeyman courses in pathogenic microbiology as a prerequisite for those going into research on genetic recombination or life synthesis.

And now a look at pathogenicity. As I mentioned at the outset of this discussion, the initial fears regarding recombinant DNA technology were that introduction of foreign genetic material into a microorganism might confer upon it pathogenic properties (or alter those present) or undesirably modify its natural history. Pathogenicity is not a simple attribute of a microorganism, but a complex function (Thomas, 1977; Schlessinger, 1979), the aggregate expression of a number of factors or properties (see, for example, Chan and co-workers, 1979; Formal and Hornick, 1978; Fretor, 1978; Israel and co-workers, 1979; Montgomerie, 1978; and Ørskov, 1978). Of the numerous families, genera and species of microorganisms known, only a few are endowed with pathogenic properties; to convert the myriad nonpathogens into disease incitants would require insertion of foreign information coding for all those attributes which in the composite make for pathogenicity or virulence (Curtis, 1978); merely putting together random new combinations of genes apparently will not suffice.

It has also been suggested (King, 1978b) that a chimeric organism may acquire new antigens which cross-react with tissue antigens of the infected host, and thus give rise to a destructive autoimmune response. Infection by a microorganism possessing antigens which cross-react with those of the host may raise antibody directed at both the foreign antigen and the self antigen (cf. Ebringer, 1978). Attachment of antibody to the tissue receptors activates an inflammatory reaction but since the antibody concentration is too low to eliminate the invading organism, a chronic disease process is established in which sporadic flare-up of infection elicits anew the production of tissue-destructive antibody. In rheumatic fever, for example, antibody raised by the causal streptococcus attacks heart muscle because of the molecular mimicry between the bacterial and the cardiac antigens (Ebringer, 1978); the cross-reacting antigens have been identified as glycoproteins. Although this is the only disease in which the mimicking molecule is known, circumstantial evidence points to a similar immunological situation in other diseases, e.g., ankylosing spondylitis and *Klebsiella pneumoniae*. Myasthenia gravis also appears to be an autoimmunity phenomenon (Anonymous, 1978b). The serum of such patients contains antibodies which attack the acetylcholine receptors on the motor end-plate of muscle tissue; the initiating cause is unknown. Under the "one-gene" or cross-tolerance theory, such diseases are linked to HLA (human lymphocyte antigen)-genes; where disease occurs, it is because the HLA gene product cross-reacts with the pathogen. A diversity of diseases is being added to the expanding list of HLA-linked diseases. With respect to manufactured, chimeric microorganisms, we might refer to Snell (cited by Ebringer, 1978), who has suggested that the broad spectrum of HLA antigens perhaps reflects an attempt by Nature to provide a wide enough coverage of self antigens "to prevent bacterial mutants from completely mimicking self structures in order to avoid immunological recognition and destruction". There is thus a built-in mechanism to protect "self" against "non-self"; not perfect, but reasonably effective.

Some five years have elapsed since the Asilomar Conference and the "Berg" letter; none of the misgivings, concerns or frightening scenarios to which these events gave rise have been realized. A host of committees and commissions -- lay, scientific, and governmental -- has sprung up to control and even to prohibit scientific application of genetic engineering, a process which occurs in Nature with or without man's inadvertent intervention. Because of its minute size, the bacterial cell can carry only a limited amount of genetic information, but this apparently interposes no constraint in coping with its environment. Sharing of genetic material occurred early between subclones of the ancestral forms (Sonea, 1971) and development of genetic transfer mechanisms gave the bacterial cell ready access to virtually a universal extra-cellular gene pool. Although the frequency of recombination diminishes the greater the taxonomic unrelatedness between the donor and recipient (Sonea, 1971), in such instances transfer between two bacterial groups may be effected through a third acting as an intermediary (Jones and Sneath, 1970, cited by Sonea, 1971). Among eukaryotes, viruses play a similar, significant role in the evolutionary process (Anderson, 1970; Gruneberg, 1970; L'Heritier, 1970; and Mourant, 1971); in Nature, they not only cross species barriers but, as is exemplified by the zoonoses, may be transmitted from individuals of one phylum to another. As Anderson (1970) states, "if the extent of this crossing were known and fully mapped, pathways of infection would probably be found which interconnect each cell through other cells to all other cells -- both plant and animal". All this suggests the myriad possible permutations of gene combinations, so vast that most have never been effected in Nature (cf. Danielli, 1974). Perhaps, as Colin Tudge (1977) has commented, we should look upon the world as "one big happy genome".

While the conjectural hazards of recombinant technology through laboratory intercession are being debated, genetic engineering is occurring naturally outside the laboratory. There is, for example, the transfer of DNA from a bacterium, *Agrobacterium tumefaciens*, to cells of dicotyledenous plants to produce crown-gall disease, a tumor (cf. Marx, 1979). In the virus field, there is some evidence that new strains of influenza A virus arise through recombination between the influenza viruses of man, animals and birds (Webster, Isachenko and Carter, 1974; cf. Beveridge, 1975) and in foot and mouth disease new subtypes may arise through recombination in animals carrying several immunotypes of the virus or in animals carrying one immunotype and inoculated with a live attenuated vaccine virus of a different immunotype (MacKenzie and Slade, 1975).

The hand of man is suggested or obvious in a number of natural situations. Thus, a recent outbreak of porcine diarrhea (Gyles, Palchaudhuri and Maas, 1977) was caused by a naturally-occurring strain of *E. coli* containing a recombinant of the gene for enterotoxin and that for antibiotic resistance. Over the past decade, outbreaks of bacillary dysentery caused by antibiotic-resistant bacteria have produced hundreds of thousands of cases and thousands of deaths in Central America and Mexico (Anderson, 1978), and epidemics of typhoid fever due to chloramphenicol resistant *Salmonella typhi* have occurred in Vietnam and in Mexico (Finland, 1978). Antibiotic resistant strains of *Haemophilus influenzae* have made their appearance within recent years (cf. Finland, 1978) and to compound our problems there has been the very recent emergence of multiply resistant pneumococci (Finland, 1978; Jacobs and co-workers, 1978).

These developments arise from abuse and overuse of antibiotics; the present situation, although ominous, is conceivably not irreversible. Pritchard (1978) states that he does not know "of a single example...of *any* bacterial species in which resistance to *any* antibiotic has become the prevalent phenotype except in an antibiotic-contaminated environment". This suggests that acquisition of resistance is of no permanent advantage to the economy of the species and a return to susceptibility is a return to the normal physiological state. In support of such a position is the observation of Levy (1978) that chickens fed antibiotics developed resistant intestinal bacteria and that similar resistant organisms appeared in the intestinal flora of exposed farm personnel. Discontinuance of the antibiotic supplemented feed led to the disappearance of the resistant forms both in birds and man. Chau, Wong and Fok (1978) reported that over a five year period the proportion of tetracycline-resistant strains of *Salmonella johannesburg* decreased markedly, concominantly with a decrease in the use of tetracycline in the general hospital and in the community at large. (The proportion of chloramphenicol resistant strains, however, showed no decrease, for a number of possible reasons suggested by the authors.) How best to attack the problem poses a dilemma. It is doubtful that aside from establishing another bureaucracy, strict controls and regulations on the use of antibiotics would be productive. Indeed, complete interdiction of the use of antibiotics as an animal feed supplement could result in undesirable economic repercussions on the meat-consuming public. A more rational objective is to revamp teaching in medical, veterinary and agricultural schools to provide the professionals involved with a knowledge and understanding of antibiotic agents.

Others at this Conference (and elsewhere (Cohen, 1977)) have discussed some of the more obvious benefits that recombinant DNA technology may produce for the medical, veterinary and agricultural sciences, and I should like at this point to touch upon some of these briefly. First off, there are the concerns relating to the constantly increasing world population and the inability of food supplies to keep abreast of needs. Increased crop production imposes a high demand on chemical fertilizers, which are expensive. Danielli (1974), several years ago, calculated that to build the required fertilizer plants would require some ten years and twenty billion dollars. An obvious and more practical approach is to incorporate nitrogen-fixation genes into crop plants. The probable cost of this approach was estimated by Danielli (1974) to be 100 million dollars or less, a considerable difference as compared with production of chemical fertilizers. Additionally, if, along with the nitrogen-fixation genes there were also inserted genes to increase or add proteins containing essential aminoacids, the nutritional value of the crop plant would be desirably elevated. Such achievements might be associated with a resultant decreased dependence on animal proteins.

Mention has been made of the successful synthesis of hormones. One important development in this area would be production of interferon, a hormone-like substance with a very potent antiviral effect (Cantell, 1978). Application of this potent antiviral has been limited in scope because of the difficulties surrounding the production of exogenous interferon; manufacture depends upon stimulation of various mammalian cells in culture by any of a variety of inducers (Cantell, 1978). At the moment, costs of a major production effort of interferon in human leukocytes are being underwritten by the American Cancer Society for a study to evaluate the therapeutic potential of interferon against cancer (Merigan, 1979). Because of its potent activity against viruses, it would undoubtedly find wide medical application if it could be produced cheaply and in bulk, something not feasible with current dependence on mammalian cell culture systems. Insertion of the appropriate genes into a host micoorganism should provide a way around this obstacle.

Antibodies with a sharp specificity are a prime requisite in immunology, and difficult to attain. In such areas as typing of tissue antigens for organ transplants or grafts, success depends considerably upon the closeness of the tissue match. Very fine antigenic distinctions are achievable through the use of monoclonal antibodies, which are now on the horizon through the use of so-called hybridomas. These are derived by fusing myeloma cells, which secret large amounts of antigenically uncommitted immunoglobulin, with cells from a hyperimmunized animal that are processing a selected antigen. Such hybridomas have a wide application (Anonymous, 1977; 1978a). Here again, as in the case of hormones, present exploitation on a commercial scale is cost-intensive, and application of recombinant DNA technology would appear to offer real rewards.

Contamination with fertilizers and pesticides from farm run-off, and with effluents from industrial operations and sewage treatment plants poses environmental problems such as eutrophication, accumulation of toxicants in fresh water or marine organisms, etc. Organics arising from industrial processes could be contained at the source through the use of appropriate degrading or converting microorganisms (as, for example, the recent construction of bacteria which utilize haloaromatics (Reineke and Knackmuss, 1979)). Sewage treatment traditionally depends upon the presence of naturally occurring organisms to break down organic constituents; organisms better suited to the digestive processes involved probably could be manufactured for specific activities or purposes in the degradative cycle. And for those concerned with the environment, it might be kept in mind that biological processes, as compared to industrial chemical processes, are not only more efficient but cleaner, since byproducts can be used as animal food or as fertilizer (Danielli, 1972).

One might wonder what directions the Asilomar Conference might have taken and what conclusions it might have reached had the organizers included among the participants practicing scientists with expertise in medical microbiology, infectious disease, immunology and epidemiology; a conference so constituted might have precluded the horrendous bureaucracy which has since arisen in the United States and elsewhere. Be that as it may, the existence of guidelines is a fact and something that must be lived with.

On a number of occasions during the course of this discussion I have mentioned laws or regulations or guidelines and perhaps I should collate some of these points.

The guidelines may have, in a way, served a useful purpose by calling attention to a number of facets of safety and containment of dangerous pathogens. None of the conjectured possibilities which gave rise to the guidelines have eventuated, and over the past several years these guidelines have been modified and relaxed. As further experience is gained, and present questions and doubts are resolved, we should see a broader and more intensive application of recombinant DNA to the solving of problems, such as those in medicine, public health and agriculture, concerned with the well-being of mankind.

The smallpox incident at Birmingham University is an unfortunate illustration of how things may go awry when appropriate safety and containment requirements are breached and when the control responsibility of an overview committee is ineffectual.

Two points seem relevant here with respect to work with potentially hazardous pathogens, naturally-occurring or laboratory-contrived. First, no set of laws, regulations or guidelines will ensure the safety of laboratory personnel or containment of the pathogen unless those who work in the field have a full comprehension and understanding of the hazards involved and are thoroughly trained and indoctrinated in the methods, procedures and technics for handling dangerous pathogens. Safety is a matter of compliance and even an extramural review committee, competent to make scientific judgements and evaluations and with authority to enforce compliance, must have a knowledgable staff to work with. The second point is the need to bring back to our professional schools and universities formal and laboratory courses which have been emasculated or abandoned in recent years and which gave the student the fundamentals of the natural history of pathogenic microorganisms and of the pathogenesis, pathophysiology and immunology of infectious diseases.

What is needed, then, is not a set of laws or guidelines which dictate the scientific investigations or experiments that may or may not be undertaken, but rather a set of operational principles that will serve to ensure the safety of the scientific staff and of the public. Enforcement and compliance, by law or regulation, may well emerge as a consequence of the smallpox incident. Whatever path we take with respect to new laws and regulations affecting recombinant DNA technology, we should proceed most judiciously and carefully, and with due regard to what impact our actions might have on the future. The emotional turmoil orginally surrounding genetic engineering has raised issues threatening the freedom of scientific inquiry in all fields of science, not only biology. As a warning and a possible prophecy I should like to refer to Marvin Harris' book (1977) entitled "Cannibals and Kings", wherein he comments on the evolution of the pristine state from simple band and village societies in which "the average human being enjoyed economic and political freedoms which only a privileged minority enjoy today" (page 101). Paraphrasing Malcolm Webb, Harris states that "tribal equalitarianism would gradually vanish even as it was being appended, without awareness of the nature of the change, and the final achievement of absolute control would at that point seem merely a minor alteration of established customs. The consolidation of governmental power would have taken place as a series of natural, beneficial....responses to current conditions, with each new acquisition of state-power representing only a small departure from contemporary practice" (page 122). Pirt (1979) quotes Keeton that "rights are not violently overturned. They are quietly and insidiously eroded so that the extent of the inroad is apparent only at a later date". My caveat, then, is to consider "the possibility that at this very moment we are again passing by slow degrees through a series of 'natural, beneficial'...changes which will transform social life in ways that few alive today would consciously wish to inflict upon future generations" (Harris, 1977, p. 122).

REFERENCES

Anonymous. (1976). Editorial. Coliform meningitis in the newborn. *Lancet*, 2, 778-779.

Anonymous. (1977). Editorial. Spin-off from cell fusion. *Lancet*, 1, 1242-1243.

Anonymous. (1978a). Hybrid cells are a good source of antibody. *New Scientist*, 79, 271.

Anonymous. (1978b). Monitor. Antibodies coat muscle receptors in myasthenia. *New Scientist*, 79, 769.

Anderson, E.S. (1978). Genetic engineering -- the manifest hazards. *New Scientist*, 79, 34-35.

Anderson, N.G. (1970). Evolutionary significance of virus infection. *Nature*, 227, 1346-1347.

Beveridge, W.I.B. (1975). The origin of influenza pandemics. *WHO Chronicle*, 29, 471-473.

Cantell, K. (1978). Towards the clinical use of interferon. *Endeavour*, 2, 27-30.

Chan, H.W., and co-workers. (1979). Molecular cloning of polyoma virus DNA in *Escherichia coli*: lambda phage vector system. *Science*, 203, 887-892.

Chau, P.Y., Wong, W.T., and Fok, Y.P. (1978). Resistance to chloramphenicol and ampicillin in *Salmonella johannesburg* in Hong Kong: Observations over a five-year period 1973-1977. *J. Hyg.*, 81, 343-351.

Cohen, S.N. (1977). Recombinant DNA: Fact and fiction. *Science*, 195, 654-657.

Curtiss, R. III. (1978). Biological containment and cloning vector transmissibility. *J. Inf. Dis.*, 137, 668-675.

Danielli, J.F. (1972). Industry, society and genetic engineering. *Hastings Center Rep.*, 2, 5-7.

Danielli, J.F. (1974). Genetic engineering and life synthesis: An introduction to the review by R. Widdus and C. Ault. *Internat. Rev. Cytol.*, 38, 1-5.

Dixon, R.E. (1978). Effect of infections on hospital care. *Ann. Int. Med.*, 89, 749-753.

Ebringer, A. (1978). The link between genes and disease. *New Scientist*, 79, 865-867.

Finland, M. (1978). Editorial. And the walls came tumbling down. More antibiotic resistance and now the pneumococcus. *New Eng. J. Med.*, 299, 770-771.

Formal, S.B., and Hornick, R.B. (1978). Invasive *Escherichia coli*. *J. Inf. Dis.*, 137, 641-644.

Freter, R. (1978). Possible effects of foreign DNA on pathogenic potential and intestinal proliferation of *Escherichia coli*. *J. Inf. Dis.*, 137, 624-629.

Gruneberg, H. (1970). Is there a viral component in the genetic background? *Nature*, 225, 39-41.

Gyles, C.L., Palchaudhuri, S., and Maas, W.K. (1977). Naturally occurring plasmid carrying genes for enterotoxin production and drug resistance. *Science*, 198, 198-199.

Harris, M. (1977). Cannibals and Kings. Random House, New York.

Hawkes, N. (1979). *Science in Europe*/ smallpox death in Britain challenges presumption of laboratory safety. *Science*, 203, 855-856.

Hewitt, W.L., and Sanford, J.P. (1974). From the National Institutes of Health. Workshop on hospital-associated infections. *J. Inf. Dis.*, 130, 680-686.

Israel, M.A., and co-workers. (1979). Molecular cloning of polyoma virus DNA in *Escherichia coli*: plasmid vector system. *Science*, 203, 883-887.

Jacobs, M.R., and co-workers. (1978). Emergence of multiply resistant pneumococci. *New Eng. J. Med.*, 299, 735-740.

King, J. (1977). A science for the people. *New Scientist*, 74, 634-636.

King, J. (1978a). New diseases in new niches. *Nature*, 276, 4-7.

King, J. (1978b). Recombinant DNA and autoimmune disease. *J. Inf. Dis.*, 137, 663-666.

Jones, D., and Sneath, P.H.A. (1970). Genetic transfer and bacterial taxonomy. *Bacteriol. Rev.*, 34, 40-81.

Levy, S.B. (1978). Emergence of antibiotic resistant bacteria in the intestinal flora of farm inhabitants. *J. Inf. Dis.*, 137, 688-690.

L'Heritier, P. (1970). *Drosophila* viruses and their role as evolutionary factors. In Evolutionary Biology, 197 , Vol. 4, edited by T. Dobzhansky, M.K. Hecht, and W.C. Steere, Plenum, New York City.

MacKenzie, J.S., and Slade, W.R. (1975). Evidence for recombination between two different immunological types of foot-and-mouth disease virus. *Austral. J. Exp. Biol. Med. Sci.*, 53, 251-256.

Maki, D.G. (1978). Control of colonization and transmission of pathogenic bacteria in the hospital. *Ann. Int. Med.*, 89, 777-780.

Marx, J.L. (1979). Research News. Crowngall disease: Nature as genetic engineer. *Science*, 203, 254-255.

McGinty, L. (1979). Smallpox laboratories, what are the risks? *New Scientist*, 81, 8-14.

Merigan, T.C. (1979). Editorial. Human interferon as a therapeutic agent. *New Eng. J. Med.*, 300, 42-43.

Mirkovic, R.R., and co-workers. (1973). Enterovirus type 70: the etiologic agent of pandemic acute hemorrhagic conjunctivitis. *Bull. Wld. Hlth. Org.*, 49, 341-346.

Montgomerie, J.Z. (1978). Factors affecting virulence in *Escherichia coli* urinary tract infections. *J. Inf. Dis.*, 137, 645-647.

Mourant, A.E. (1971). Transduction and skeletal evolution. *Nature*, 231, 466-467.

Neu, H.C. (1978). How is the medical student being trained in microbiology and infections? *Ann. Int. Med.*, 89, 818-820.

Ørskov, F. (1978). Virulence factors of the bacterial cell surface. *J. Inf. Dis.*, 137, 630-633.

Pirt, S.J. (1979). Letters. 1984 beckons? *New Scientist*, 81, 114.

Polk, H.C., Jr. (1978). Prevention of hospital wound infection. *Ann. Int. Med.*, 89, 770-773.

Pritchard, R.H. (1978). Recombinant DNA is safe. *Nature*, 273, 696.

Reineke, W., and Knackmuss, H.J. (1979). Construction of haloaromatics-utilising bacteria. *Nature*, 277, 385-386.

Robbins, J.B., and co-workers. (1974). *Escherichia coli* K-1 capsular polysaccharide associated with neonatal meningitis. *New Eng. J. Med.*, 290, 1216-1220.

Schlessinger, D. (Ed.). (1979). Microbiology - 1979. Am. Soc. Microbiol., Washington, D.C., 356 pp.

Sonea, S. (1971). A tentative unifying view of bacteria. *Rev. Can. Biol.*, 30, 239-244.

Thomas, L. (1977). The hazards of science. *New Eng. J. Med.*, 296, 324-328.

Tudge, C. (1977). Forum. One big happy genome. *New Scientist*, 73, 659-660.

Webster, R.G., Isachenko, V.A., and Carter, M. (1974). A new avian influenza virus from feral birds in the USSR: recombination in Nature? *Bull. Wld. Hlth. Org.*, 51, 325-332.

(This paper was given by N.D. Zinder, The Rockefeller University, New York, USA.)

DISCUSSION

W.F. BODMER: I want to comment on a point of science in the talk which I don't expect Norton to answer. This is in relation to his comments on auto-immunity and HLA association with disease. I believe his interpretation of those associations is not correct. It doesn't fit in with that which is widely discussed by those who work in the field. There is no evidence, apart from one source of work which I think certainly needs confirmation that any of these associations are due to cross reactions with HLA antigens. The work that he is quoting relates to possible cross-reaction between *Klebsiella* and B27 in cases where, in any case, the associated diseases are probably due to humoral rather than cellular responses. I think there is very little case for assuming that antibodies to histocompatability antigens themselves are disease causing, because some 20%-30% of all women who have had children form those antibodies in response to fetal stimulation. I think, therefore, we must take with great caution such interpretations of the HLA and disease associations and that the case for assuming that antigens picked up in that way could cause diseases is very poor.

N.D. ZINDER: I should say in response to Walter's statement that when I pseudo-edited this for presentation - it's a very long paper - I would have eliminated that section since I feel about it as Walter does. However, a message from Dr. Lennette came from Geneva and there were three points that he particularly wanted stressed and they are the Birmingham incident, the nosocomial infections and also the auto-immune diseases. I did not feel that I should take it upon myself to eliminate any of them.

I.S. JOHNSOŃ: Earlier in her talk Dr. Weiss spoke of the risk assessment of the space programme and in support of Dr. Lennette's "astonishment", I think that is the term, at Asilomar - I couldn't help but wonder if there wasn't an analogy that might be made. Here, again, without impuning anyone's motives, I think we had a risk assessment made by a geneticist and molecular biologist. The environment on the moon was certainly hostile i.e. cold, no water, radiation, etc. - and the probability of life had to be slight. If you reflect for a moment on what actually occurred to protect us from this unknown risk, the space capsule came down through the atmosphere, through intense heat which probably sterilised the outside. Inside the astronauts removed their helmets so that their respiration was no longer self-contained. Frogmen would swim over, open the hatch, inhale deeply and say "Are you okay?". They would then take the astronauts back to the flight-deck of a ship where they shook hands, put arms around each other's shoulders and *then* they put them into isolation!

M.G.P. STOKER: Well, Dr. Lennette has given us a long and complicated statement and I think it's rather difficult to analyse it immediately. Obviously, we must read it in full. Are there any more general points on the public health side that anyone wants to make?

N.D. ZINDER: I'd just like to say a few words - I am not going to go over at this time either Berg et al or Asilomar but I have been somewhat concerned and I heard an echo of this in Dr. Frederickson's paper the other day that, because of the untoward events that followed the statement by Berg et al, scientists in the future will be afraid to speak out on significant public issues especially if they feel there might be some ultimate regulation involved. Now, in some ways, if this turns out to be true, and I believe it's going to be true, it must be the most tragic of the consequences of these events. I felt and I still feel that we scientists have special responsibilities. We are aware before anyone else is of changes in knowledge - and it's not all right to say that we cried wolf and therefore got everybody into trouble and one should not do this again. I guess what I am saying is that I hope that the lesson of these last years is that we scientists should always try to act as responsibly as we can whatever the ultimate consequences may be.

THE VIEW OF A SCIENCE JOURNALIST

R. Lewin

Association of British Science Writers, New Scientist, King's Reach Tower, Stamford Street, London SE1, U.K.

I will make some comments that derive specifically from my experience in reporting on genetic engineering over the past few years and some more general observations to do with the way I see the relationship between scientists and journalists.

In a recent conversation with Stanley Cohen he told me about his baptism to the world of science journalism. Some few years ago a reporter from the *San Jose Mercury* came to his laboratory for half a day to discover what this new science of genetic engineering was all about. Stan spent a lot of time with the journalist explaining things as he saw them. He had the impression that the journalist understood the issues relatively clearly. But when Stan saw the paper the following day he was appalled to read the headline: "Man-made bugs ravage the earth". The story was pretty close to the headline - and far away from reality.

This sort of experience is - unfortunately - not uncommon, and I know that many people at this conference have been misrepresented or their views distorted in the process between explanation and publication. Sensationalism is a frequent criticism of the press by scientists, and, as Chairman of the Association of British Science Writers, it is something I very much regret. There are practical reasons why this sometimes occurs, and I will mention a few later.

Genetic engineering happens to be one of the few areas of science that the public apparently finds interesting. Perhaps I should say that it produces stories that journalists are able to persuade their news editors are worth printing. Media interest in a scientific subject is, scientists might think, a mixed blessing. No one likes to be ignored, and responsible public exposure can be very helpful in generating public support for worthwhile research programmes. But there can be problems, both if reports are distorted or sensationalised and if research is "over-sold".

For instance, the war on cancer of five or six years ago turned out to be a pretty unrewarding offensive. And nuclear energy failed to produce the promised solution to energy problems on the time-scale implied by the more enthusiastic proponents. Indeed, nuclear power offers hazards that were totally played down in the early days. Now, the media was a willing partner in generating far-reaching hopes in both these areas, and many unreflective newspaper and magazine articles, and TV programmes blazed with hopes shortly to be fulfilled. When, as time passed, those hopes remained unfulfilled, however, more and more antagonistic reports began to appear, especially in the case of nuclear energy. It is therefore not surprising that the public is much more cynical now towards these areas of endeavour: the scientific case was over-played in the first place and an uncritical press swallowed; the scientific community is now paying for this earlier lack of judgement - both by the scientists and the journalists - in having to face a public that is now more suspicious of science than it used to be.

The point I want to make is that scientists should try to be absolutely honest with the press. I know there is a temptation to make things seem simpler or more straightforward than they are, and the promise of a new route for synthesising human insulin is more likely to be thought newsworthy than discoveries about the unusual structure of eukaryotic genes. But if you make promises simply to catch the eye of the news editor, everyone falls into the trap I have just outlined.

Until recently science had the reputation for having answers to everything - uncertainty was not part of the vocabulary of the scientist on the screen or in popular print. But science *is* full of uncertainties, as we all know. It is also full of other human things, such as competition and jealousies. I always find it distressing when I encounter scientists who try to present their world as if it were some special kind of system untouched by the features known to the world of ordinary mortals. So I would encourage you to be more honest, more straightforward, when telling the media about your work.

I know there is a danger of being misquoted or having your words used out of context, and there are many reasons for this - the pressures on a daily journalist are enormous: fighting for space in the paper, looking for something catchy, the threat of deadlines looming, all are occupational hazards. The question of honesty I am referring to has to do with the relationship between the scientist and the reporter who has to write a story by 8.00p.m. that evening. The 11 people who signed the "Berg" letter were being honest and open - some may think naive too. But the letter and all that followed it have generated a climate of public concern with which the scientific community is going to have to cope. And the best way to handle suspicion is to be absolutely open and honest, to admit limitations and uncertainties. I will return to this later.

One can ask, what is the role of the science journalist? Is he or she a simple mirror for the events in the scientific arena, or is the profession more active than that, more interpretative if you like. I certainly see my role as something akin to that of a critic, not quite involved in events, but knowledgeable enough to be more than a passive spectator.

A number of people at this conference have mentioned the "trauma" of Asilomar. I was not there, but I have read a good deal about it. I have to admit that I find it difficult to understand when people complain that the presence of the press there did a disservice to science. No doubt some participants were misquoted or quoted out of context - the usual bogeys. But, overall, the reporting seemed to me to be very fair. In some instances it was inspired. For instance, Michael Rogers' report in *Rolling Stone* was undoubtedly the best, not because it gave the clearest account of the meeting's conclusion but because he managed to convey a real *feeling* of the anxieties, concerns and motivations of people there. It was good interpretative journalism. As a reflection of the world of scientific debate as it really operates, Rogers' article is unmatched. I believe he did a great service to science, by being honest about what he saw and heard, and by revealing science as a human activity, rather than some form of magic performed by boffins in a back room.

If I can move from the far-off days of Asilomar to the present, I will remark on what I see as a tangible shift in public interest in science. First, there appears to be a greater appetite for science as a whole, and this is reflected in the increasing coverage of science in many newspapers and magazines, particularly in America, and in the number of new science magazines that are on the launching pad. Secondly, that interest is not confined to the gee whiz type of science, the wonder-drugs and new inventions. Many of the most popular articles and TV programmes have been on somewhat abstract subjects, such as paleoanthropology and cosmology. In moving into this new era of public interest in science, I strongly urge scientists not to "sell" genetic engineering on the grounds of insulin or other medical and industrial benefits. Yes, there will be benefits, but the science is exciting in itself, and I am sure the public is ready to hear about it. As the benefits move closer to reality, that will be the time to talk about them.

One particularly interesting aspect of this whole genetic engineering issue is the difference in the nature of the debate on the two sides of the Atlantic. In the USA, pressure groups and the general public were very interested in the problem, and they frequently contributed to the debate. Often, what they said seemed foreign to the experience of most scientists - issues such as public participation in science policy, the local community demanding to know what is going on in the laboratory down the road. This sort of thing. I once commented in print that the recombinant DNA issue was being used as a vehicle for expanding democracy into science. I do not think the vehicle moved very far, but it did make some progress.

The public debate in America has without question increased interest in and awareness of science - the public's understanding of the nature of the process of science is enhanced too, I believe. You cannot have countless newspaper articles and magazine stories, five popular books, and several major television programmes without someone noticing it all.

In the UK, however, the debate has been much more low key - so low key as to be almost inaudible. Surely, there are differences in the British and American character. Americans are much more openly responsive to issues: they pick up a topic and pursue it enthusiastically for a couple of years, and then move on to the next one. The Vietnam war was over, so the public conscience turned to recombinant DNA. We in Britain are much more reserved, much more cautious about championing topics in public. But it did surprise me that the British Society for Social Responsibility in Science - the counterpart to the Boston Science for the People group that had such an impact on the debate in the USA - were so slow in raising any interest in the issue. So, without a focus of dissent, public debate here has been conspicuous by its almost total absence. Some unions have taken up the challenge, mainly on the question of laboratory safety for individuals. But, there have been few public meetings.

There are differences too in the press between the two countries. We pride ourselves in having a good press in Britain - the "quality" newspapers at least. But science is poorly served. For instance, the London *Times* - when it was still publishing - has one science correspondent and one medical correspondent. By contrast, the *New York Times* has eight people doing the same job. And there is a difference in access by journalists to scientists and politicians and their aides. The USA, with its Freedom of Information Act, allows responsible and diligent journalists to present a fuller picture of events to the public. In the UK, with its Official Secrets Act and reticent civil servants, one has to fight hard to get anything like a clear image of what is going on. It is a system that is certain to induce premature ageing in British journalists and one that ensures that the public are inadequately informed.

The period between Asilomar and the present was filled with many things that alarmed scientists, such as the events at Cambridge, Massachusetts, and the vigorous response from the Friends of the Earth. But what one must recognize is that the "Berg" letter catapulted what previously was an obscure branch of molecular biology into the political arena. The letter, and all that followed it, alarmed the public. The environmentalists jumped in and organised their protest, just as they have a right to do so under the First Amendment. The scientific community was shocked to see its genuine concern treated so sorely. Meanwhile, time passed and you decided that the conjectured hazards are not as bad as was first supposed, a change of mind based partly on some small amount of evidence, but mainly on more cerebral processes. You changed your mind. What do you do about it?

I tell you what you do not do. You do not organise a conference like this, in the remote Kentish countryside and say: "Sorry, no journalists". Nothing excites the interest of a reporter more than telling him he is excluded. Given the nature of the subject and its past political history, a meeting that appears to be closed to reporting is certain to arouse suspicion. I doubt whether the *Guardian* would have published its highly critical story about the meeting if Anthony Tucker had not been given the impression that participants would be saying things they did not want immediately reported. It was a blunder on the organisers' part to give that impression, and I am very pleased that, eventually, reporting restrictions were lifted. Better to risk the danger of being misquoted than generate suspicion by being unwilling to go on record. Politicians live in that world all the time. And when a scientific topic is as political as this one undoubtedly is, then scientists must put up with it too. In the end public confidence in science and trust of scientists will be enhanced.

As I said earlier, the public concern and interest in the science of genetic engineering - its benefits and possible hazards - demands complete honesty from you. If you really feel you made a mistake, then say so openly. But don't expect the *mea culpa* to be accepted readily. You must allow the public to be a little wary at such a remarkable turn-around. You cannot expect the restrictions you now face - the NIH Guidelines, GMAG, etc. - to be dropped just because you claim it was all a dreadful mistake. The issue developed through a highly political process, and your "re-entry", as Professor Pritchard calls it, will have to be through the political process too. That takes time and public confidence. And I see the GMAG arrangement, where technical problems are considered jointly by experts and by other interested groups, as one very promising way of achieving that confidence.

In conclusion I should like to reiterate one or two points. First, public understanding of science is better served if scientists are more open - more honest if you like - about the world in which they work. If mistakes are made, admit them. If doubts exist, spell them out. Second, do not over-sell the promise of your endeavours; in the end it is a dangerous occupation given to backfiring. Third, try to understand the limitations under which many journalists operate, particularly those on daily newspapers. Last, the growth of public understanding in science is essential for progress, but it is a slow process. Like politicians, you must learn to tolerate the hazards of being reported, because, ultimately, the relationship between you and the media will improve.

DISCUSSION

A. CAMPBELL: At least two speakers, one of them a journalist, have described the Berg Committee Report as a "very honest" document and have asserted that they are able to infer the honesty of the report from its contents. I was told by a colleague over coffee that this inference is proper because the document is apparently not self-serving, but rather the reverse. I strongly disagree with the proposition that one can judge the honesty, candour or sincerity of the authors of this or any other statement from internal evidence alone. As I consider this question basic to some highly unfortunate misunderstandings between scientists and other members of society, I would like to elaborate.

The first, and less important, point is that an action may be self-serving for reasons not obvious to the observer. In this regard it may help to suggest a plausible though hypothetical sense in which the Berg report might have served the interests of some of its authors. A widely disseminated public announcement to the effect that "We fear that our revolutionary new accomplishments might endanger mankind" not only warns society, but also publicizes and promotes the importance of the discovery itself. I do not intend to imply that the actual purpose of the "Berg" letter was publicity rather than safety. Suffice it to say that however sincere the authors' concern for safety, I doubt that all of them were totally unaware of the promotional aspect. But the point is that I cannot ascertain, merely by reading the document itself, the relative influence of these two goals on the authors.

With this example, we can turn to the more important point. The characterization of the Berg report as "very honest", especially in a talk where scientists are admonished a dozen times to be honest with the press and public, seems to indicate that other statements on the same subject are perceived as less than completely candid. If these are also to be identified by the degree to which they appear to be self-serving, one may easily fall victim to a circular logic.

Several of the non-scientists at this meeting have summarized the current situation somewhat as follows: "A few years ago biologists thought that recombinant DNA research might be hazardous. Now, without extensive new data, they are saying that it is probably safe". None of them said instead: "Most biologists believe that the research is safe. But a few years ago, on the basis of no evidence and superficial theoretical analysis, they were saying it might be dangerous". The difference is significant, if the only informational base is what was said on both occasions. I submit that, without independent evidence from some other source, there is no way of inferring whether scientists were more honest or open at Asilomar or at Wye.

The news media function is a highly selective filter between the input of scientists and the output that reaches the public. One component of this filter is the journalist's appraisal of the honesty of various inputs. There is a danger that journalists may measure honesty, not by objective standards, but rather by their own preconceptions of what an honest man should say. If that is indeed what they are doing, honesty should require them to make that fact known to their readers.[1]

[1]For clarification in the published version, the above comments were expanded and rewritten by A. Campbell.

R. LEWIN: The question of honesty to which I referred has to do more with openness than with the business of telling the truth or not. The public reacted strongly to the "Berg" letter, and the scientific community, I believe, withdrew even more into its shell. Many scientists edit what they say to the media, not just making things simple to understand, but so as not to be totally open. That's what I mean by honesty.

A. CAMPBELL: Your use of the word honest, I think, is different from mine. When I ask if someone is honest, I am thinking does he mean what he is saying and I don't see how you judge that by internal criteria.

R. LEWIN: You can mean what you're saying without saying everything that you could say about a subject. It's a question of degree. It's a question of really being absolutely open with everything you have to say about it. It's a question of selecting data if you like.

W.F. BODMER:You comment, in my view rightly, about the problems of whether the press were here or not. But I do wonder whether, if this had been declared right from the start an open meeting, it really is the sort of meeting that is appropriate for the press to be present at and for education of the general public. I'd like to know your views on that and, indeed, your views on what sort of a meeting would be the right one for the proper communication of the sort of thing we have been dealing with here to the science journalists.

R. LEWIN: I think the general format of this meeting is one which is appropriate for having a contingent of the press. You, after all, on Monday had a review of most of the areas of science. You had historical perspectives. You had this morning the legal aspects. It's a very general overall picture. I think, had this meeting been more accessible to people who can't necessarily spend three days away in the country - someone on a daily newspaper perhaps - you would have got some responsible attendance from the daily press if it were in London. So I think this format is fine. It doesn't present any great difficulties as far as I can see.

M. SINGER: I am going to start out exactly the way Allan Campbell did and point out that you urged us to be honest. You indicated that you've heard many complaints from the scientific community over the years but that they really didn't worry you very much. They worry me very deeply. You are our conduit to the public and therefore when we talk to you we are doing so in an effort to carry out our responsibility to inform the public. We think that responsibility is important and we agree that we must be honest. But the presentation to the public is then dependent on you. It's true that the reporting of the Asilomar Conference was very fine in quality and accuracy. It was very important to have that reporting because no published proceedings of that Conference were planned as there will be for this meeting at Wye. But, after Asilomar, at least in the United States, the situation deteriorated. I've heard from many people besides yourself that editors impose severe pressure on reporters and I have no doubt but that it's true. I also have no doubt that people other than reporters write the headlines. The reporter may struggle very honestly with the story and then the hand of the editor and the headline writer can completely alter the story's impact. But, I don't need simply to accept that and continue to deal with it. I have to look for other options. You may feel that you are not in a position to do anything about it but that doesn't solve my moral conflict when erroneous or misleading stories appear in the press. Everyone in this room probably has some terrible tale to tell. For example, I had a debate on TV with George Wald. The debate never actually occurred: it was an electronic fabrication. I had spent two hours doing exactly what you asked - trying to describe science in front of a TV camera in my laboratory. What appeared on TV was a five minute debate where bits of sentences that had been set in a scientific context were pulled out and put into a political context. The public was completely misled. I felt after that experience and others like it that it was probably immoral for me to participate in such ventures and that's the problem that many of us have and which I don't think you really addressed.

W.J. WHELAN: On the original question of press reporting and whether or not this was a closed conference - it never was and never has been a closed conference. I don't think you said that, but I do want to make this clear because you have told me in private conversation that there has been a belief to the contrary. We did advertise the conference quite widely. We took half a page in *Nature* in which we invited anyone to come along. Of course there was a limitation on the accommodation here but, in fact, that limit was never reached. We could still have taken ten more people and as far as I know everyone who applied to come did, in fact, attend. As far as the question of press reporting was concerned, it was clear that the reasons why it was thought that it would be a good idea to have it off the record had become misunderstood. Once it was clear they were misunderstood then the correct thing to do - which is what COGENE and the Organising Committee did - was to open the conference to reporting and I am glad to see Dr. Lewin here and to have an opportunity to correct any misapprehensions.

R. LEWIN: I would like to make a comment about that because Stan Cohen also said to me it was never a closed meeting in that sense and, yes, I did see the advert in *Nature*. But some months ago I was told that it was a closed meeting, which sounded odd. So I called John Tooze whose comments seemed to confirm the point and we discussed how best I should challenge it if I wanted to. I wrote to several of the organisers, and eventually received a reply which said "As a compromise over this difficulty we will allow three nominations to the Association of British Science Writers". This surely implied that, whereas three science writers were now to be allowed to attend, none was previously.

To return to Maxine Singer's points. You have obviously had a bad time, and this is unfortunate. Now, if it were the case that more than 50% of press reports on genetic engineering - and by press reports I mean TV, radio and everything - were irresponsible then I would be seriously worried. But my impression is that it's a smallish percentage that one complains about. It is very irritating to go through a two-hour TV debate and then have the whole thing chopped up. Everything like that is just very distressing. If you are not offered an opportunity to see what was produced in the end, then that's somebody's malpractice. The thing I would suggest to improve things, and it's not something that would work at a stroke, is that if scientists spend more time with journalists, more mutual confidence will develop. Better stories will be produced in the end. I can see this process each time that I move into a new area. One goes in and has an initial impression of it and the people but as you get to know more and more people, you get to know more and more of the nuances. Now the problem with all this type of journalism is that science correspondents on newspapers have to cover a very wide range of topics. It's very difficult for them to have time enough to explore every nuance of every topic.

M.G.P. STOKER: I really think it is time to stop before somebody proposes there should be a moratorium on press reports.

HISTORICAL PERSPECTIVES ON THE RECOMBINANT DNA CONTROVERSY

C. Weiner

Program in Science, Technology, and Society, Massachusetts Institute of Technology, Cambridge, Massachusetts 02139, U.S.A.

I have been observing what has become known as "the recombinant DNA controversy" since shortly after the February 1975 International Conference on Recombinant DNA Molecules at Asilomar. My attention had first been attracted a few months earlier when David Baltimore, one of those involved in the so-called "moratorium letter," spoke at a colloquium at MIT about the issues involved and the planned conference at Asilomar. Here, clearly, was a new technique for fruitful research on problems at the forefront of fundamental knowledge, with possible applications that would have a major impact on society. At the same time scientists involved in this work were making an unparalleled effort to inform the larger scientific community, and indirectly the public, not only of the potential benefits of their research but also of their concerns about its possible hazards. The extent of their concern was expressed by their decision to call on fellow scientists to voluntarily refrain from doing certain experiments until the likelihood of hazards was assessed and minimized. I was especially fascinated as a historian of science because of my studies in the emergence and growth of new research fields and my interest in the ways scientists have perceived and defined their social responsibilities. On all counts, these were events of considerable significance in the history of science.

Therefore, I made sure that I was present when four MIT biologists who had been at Asilomar reported on their experiences before about one hundred of their colleagues at an informal meeting in the biology department. The tape I made of that session was the beginning of an experiment to document a significant scientific development as it was unfolding. The recombinant DNA history project at MIT was initiated in collaboration with my colleague Rae Goodell; its aim was to make sure that the full and detailed record of these events would be available for future study.[1] From Spring 1975 until the beginning of 1979 we collected letters, minutes, notes, memoranda, press clippings, reports--about 5000 items in all. We observed and recorded on both audio and video tapes the proceedings of meetings and legislative hearings; and we conducted and transcribed oral history interviews with one hundred and twenty

---

[1]The project was supported by the MIT Oral History Program, the Ethics and Values in Science and Technology Program of the National Science Foundation, and the National Endowment for the Humanities Program of Science, Technology and Human Values. For a report on the project see: Weiner, Charles. The rDNA Controversy: Archival and Oral History Resources. Science, Technology and Human Values, Winter 1979, pp. 17-19.

participants in the United States and Europe. These materials were catalogued and deposited in the Recombinant DNA History Collection at the MIT Archives where they are now available, and are being used, for research. The early establishment of this project enabled us to collect materials before they were lost or discarded and to record recollections while they were still fresh.

The project staff interviewed a broad range of persons involved with the recombinant DNA research issue. They included thirty-six participants in the Asilomar Conference, many of them scientists directly involved in recombinant DNA work; critics of the research as well as opponents of efforts to regulate it; policy makers and their advisors involved in regulation and legislation on national and local levels; citizens who served on local review boards and the representatives of universities and industry who interacted with them; journalists and science writers who covered some of the major events. The interviewees were unusually open and cooperative, supplementing their recollections with extensive written documentation. The most active stage of the project is completed, but it is evident that the story is not over yet. Although there are inevitable gaps, the collection is a good starting place for serious study of what happened. It will still be a valuable resource when all of these recent events may seem like ancient history. Additions to the collection that will help to complete the historical record are welcome.

I would like to offer a few observations on aspects of the controversy surrounding recombinant DNA research that I find especially interesting and relevant to the future.

Most accounts of the controversy begin with the quasi-biblical statement, "In the beginning there were the letters," the one from the Gordon Conference in the summer of 1973 and the one signed by the eleven scientists in 1974, the so-called moratorium letter. Now, as we approach the fifth anniversary of the moratorium letter, and especially in light of current attitudes, it is appropriate and important to look at archival documents and recollections of people who participated in the meeting at MIT in April 1974 from which the letter emerged and of those who subsequently added their names to it.

The purpose of that meeting was clearly stated in the letter sent out by Paul Berg to the participants: "[Our] agenda is to consider whether or not there is a serious problem growing out of present and projected experiments involving the construction of hybrid DNA molecules *in vitro*. If a problem exists, then what can be done about it; both short and long term actions."[2]

The first decision at that April 17, 1974, meeting was to hold a conference; the idea for the letter was a consequence of that. Paul Berg described these actions in a letter to Hans Kornberg:

> We met at MIT for a day and settled on the idea of calling a conference next February of those scientists working on methods of joining DNA molecules and particularly those involved in constructing hybrid DNAs. It was our plan that one of the major purposes of the Conference, besides a report on the scientific progress, would be a wide ranging discussion of potential hazards growing out of these types of experiments. Were there any experiments that should not be done? How could such a moratorium be proposed or enforced? In short, we expected a frank and searching review of what people were doing or wanted to do, particularly from the point of view of whether they should be done. But as we talked we realized that the pace of events might not wait for February

---

[2] Letter from Paul Berg to all participants in MIT meeting on biohazards, April 8, 1974. MIT Archives, Recombinant DNA Collection.

> and that some of the experiments many people would agree could be hazardous would be done by then (e.g., attempts to fuse portions of Herpes DNA to appropriate plasmids for cloning in *E. coli* were imminent). Since the technology for constructing hybrids has become ridiculously simple, that fear was well founded.
>
> Consequently we decided to devise a letter to be submitted to *Science* and *Nature* calling on scientists to defer certain kinds of experiments until these potential hazards could be better evaluated and certainly until there was an opportunity to discuss the issues at the February meeting.[3]

Further evidence of the group's concern is provided by this excerpt from Richard Roblin's letter to Maxine Singer who was then on the editorial board of *Science*: "We feel a sense of urgency about rapid publication, since research on construction of recombinant DNA molecules is currently proceeding at a fast pace and since the most important feature of the letter is its call for a voluntary moratorium on certain types of experiments until an evaluation of the potential biohazards is made."[4]

It is clear that the letter and the Asilomar conference were planned as a package and were based on the recognition that the new recombinant DNA technique was going to be an extraordinarily powerful research tool whose use would grow by leaps and bounds. It was evident that many scientists would be attracted to the field, and that most of them had no experience working with potentially hazardous materials. The quickest and best way to alert these scientists was to publish a letter in an appropriate scientific journal.

The letter was not a hastily written, relatively private enterprise. Between the April 17 meeting and the date of its publication in *Nature* and *Science* three months later, the letter went through four drafts which were widely circulated in the United States and Europe. For example, two of the original signers discussed it with participants at the European Molecular Biology Organization restriction enzyme meeting in Belgium in May, and a third talked about it at the June Cold Spring Harbor meeting on tumor viruses. The relevant scientific community knew what was up.

When the letter was about to be published in *Science* and in *Nature*, the U. S. National Academy of Sciences held a press conference. Why did they go public? The group of scientists who had to be alerted was potentially large and the fastest and most effective way to reach them was to publish the letter in major scientific journals. Such an action was sure to attract wider attention. An orderly release of information to the press with proper scientific explanation was preferable to a random sensationalized approach.

The response to the letter by the relevant scientific community was generally favorable. When the draft was read at the Cold Spring Harbor meeting in June, twelve of the European scientists in attendance immediately drafted a letter to John Kendrew recommending that the European Molecular Biology Organization give urgent consideration to the matter. They felt that it was essential that research utilizing the

---

[3]Letter from Paul Berg to Hans L. Kornberg, June 18, 1974. MIT Archives, Recombinant DNA Collection. Present at the meeting were P. Berg, D. Baltimore, J. Watson, D. Nathans, Sherman Weissman, N. Zinder, R. Roblin, and Herman Lewis.

[4]Letter from Richard Roblin to Maxine Singer, July 2, 1974. MIT Archives, Recombinant DNA Collection.

new recombinant DNA techniques be made possible in Europe and that urgent and careful consideration be given to providing appropriate special risk laboratories.[5]

When the Berg _et al_ "moratorium letter" was published in _Nature_, it was followed by responses from Michael Stoker and Kenneth Murray which included these statements:

> No doubt a good many dirty tricks have been attempted and discarded by nature in the course of evolution, but the disquiet arises from the utterly novel associations of genetic material which are now possible. The potential benefits should, therefore, be delayed, not for ever, but until consequences can be assessed, and preliminary experiments carried out under conditions of maximum security. . . . it is encouraging that the very leaders in the field have taken the initiative and have been supported by the [National Academy of Sciences]. It is now to be hoped that academics and learned societies in other countries will add their weight, and that international organisations such as the European Molecular Biology Organisation will lend support. . . . For many it will be a test of self denial and social responsibility in the face of strong intellectual temptation.
>
> _Michael Stoker, ICRF_[6]

> The NAS request is both reasonable and responsible and deserves to be universally respected. It recognises both the difficulty in evaluating real or potential hazards that may be involved in such work, as well as the obvious criticism that these will remain obscure in the absence of experimental study; urgent consideration of the latter is explicitly recommended.
>
> Fears that the proposed limitations to experiments will seriously obstruct research in vital areas of biology seem unfounded. The NAS initiative, by focussing attention on the hazards involved, could well promote rather than hinder work on _in vitro_ recombination in animal viral systems, an area believed by many to hold the key to gene therapy in its broadest terms. . . . if we follow the moderate tone set by the NAS we shall be careful not to oversell the social benefits devolving from recent experiments.
>
> _Ken Murray, Edinburgh_[7]

The materials in the historical archives demonstrate that the "moratorium letter" was seen by the public and accepted by a significant part of the relevant scientific community as a commendable, responsible action. The subsequent story is well known, including the February 1975 Asilomar conference and the efforts that followed to develop workable guidelines on national levels.

I would like to offer some observations on the process. The issue was defined by the scientists as a problem of the scientific community, a limited technical problem to be solved by technical means. The possible abuse or misuse of the research was occasionally mentioned but was excluded from consideration as were the social implications of the genetic engineering developments that it might lead to. The discussions were narrowed to the questions: "Is there any danger in what we're doing? If so, how can we avoid that danger and still proceed?"

---

[5] Letter from L. Crawford, F. Cuzin, R. Dulbecco, W. Fiers, B. Hirt, L. Philipson, J. Subak-Sharpe, K. Weber, R. Weil, Charles Weissmann, E. Winocour, P. F. Spahr to John Kendrew, June 7, 1974. MIT Archives, Recombinant DNA Collection.

[6] Stoker, Michael. Molecular dirty tricks ban. _Nature_ (July 26, 1974): _250_, 278.

[7] Murray, Ken. Alternative experiments? _Nature_ (July 26, 1974): _250_, 279.

The process of developing guidelines was influenced by the realization that a popular, exciting new field was emerging and by the conviction that it should be encouraged to grow. Scientists were eager to get involved in recombinant DNA research and the developers of the guidelines felt that artificial restraints should interfere as little as possible. In the United States the National Institutes of Health was in the ambiguous position of encouraging and facilitating the growth of research using recombinant DNA techniques while at the same time it was responsible for establishing and enforcing safety regulations that would restrict it. While all of this was going on scientists were tooling up to use the new technique and were waiting for the green light that would allow them to proceed as rapidly as possible. Those charged with developing guidelines felt that they had to move as quickly as possible, despite the lack of needed information.

From the start, the scientists involved believed that there was a need to demonstrate to the public that scientists, on their own, could act responsibly to protect the public. They felt that, "If we don't do it on our own, someone else will do it for us." When pressures exist, from scientists and the public, and a large degree of uncertainty prevails, there is bound to be disagreement on the scientific basis of risks and the weight attached to them. As a result, the process of establishing rules involved a series of compromises aimed at achieving a consensus within that portion of the scientific community affected by the regulations while at the same time providing assurance to the public that they would be protected against possible hazards.

These events were set in motion by the letter in 1973 from the participants in the Gordon Research Conference on Nucleic Acids. But there was another story coming out of the June 1977 Gordon Conference on Nucleic Acids. There the assembled biologists also endorsed a public letter about the hazards of recombinant DNA research, but with two significant differences, the content and the intended audience.[8] In this letter they said that the hazards had been exaggerated and that, in fact, the experience of biologists since the 1973 Gordon Conference, when they sounded the alarm, had convinced them that much of the concern was groundless. This letter was not addressed to the scientific community but to Congress. The timing was no accident. Just that week a bill to regulate recombinant DNA research in the United States was expected to be reported out of the Senate Subcommittee on Health and Scientific Research, chaired by Senator Edward Kennedy.

A major purpose of the voluntary moratorium and the subsequent guidelines--and this was declared repeatedly at Asilomar--was to demonstrate that scientists could regulate themselves and to assure the public that legislation was not needed. It is clear that the recent experiences of the biologists referred to in the 1977 Gordon Conference statement were not only scientific, but social and political as well.

Public interest was relatively limited until 1976. Then concern surfaced in a few academic communities as universities tooled up for research in this exciting new field. Some of this public awareness was sparked by scientists who were critical, on scientific grounds, of the guidelines. But it also soon became clear to many non-scientists that a good deal of the scientific information essential for evaluating the probability and extent of potential mishaps and, thus, the adequacy of the guidelines was just not available. The recombinant DNA history project interviews with laypersons who assumed various community responsibilities reveal their concerns; NIH documents also reflect increased public anxiety. There were many unknowns. Was it a matter of faith? Which scientist should you trust under such circumstances? Were the scientists who advocated the safety of the research acting

---

[8]Letter from 79 participants at Gordon Conference on Nucleic Acids to members of Congress. June 29, 1977. MIT Archives, Recombinant DNA Collection.

out of self-interest? What public health benefits might be lost or delayed by slowing down the research until more was known? People were asking whether the guidelines were enforceable and whether they should be left to the universities to implement. Should non-scientists play a role in enforcement? What was going on in industry? The debate continues even now and the process is still in motion.

Where do things stand now, five years after the "moratorium letter" was written? The drafters of the letter showed good scientific judgement in their assumptions about the expected rapid growth of the field. While the regulation of the research was being debated in committees and in public hearings during the past several years, the laboratory use of recombinant DNA techniques has been booming, as its potential is being explored in one subfield of biology after another. It is rapidly becoming a central tool in research that was previously impractical or impossible. Combined with other new developments such as gene sequencing it has created great intellectual excitement in biology. As predicted in 1974, people are flocking to the field. This is something every biologist knows and it is borne out by rough grant statistics.

In the United States alone, the National Institutes of Health, in May 1977, supported 207 projects that made use of the recombinant DNA technique. By September 1978 the number had increased to 560. Excluding the projects solely within the National Institutes of Health, the number increased steadily from 498 in September 1978 to 594 in February 1979. Projects supported by the National Science Foundation (there is some overlap with the National Institutes of Health) increased from 53 in 1977 to about 250 in February 1979. A greater increase will follow as a result of the recent relaxation of the National Institutes of Health guidelines. The NIH Office of Recombinant DNA Activities already processes about 1000 applications per year, and other projects are supported by the Department of Agriculture and the American Cancer Society, and, of course, by industry.[9]

Recombinant DNA is rapidly becoming a "required" technique in biology laboratories. There is considerable competitive pressure to use it because of its simplicity and the shortcuts it provides. At the 1975 Asilomar conference only a handful of the participants was involved in recombinant DNA research. At recent biology conferences the impact of the technique is quite visible. For example, a participant at the Keystone, Colorado, meeting on eukaryotic gene regulation in March 1979 estimated that recombinant DNA was essential to at least three quarters of the papers presented there. Young scientists have generally not been discouraged from entering the field, despite earlier fears that they would avoid research involving recombinant DNA techniques. The intellectual appeal, the power of the technique, and the scientific incentives have been too powerful.

The rapid growth of the field has also been reflected, and enhanced, by the emergence of new journals and newsletters, special conferences, and training workshops devoted to recombinant DNA research. The competition that is evident among academic researchers and laboratories is accompanied by competition among industrial firms with interests in utilizing the technique. Potential applications of recombinant DNA techniques have influenced a number of university researchers never before involved with industry to become industrial consultants or to form their own companies. A recent international conference on the patentability of biological organisms, involving scientists and lawyers, is but one example of this change, a new development for relatively esoteric molecular biology.[10]

---

[9] Data supplied by National Institutes of Health and National Science Foundation.

[10] New Scientist (1979). What's bugging the European Patent office? New Scientist, 81:1146, 845.

All of this has occurred despite the many difficulties: the trauma, the feeling of being misunderstood and distrusted, the endless hours of committee work and public explanations, the lobbying and politicking, the immersion in forms, applications and memorandums of understanding required in order to embark on experiments and to vary them while they are in progress, and the delays some face because of lack of appropriate facilities. We might speculate whether, on balance, the controversy has accelerated rather than retarded the growth of the field. At Asilomar, the press representatives were not allowed to write about the meeting until it was over. Instead, the scientists were the ones rushing to the telephones to tell their colleagues back at their own laboratories what was going on and to suggest ways to get involved in using the new technique. The perceived need for control of the research resulted in exchange of information in a way that went beyond the normal channels of scientific communication.

In conclusion, I would like to offer some thoughts on options for the future. I disagree with the statements made by some biologists that now they are acting "scientifically" when they maintain that the conjectured risks were overstated, and that they were acting "politically" five years ago when they voiced concern about such risks. Such statements are bound to be seen by the public as self-serving. In any case, it is naive to think that science and politics can be neatly separated. Science is imbedded in a social context, and recombinant DNA research, like other fields, will continue to be subject to pressures and questioning, especially as it becomes more successful and has greater impact on society.

Nothing will be solved by facing the future with a "scientists versus the public" attitude. It is far more constructive to attempt to understand the basis of the public's perception of risks than to belittle and ridicule their concerns. In addition, confidence in the credibility of technical experts could be enhanced rather than diminished if they are open and explicit about their professional and financial interests in the outcome of the issues being discussed; and if they admit when they do not have the answers, or when the available data are ambiguous, or when uncertainty prevails, or when there is disagreement among experts.

Just as the recombinant DNA technique has proliferated so has the demand for increased public participation in scientific affairs which have an important social impact. The question now is not whether the public should participate but how they can do it in an effective, intelligent manner. Scientists must learn how they can help in these new approaches, and how they can contribute to efforts to increase public understanding of science in order to enhance public participation.

Scientists have a special responsibility, because of their special knowledge, to identify early and to point out to others possible problems related to their research and to participate with other groups in considering all appropriate solutions. With the continuing success of genetic manipulation, situations will certainly arise that will go beyond the laboratory safety problems. They will involve social and ethical issues associated with application of the new knowledge. When is the time to consider such matters? Almost all the people in the field that I have talked with agree that it is important to confront these issues and that they should help initiate such discussions when the time comes. Perhaps that time is now.

Finally, it is important for scientists to counteract the impression that many of their younger colleagues may be carrying away from the events of the past few years, that is, that one should "Shut up or be shut down." There is a fine tradition of responsible action among biologists, physicists, and other scientists who have spoken out on a number of issues in the public interest. Despite the frustration of the current situation, I am sure that this tradition will continue.

THE PUBLIC PERCEPTION OF RISK

Sir Frederick Warner

140 Buckingham Palace Road, London S.W.1, U.K.

Many new developments in science and engineering meet with some form of public response which questions the need for the development because of its potential social and environmental risks. As the argument develops, the ground shifts from one side to another - at one time to the acceptability of the risk and at another to the view that growth as an objective for society should be abandoned.

In the television lecture given by Lord Rothschild (1978) on "Risk", he attempted to make the concepts of risk and its assessment more understandable to the public in general. For this he used British figures for occupational mortality and the figures prepared by Inhaber (1978) on "The Risks of Energy Production". In the public discussion which followed, Professor S. Cotgrove pointed out that the arguments were about values not facts. His 1976 paper on "Environmentalism and Utopia" discusses this more fully (Cotgrove, 1976).

The public in perceiving risks is at a disadvantage. People are used to making value judgments on a development such as a new airport, industrial expansion or a gas storage installation particularly when they are involved as inhabitants of housing close by. The associated risk is something they see aimed at them personally rather than a figure to be weighed against increased prosperity or a reduction of risk elsewhere. Increased prosperity is in any case criticized as an illusion fostered by professional and establishment interests, to guard their own status (Illich and co-workers, 1977).

It is a matter for note that perception of risk varies with location, prevailing socio-economic development, and individual standards of living, education and culture. It is also strongly influenced by familiarity and the ability to make direct observations - say an experience factor. There are probably others but these factors will do to begin a discussion.

LOCATION

Does the perception of risk vary inversely with distance from source, or as the square or the cube? The actual risks result in effects which follow an inverse square or cube law because they are attenuated depending on the area over which they can spread or the volume which takes them up. An explosion can cause a pressure wave effect on individuals or houses inversely as the square of the distance from the epicentre. The accompanying noise is reduced in a similar way but is it not perceived more nearly as if the reduction were linear?

The spreading of chemicals or bacteria in the environment cannot usually be directly perceived except that some chemicals have characteristic smells. The study of dispersion of bacteria released deliberately for purposes of warfare was begun nearly 50 years ago by Sutton (1932) in a series of papers providing the basis of all methods for predicting maximum ground level concentration of pollutants released from a point source. The maximum concentration varies as the inverse square of the height of the release point and, after reaching the ground, is similarly reduced with distance. Multiple sources in large conurbations give a different result and pollution concentration becomes proportional to the town diameter to the 0.4 power (Smith, 1973).

SOCIO-ECONOMIC DEVELOPMENT

In 19th century England it was thought that "where there is muck there is money". The smoke-pall over towns was regarded as a satisfactory indicator of thriving industry, providing work for a dispossessed peasantry who could add to the gloom by cooking and heating with open coal fires. The risk of being killed by a combination of sulphur dioxide and smoke was not perceived against the background of better nutrition, housing and sanitation.

In the 20th century, a greatly increased life span has been achieved by adequate nutrition and the elimination of major diseases through chemotherapeutic and antibiotic agents. The chronic damage caused to the respiratory system over a lifetime by smoke, sulphur dioxide and cigarette smoking then became revealed at times of acute pollution, as in the famous London fog of 1952 when an estimated 4000 extra deaths occurred. Since that time, excess deaths have fallen to zero during comparable episodes with high SO_2 but the perception of risk has risen. The greatest risk, from the smoking of cigarettes, is still not perceived and additional targets are presented as more young people and women become addicted.

INDIVIDUAL STANDARDS

The example of smoking shows that the perception of risk is idiosyncratic and apparently independent of age, sex and education, although perhaps slightly influenced by scientific education and socio-economic group. The influence of occupation and social class on mortality statistics in general throughout Britain have been reviewed by Fox and Adelstein (1978). They arrive at a high correlation (+0.72) between standardized mortality rates for lung cancer against smoking score (the ratio of number of men smoking to the number expected on the basis of national proportions). However, when the lung cancer mortality rates are standardized for social class, the correlation is only +0.16. They suggest that this residual correlation might reflect an association between smoking and occupation over and above the association between smoking and social class. There is no sign that these findings have been widely recognised.

EXPERIENCE FACTOR

Familiarity is dependent on time elapsed and the statistics of accidents or harm over that period. Perception decreases with time but when incidents involving multiple deaths occur it can be revived. Perception does not relate directly to acceptability of risk which itself depends, according to one viewpoint, on the voluntary or involuntary nature of the risk. The Royal Commission on Environmental Pollution (1976) stated "The risk of death or injury through accident is a condition of living and many human activities imply a judgment that the benefits outweigh the related risks. The following table gives some examples:

Probability of Death for an Individual per Year of Exposure

Risk	Activity
1/400	Smoking (ten cigarettes a day)
1/2,000	All accidents
1/8,000	Traffic accidents
1/20,000	Leukaemia from natural causes
1/30,000	Drowning
1/30,000	Work in industry
1/100,000	Poisoning
1/500,000	Natural disasters
1/1,000,000	Rock climbing for 90 seconds driving 50 miles*
1/2,000,000	Being struck by lightning

*These risks are conveniently expressed in the form indicated rather than in terms of a year exposure.

The table is intended simply to give an appreciation of the scale of risks associated with a range of activities. Some of these risks, such as those arising from smoking or from rock climbing, are accepted voluntarily by individuals. Others are concomitants of our everyday lives, such as risks at work or from aircraft or car accidents."

More complete tables, based on the International Classification of Diseases (1976), are given in Tables 1.2 and 1.5 of a publication by Grist (1978) on recent British data (modified in Tables 2 and 1 respectively).

The criteria of familiarity and possibility of direct observation can be applied to activities in research on recombinant DNA. Hardly anything would be more unfamiliar to the general public or less likely to come under direct observation. It is, therefore, a firm candidate for being categorized as bearing an unacceptable risk even though the risk has not been defined or perceived.

PARALLELS IN PERCEPTION

In advanced societies which consume large amounts of resources achieving high standards of living, extension of lifespan and family limitation, attitudes to risk continually change and have been partly discussed under "Location". In such activities, the risks of cancer, foetal damage and mutation are perceived out of proportion to the effects which have been measured. Recombinant DNA is on the way to joining radioactivity, chemicals and new drugs as a subject for public question and suspicion. It may replace radioactivity if the increasing pressures of energy demand result in the building of more nuclear power stations. The changing attitudes are shown by the narrow vote in Austria against commissioning the Zwentendorf nuclear station, followed by the Swiss referendum in favour of nuclear power. The public hearings now taking place in Lower Saxony over proposals to develop the reprocessing of nuclear fuel and storage of highly-active waste at Gorleben will again bring out the arguments heard nearly two years ago at the Windscale Inquiry into similar British proposals. It is doubtful whether these public debates result in greater public perception. At the Windscale Inquiry, the local population most likely to form a critical group, whether from eating fish with ^{137}Cs or breathing air containing minute quantities of transuranic elements showed little concern. Inhabitants over 50 km. away showed great concern particularly about possible genetic effects. In France,

TABLE 2 Individual Risk Estimates (Accidents & Violence)

Description of Cause	Individual Risk (per year, x 10^6)	Notes	0.95 Probability Confidence Interval
Railway Accidents	3.3	b	
Motor Vehicle Traffic Accidents & Motor Vehicle Non-Traffic Accidents	122	a	
Motor Vehicle Traffic Accidents Involving Collision with Pedestrian	45	a	
Water Transport Accidents	2.9	b	
Air & Space Transport Accidents	1.4	b	
Accident to Unpowered Aircraft	7.4×10^{-2}	c	$5.0 \times 10^{-2} - 10.4 \times 10^{-2}$
Accidental Poisoning by Drugs & Medicaments	11	a	
See text (Section 7.6)	1.9	a	
Accidental Poisoning by Other Liquid & Solid Substances	2.2	a	
Accidental Poisoning by Noxious Foodstuffs & Poisonous Plants	9.2×10^{-3}	c	$2.3 \times 10^{-3} - 23.5 \times 10^{-3}$
Accidental Poisoning by Gases & Vapours	3.2	a	
Accidental Falls	111	a	
Falls on or From Ladders or Scaffolding	2.1	a	
Accidents Caused by Fire & Flames	16	a	
Accidents Due to Natural & Environmental Factors	4.2	c	
Excessive Heat	4.6×10^{-2}	d	$3.4 \times 10^{-2} - 6.2 \times 10^{-2}$
Excessive Cold	1.2	c	
Hunger, Thirst, Exposure & Neglect	2.3	c	
Bites & Stings of Venomous Animals & Insects	8.5×10^{-2}	d	
Other Accidents Caused by Animals	4.0×10^{-1}	d	

TABLE 2 (contd.) Individual Risk Estimates (Accidents & Violence)

Description of Cause	Individual Risk (per year, x 10^6)	Notes	0.95 Probability Confidence Interval
Lightning	1.0×10^{-1}	d	
Cataclysm	4.3×10^{-2}	d	3.1×10^{-2} - 5.8×10^{-2}
Accident Due to Other Natural & Environmental Factors	9.2×10^{-3}	c	2.3×10^{-3} - 23.5×10^{-3}
Other Accidents	43	a	
Accidental Drowning & Submersion	12	a	
Inhalation & Ingestion of Food Causing Obstruction or Suffocation	11	a	
Hit by Falling Object	3.4	b	
Accident Caused by Fire-arm Missiles	6.2×10^{-1}	c	
Accident Caused by Explosive Material	8.3×10^{-1}	c	
Accident Caused by Electric Current	2.5	b	
Surgical & Medical Complications & Misadventures	1.5	b	
Late Effects of Accidental Injury	2.2	b	
Suicide & Self-Inflicted Injury	76	a	
Homicide & Injury Purposely Inflicted by Other Persons	10	a	
Injury Undetermined whether Accidentally or Purposely Inflicted	30	a	
All Deaths due to Violence	442	a	
Deaths from All Accidents	325	a	
Deaths from all Non-Transport Accidents	195	a	

Notes: (a) based on 1975 data
(b) based on 1975-71 data
(c) based on 1975-1969 data
(d) based on 1975-1958 data

0.95 probability confidence intervals are quoted where the risk calculation employed a total of less than 50 deaths recorded in the appropriate time period.

TABLE 1. Risk of Death as a Function of Age 1975-1971 mean

Cause of Death	Individual Risk per Year x 10^6												
	All Ages	0-4	5-9	10-14	15-19	20-24	25-34	35-44	45-54	55-64	65-74	75-84	85+
All Causes	11910	3693	334	282	637	718	791	1945	5897	15261	37642	90023	207405
Natural Causes	11453	3427	203	174	250	310	483	1622	5504	14772	36934	88188	202378
Neoplasms	2475	79	68	55	71	95	171	583	2005	4983	9505	14143	17025
Diseases of the Circulatory System	6177	27	7	15	27	45	121	654	2535	7180	19879	52558	127752
Ischaemic Heart Disease	3120	**	**	**	**	4	39	391	1711	4655	11278	23617	45983
Pneumonia	887	320	16	13	21	21	22	51	142	466	1976	8631	29720
Diseases of the Digestive System	296	67	6	7	11	17	26	69	178	391	926	2239	4522
All Deaths due to Accidents and Violence	457	267	131	108	387	408	308	324	393	489	708	1835	5027
Suicides	78	No deaths recorded		1*	25	64	76	96	125	138	150	138	94
Homicides, Legal, Undetermined, War	36	26	6	5	27	38	33	41	48	52	51	57	54
All Accidents	343	240	125	101	336	306	199	186	219	299	506	1639	4879
Road Accidents	138	54	78	57	259	215	117	92	98	139	203	363	340
All Non-Transport Accidents	196	184	44	39	65	78	70	84	111	150	297	1268	4530
All Deaths due to Accidents & Violence as % of Total Deaths from All Causes	2.88	6.49	37.43	35.79	52.69	42.67	25.17	9.56	3.71	1.96	1.34	1.82	2.35

* fewer than 50 but more than 20 deaths registered in the period 1971-1975

** fewer than 20 deaths registered in 1971-1975

Data presented were obtained from mortality & total population statistics for Great Britain apart from one violent demonstration at Creys-Manville, there seems to be little public perception of any risks with a major expansion of nuclear power now being undertaken.

CONCLUSIONS

With old technologies such as rail, road or air transport the risk perceived of over 100 deaths in a single accident in the U.K. every 10 years for rail alone does not halt travel. The occurrence of one death in a smallpox research laboratory in Birmingham had little public effect but resulted in the questioning of any experimental work involving pathogens, in spite of the historical record. This will lead to obstacles being put in the way of work on recombinant DNA until enough experiments have been made and enough time has elapsed for the small risk to be demonstrated. During this time, the risk can be kept low by correct classification and low inventories. What will happen when large-scale application is proposed remains to be seen. Some comfort might be drawn from the health records of factories making citric acid with *Aspergillus niger*, if these records are available. It seems unlikely that a methodology like that for radiation protection will become necessary as public perception grows.

REFERENCES

Cotgrove, S. (1976). Environmentalism and Utopia. *Sociological Review*, 24, No. 1.

Grist, D.R. (1978). Individual Risk - Recent British Data. HMSO, London.

Fox, A.J. and Adelstein, A.M. (1978). Occupational Mortality. *Jnl. Epid. and Community Health*, 32 (2).

Illich, I. and co-workers (1977). *Disabling Professions*. Boyars, London.

Inhaber, H. (1978). The Risk of Energy Production. Atomic Energy Control Board, Ottawa.

Rothschild, Lord (1978). Risk. *The Listener*. 30 Nov.

Royal Commission on Environmental Pollution (1976). 6th Report, HSMO, London.

Smith, F.B. (1973). Unpublished note. Meteorological Office.

Sutton, O.G. (1932). A theory of eddy diffusion in the atmosphere. *Proc. Roy. Soc.* A, 135, 143-165.

World Health Organisation (1976). International Classification of Diseases. Geneva.

DISCUSSION

S.N. COHEN: I have a comment on Dr. Weiner's paper. He and several of the speakers that we have heard, both yesterday and today, have expressed the view that the signing of the "Berg" letter was a responsible act and that we should not allow the subsequent events to discourage us from showing such responsibility in the future. However, I think it is important to distinguish between perception of responsibility and actual responsibility. It's fair to say that the seven individuals that met at MIT and the four of us who contributed to the later revisions of the letter believed that we were acting responsibly, and at the time our action was considered by the scientific community at large as being highly responsible. Yet, in retrospect, it seems to me that while the letter was *perceived* as responsible, it was not really responsible at all. The most incriminating thing that any of us could have said at the time about recombinant DNA research was, not that there was any indication of hazard, not that there was even any valid scientific basis for anticipating a hazard but simply that we could not say with certainty that there was not a hazard. The same thing could have been said about virtually any other kind of experimental endeavour. None of us who signed the "Berg" letter would have considered publishing a scientific paper without valid data, and yet simply on a basis of a lack of certainty that there was not a hazard, and in the absence of any real epidemiological input, a public statement was released.

In doing this, I believe we showed poor scientific judgement, quite apart from any question of political judgement. However, because of the collective scientific credentials of the group, the recommendations of the letter were given credence out of proportion to their intrinsic merit. I just want to say that I hope that in the future scientists faced with a similar situation will not be deterred by the outcome and will show the good intentions that those of us who signed the letter had at the time. But I would also hope that they would demonstrate much better judgement in distinguishing between actual responsibility and perceived responsibility.

S.D. EHRLICH: I have a question I would like to address to Dr. Weiner. It has been stressed several times here that there is a reluctance on the part of the public in general, translated also by the press, to accept change in the position of the scientific community, a position which concerns recombinant DNA research. The reluctance is due to the fact that the public does not have a clear perception that there is sufficient new information available now that would justify such a change. It may be relevant to stress that it is difficult to have access to and assess the knowledge that exists. Everybody, I think, would agree with that and we are all faced with that sort of situation and when we want to gather information, even from our own field, it takes a lot of effort. It is not easy.

It has been mentioned several times, and I would like to stress it again, that during the last five years the existing relevant knowledge has been reviewed and evaluated and this process has greatly influenced our present position, which I think is that there is no danger whatsoever in the use of recombinant DNA technology which would justify maintaining regulations in their present form. My question to Dr. Weiner would be - does your research indicate that, indeed, experts were not consulted at first and that a sort of consultation was taking place during the five year period?

C. WEINER: It is my impression that expert opinion was sought from the start, and at every stage when there were questions raised there were enquiries made within the relevant community. Recombinant DNA research was a relatively new field for a number of people, and there could be criticisms that the proper people were not invited to Asilomar. I found, however, that experts were continually, and still are, being consulted; and I have also seen some difference of opinion among the experts. Sitting in on meetings, I have observed very significant arguments on scientific grounds within the framework of the regulations about the relative risk or lack of risk of a particular experiment. I don't think by any means that there has been scientific evidence that can state that there is no risk. But the immediate answer to your question is that there have been experts consulted from the start. More and more knowledge has been gained and there are challenges to this knowledge. It is a process that has gone on in many places quite openly and as that process goes on and is explained to people and there are legitimate challenges and responses, then the issues can be worked out in a way that I think would be appropriate, that is in public with all the information on the table.

The earlier part of your question referred to the public. I guess I used the word "public" loosely; we all use it loosely. I can't say that large numbers of people are up in arms about this research; I don't know of any really large public group lobbying against it. In general, the public is not anti-science; the prestige of science is still very high and always has been. I think people, in fact, support research and the benefits of research. When there were issues raised which seemed to pose a public health hazard, people were concerned. In some cases they were reassured and in some cases not. The public does want to be informed and involved in the scrutiny of new frontiers of research. As it is now, the argument that you have is not really with the public but with colleagues, guidelines committees, and so forth in specific countries. I think the response to questions that continue to arise cannot be to say, "There is absolutely no risk, we were crazy before, and we can reassure you now". I don't think that that adds to the credibility of the scientists involved.

I am much interested to hear Stanley Cohen's view, and I am not trying to argue with him. I was giving an interpretation from views of several of the people who, like Stanley Cohen, had been involved in the signing of the "Berg" letter, and I'd like to hear from others who are here. I think, though, that several of the people who participated at that early stage have stated that, based on the information they had, not to have acted would have been irresponsible, and, secondly, they were talking about the *possibility* of a hazard. They called for a scientific study to see if, in fact, the hazard existed and, if it did, how it could be minimised. The effective way they chose to accomplish their goal was to assemble a scientific group from the relevant community to consider the problem. I thought that that was a commendable thing to do, and I still think so.

M.G.P. STOKER: You raised a point that I would also like to put to Fred Warner. You say that really the public isn't much concerned. Surely it isn't just a question of degree of risk perceived, there is also the question of how many people perceive it. I wonder if, in your experience of risk evaluation, how you can tell what proportion of the population are really worrying?

F. WARNER: This is a thing on which there are no figures and we have actually got a study group going on risk in The Royal Society. It is not only a question of how many people out of a population perceive a risk, but how do they actually perceive it? What is the actual mechanism by which they recognise it? I think we are going to have to get some work done in the field of experimental psychology because it is so unclear that we need experimental work.

F.S. ROLLESTON: As I listened to the talks this morning and yesterday evening, it seemed to me the crucial point is a relationship between the scientific background of information and the political perception of that information. The conclusion that I tend to draw from that is the process of establishing guidelines by the scientists for the scientists was too centralised. I wonder whether the process should have been, with all its dangers, one that allowed the public political process to establish guidelines on the basis of facts provided by the scientist. I don't see that distinction in the way that all of us have gone about this process in this case.

H.D. STETTEN: I have a question that I would like to address to Sir Frederick. He made reference to what I have referred to on occasions as the Book of Genesis, that is carcinogenesis, teratogenesis and mutagenesis, as the three major anxieties. Each of these in a favourable situation can be approached, I think, in an actuarial fashion, and one has ample data on which to base a risk assessment. Only when the increment in hazard is small compared with the noise, as Sir Frederick pointed out, do we start to have real problems. But, with recombinant DNA, we have a situation which is different. The first mouse, as far as I am aware, has yet to be killed; the first human has yet to be damaged by this technique. Can you give us any help as to how we assess risk when we have zero actuarial basis on which to work?

F. WARNER: There is no help available and somebody has said you got yourself into this position and I think you have to live with it. Once you have sounded the alarm about risk then you have put into people's minds the fact that there is a risk. You did that presumably on the basis that one was suspected without perhaps taking time to get together all the evidence which there might have been from related fields to see if you could put into perspective what an actual risk was. This is what has puzzled me throughout all this conference. Whenever you have talked about risk assessment you really have been trying to decide whether there is, in fact, a risk at all, whereas in every other field where we talk about risk assessment we have the historical data, we know the risks, we have gone in with our eyes open. As I say, when you do engineering you go into engineering as you go into any form of economic activity, and if you spend a million pounds you are going to kill somebody. I am sorry this is the brute fact of life. You can take the statistics, and if you look at the publication by ICRP called Publication No. 27 ("Towards an Index of Harm"), you can see an attempt to arrive there at how much money is spent in various fields and how inexorably it results in so many occupational deaths which are the ones on which there are firm statistics. On population deaths we have very few statistics and they cannot be distinguished at the moment by epidemiological studies.

E.L. WOLLMAN: There are a few comments I would like to make. I was particularly struck by the public statements made here by Jim Watson yesterday and by Stanley Cohen a few minutes ago, that is, by two of the authors of the "Berg" letter. I was especially sensitive to the manner in which Stanley Cohen looked back at what happened five years ago. There is no doubt in my mind that if scrupulous molecular biologists, unfamiliar with the factors involved in disease and epidemics, had kept between themselves their forgivable apprehensions, no public concern would have ever appeared, since there was not the slightest evidence for actual risk. This would have not deterred experimenters from taking any precautions they felt desirable nor from exercising their critical sense as to the use of the newly developed technology. Time, money, fears, passions would have been economised while emphasis would have been put on the previously unforeseen possibilities offered by this approach for the analysis of genetic material, especially from eukaryotes.

Simplifying Sir Frederick Warner's previous statement, the notion of risk is strongly dependent on what is - or has been put - in the mind of the public. Thus, the ghosts are amongst us and everybody knows that it is even more difficult to disprove the existence of ghosts than to prove their reality. This is why it seems to me, that, in the light of the experience gained and of what we have heard at this meeting, the time has come to adopt clear and reasonable positions. A great step towards this goal would be made if those eleven scientists, who originally signed the "Berg" letter, would issue a joint statement on how they see the situation now, after five years of thinking and experience. As long as those who were at the origin of public anxiety do not reveal their present feelings, a doubt will remain in the public consciousness. If a common text is not acceptable by the eleven original signatories, let them act as the judges in the Supreme Court of the United States and issue a majority and a minority statement. This seems to me, in view of the worldwide consequences of the original letter, a responsibility that they should not escape.

My last point is the following. In most of the countries in which DNA hybridization research had started or was contemplated, the Asilomar conclusions were passively accepted and the endless game of guidelines, regulations, etc. started There was great pressure for getting a universal set of guidelines and this point of view was again advocated here by some participants. On the contrary, I would like to stress that on matters such as this one, in which the basis for guidelines were so weak, diversity of appreciation and variability in rules, which at first caused certain irritations, rapidly favoured a critical attitude towards some previous excesses and became the main factor of adaptation and evolution, not only of the guidelines, but also in the mind. This is another lesson that the DNA hybridization debate teaches us.

R. CAPE: I'll be very brief, but I want to address a remark which was included in Dr. Gartland's presentation of Dr. Fredrickson's address, a remark that seems to have escaped attention. We are involved here in a process of disengagement and there are two aspects of that disengagement that are important. The first one was mentioned by Dr. Frederickson. It is that there has been a shifting of the burden of proof. If that is true, then when people describe a very remote conjectural hazard, the burden of proof that the hazard is real should be on the people who describe the scenario. The second point has to do with disengagement from what I would call the "where there's smoke, there's fire" analogies. This is not Birmingham. This is not Harrisburg. There is no fire. However, there is, unfortunately, still a great deal of smoke and we are all going to be breathing it for a long time.

M.G.P. STOKER: We have sat and talked and talked for nearly three days and I think we should ask Congressman Thornton if he would give his view as a legislator and complete the meeting for us.

THE VIEW OF A LEGISLATER AND A SUMMARY OF THE CONFERENCE

R. Thornton

Former Chairman, Subcommittee on Science Research and Technology, U.S. House of Representatives Joint Educational Consortium, Ovachita Baptist University and Henderson State University, P.O. Box 499, Arkadelphia, Arkansas 71923, U.S.A.

I would like to begin by thanking all of those who are responsible for this conference being held - The Royal Society and COGENE. I would also like to express our thanks for the hospitality shown us here at Wye by the Conference Officer, Mr. Austen, and by the very competent people who have transcribed the discussions so that they may be edited soon after they have been delivered in order that an early publication may occur. I think it has been a very fine and thorough conference. I would like to add congratulations to the Chairman of each of the seven sessions for conducting the activities in such a way that we have yet to miss a meal and I assure you that we will not miss a meal on this day! I also appreciate, since I am a non-scientist, the occasional references to other disciplines - the splendid lessons in Greek mythology and English literature, ranging from Zeus, and Pandora, to Mary Shelley's Frankenstein. Somewhat in the same vein, I would like to refer to a statement made by the great English philosopher, Francis Bacon, writing in *Novum Organum*, who said: "It would be unsound and contradictory to suppose that that which has never been accomplished can be accomplished except by means which have not yet been tried".

In a very real sense that is what science is about - how to discover ways to accomplish things which have never been accomplished. During this session we have witnessed significant accomplishments. We have learned from Dr. Waclaw Szybalski how to ask the question: "What degree of containment was required and why?". That question is one which is much on everyone's mind, because, in many instances, there seems to be no rational connection between the level of containment required and the experiment which is being conducted.

I have accepted today a double assignment, that of combining with my own report expressing the views of a legislator, a summary of this scientific conference with all of its legal, social and political overtones. Recognising I suppose the value of Bacon's suggestion of new means of accomplishing difficult tasks, you have charged me with this responsibility after first cleaving my mental storage banks at perhaps base pair 63 and inserting a rather complicated fragment of scientific data. Now you are asking me to find the path between Scylla and Charybdis of science and society interacting with each other. So perhaps I stand here as something of a recombinant trying to address the combined issues of science and public policy which have been raised during this meeting.

At the outset, I would like to say that the science in this conference has been outstanding. I am not a scientist and I am not an expert. I am not even nearly an expert, but I have had the opportunity during the past few years to be near a lot of experts, and I can recognise something new and challenging and interesting. Much which merits that description has been presented during these days. Dr. Michael Fried and Dr. Malcolm Martin, Dr. Anna Marie Skalka and many others have demonstrated that many of the hypothetical risks which were outlined at Asilomar lack substance. We have heard an excellent and vital presentation by Professor Charles Weissmann as to how one could choose specific alternate sites for insertion of genetic material - a technique requiring an enormous amount of skill and perhaps a little luck as well as good laboratory facilities - in order to gain knowledge from the different locations of the insertions. Dr. Bill Rutter gave us insight as to practical benefits which could be obtained from DNA research. Very carefully drawn and well-presented papers by Professor Jeff Schell and Dr. Ralph Riley offered insights into the background of molecular genetics affecting a field in which I have a great interest - agriculture - fungi and plants. I am not going to try to give credit to all of the people who have done outstanding jobs in scientific papers, but I would like to say that the conference has been an excellent conference and that it has brought together a great deal of useful information on scientific developments.

All of the presentations have re-enforced the suggestions that as more and more is learned from experimental data, early concerns about DNA research are being dissipated. However, I have not heard anyone suggest that the science has reached a plateau in which no exciting discoveries may still be expected. Rather I hear people saying that we are looking at the alphabet of life, studying the function of individual words - codons - reorganising some sentences by the insertion of a few words. And yet, understanding the complexities of the organisation and function of higher life through an analysis of DNA molecules is at least as far removed from the present level of understanding as Shakespear and T.S. Elliott are separated from kindergarten exercises. It would be naive to suppose that a problem as complex and interwoven as the present mix of society's interest in this area of scientific inquiry can be resolved with any simple answer, no matter how appealing such an answer may sound or how eagerly such an answer may be sought. H.L. Mencken once remarked that: ".... to every complicated question, there's a simple answer and it's wrong".

The interest of society in this field of inquiry is amply demonstrated by the complicated problems that have been discussed - problems such as those of international law which have been described by Dr. Edith Weiss this morning. The Chairman, Dr. Michael Stoker, aptly remarked that perhaps the scientists may have frightened the public and are now alarmed at the response. In my view, the responsible actions of the scientists in expressing concern about the research are commendable. I cannot agree that this original action was mistaken. It was useful to be reminded by Dr. Norton Zinder that at the outset those people who were consulted about infectious diseases were concerned about the possibility of creating new portals of infection. As Professor Mark Richmond commented in describing the relationship of science and the public: "Asilomar was a political act. Now science has lost its innocence".

So we are faced, I think, with a problem addressing both the fields of science and public policy. Dr. Edwin Lennette's report, which was presented by Dr. Norton Zinder, spent a good deal of time advocating the desirability of overview and extra-mural review committees to check operations of microbiological laboratories. The paper contained a very strong and positive statement that no set of laws will provide safety without competent and careful laboratory technology. What is needed is not more laws but a dedication to proceed most judiciously and carefully. I think that is true both in science and in government.

In my own role as Chairman of the Committee on Science and Technology, I must confess to a bit of surprise when shortly after our Committee began hearings, I found a number of scientists coming to my office urging me to support the enactment of Federal legislation in order to pre-empt state and local legislation and to avoid the Balkanisation which would result from each city, state or county, making its own rules and regulations. In those instances I suggested to the scientists that the question was more basic - not who should regulate - but whether regulation itself was required. In that regard, I want to mention something else that was most important to this session: I was proud of and impressed by the statement which Dr. Donald Frederickson sent to this meeting reflecting, I truly believe, an understanding of the science involved and a dedication to the reduction of guidelines based on new data as rapidly as that can be accomplished. I think in the situation in which we find ourselves, that kind of statement - frank and candid - from a major public official is one which deserves to be remembered and commented favourably upon.

Dr. Roger Lewin gave us insights as to the reasons that editors select dramatic articles for the newspapers relating to recombinant DNA because that is what the public is interested in and by publishing the articles public interest is rekindled. I believe that is called "feedback". This tendency calls, I believe, for a great deal of self-discipline upon the part of those who report. I think the freedom of the press and the opportunity to attend meetings is vital. I would join in saying that this meeting should have been open fully to the press for comment. Of even greater importance, the press has a strong burden of accountability to accept for itself, since in our society no one else exercises that accountability for them.

Dr. Lewin said the "Berg" letter was very honest. His statement was questioned from the floor as to how one person could say someone else was honest. The way we do that in law - we test credibility all the time - we test it according to the motivation, according to the desire for truth that is evidenced by the documents. And in that context I cannot help but conclude that the "Berg" letter was an honest expression of a concern and a disclosure that a new field of science was opening and ought to be approached cautiously. Let me just say that as a legislator this cautious approach was of great value. We had a lot of pressure for legislation to go roaring through the Congress, to enact statutory guidelines and rules which could be amended only very slowly and with difficulty, if at all. In slowing the rush toward legislation it was helpful to be able to point out that just as recombinant DNA was an important and vital issue for science, so the question of government regulation of science was an important vital issue for government. It was pointed out that just as scientists were proceeding carefully and cautiously into an exploration into this unknown arena so should we in government proceed carefully and cautiously before regulating research. If you want a frightening scenario, let me give you one, not about DNA but about what kind of government action might have occurred had the research gone on behind closed doors without public expression of concern by the scientists involved. Such a scenario would pave the way for an alarmist to create a real hysteria about the work going on without proper restraint, or caution by the professional people involved in the work. And I truly believe that such a scenario would have led to far more onerous and restrictive conditions being imposed on scientific research than those from which we are now emerging. I further believe that one reason we are emerging from the conditions which have been imposed upon science is because government and the public have decided upon reflection that scientists were speaking cautiously and caringly in giving warning and that, therefore, they can be trusted to make decisions in their laboratories.

This morning we have had, I think, some really splendid presentations. Sir Frederick Warner pointed out to us how difficult it is to analyse hypothetical risks and I agree with him. I do not know how you go about analysing a risk for which there are no factual data. However, you can properly say that there are some things that should be approached with caution. In fact, such caution is very much a part of normal living. One of the greatest fears of risk is the fear of the unknown. With regard to the risk of the unknown there is an Arkansas saying: "Don't prod the unknown, it may prod you back" - and it is not irrational to have a concern about the unknown and to proceed cautiously. I think proceeding cautiously is a sign of good intellect and judgement. I think it is easy for us to stand on the safe side of a danger which has not materialised and look back and say that nobody should have fussed about that; there was no reason to worry about that at all. It would have been quite difficult had the research not been approached with caution, if the unknown had prodded us back.

I really appreciated Dr. Charles Weiner's splendid outline of the historic background of developments in this area. His conversations with Dr. David Baltimore and others convinced me that the "Berg" letter and the Asilomar Conference were not hastily called for, or signed, without much attention. He made me feel that the actions of the scientists were rational and responsible and that it might reasonably be expected that an orderly release to the press with a scientific explanation might give better results than a haphazard sensational release later on. Just because - notwithstanding this approach - other things developed that caused complications far beyond the expectations of those people involved in the planning of those conferences should not change our evaluation of the reasonableness of the decision at the time that it was made.

Now I do respect the views of Dr. Stanley Cohen and others who now consider that the action was a mistake because I understand that they are deeply concerned about the scientific and political consequences which have followed that action. While I do appreciate that concern and while that concern shows a great deal of reflective consideration of the matter, I would like to point out again that there are other scenarios which might well have happened which could make the action taken seem awfully good by comparison with another course. What would have happened if the alarm of secret research on new combinations of life had been raised by others?

Well, as you all may have surmised, I have been an onlooker for most of the recombinant DNA debate in recent years in the United States. I had an opportunity during my tenure as the Chairman of the Subcommittee on Science, Research and Technology of the U.S. House of Representatives Committee on Science and Technology to spend quite a bit of time considering the public policy implications of the debate on the recombinant DNA technique. At that time, I believed, as I do now, that true science always stands upon a frontier. It probes at the edges of our knowledge and our ignorance, and we accept its contributions as valuable, its continuation as a necessity. Perceived as a gradual extension of the sphere of knowledge, science is accepted and praised as both our benefactor and our servant. This is the science with which we are most comfortable, the science which explains how things work, which promises health, physical well-being, and material progress. But the boundaries of the physical and biological sciences are not so easily contained. From time to time we come upon a field of inquiry which fundamentally challenges our concepts of life and Nature, which confronts us too directly for our collective comfort or convenience, and yet intrigues us too greatly to ignore.

It is on this meeting ground of science and philosophy where man has made his greatest scientific advances. It is also here that science has caused its greatest strains upon our social, political and religious institutions.

When Galileo offered the theory that the Earth revolves around the Sun, it was bad enough to his contemporaries that he committed scientific error. It was worse that he committed heresy as well. Galileo was probing the physical universe. As science has progressed and transformed our lives, we have rejected many of the dogmas of an earlier day. And yet I suspect that many of us have harboured a feeling of security that there is at least one element of existence, the nature of life itself, that eludes scientific inquiry and control.

Recombinant DNA research challenges that presumption as profoundly as Galileo challenged the science and religion of his day. It poses for the scientific community fundamental questions of its role in society. It poses for government fundamental questions of its role in science.

The scientific community often resolves its own conflicts more easily than our political community can even understand them and has been doing so with some of the concerns which have been central to the recombinant DNA debate. But the scientific community cannot resolve by the same techniques the public policy concerns of its larger public constituency, and government cannot isolate itself from the science that it has encouraged and supported.

Consideration of these questions brings us once again face to face with the same fundemental issue which confronted Galileo in his time; the issue of society interacting with science and the place of society in the decision-making process in science. Although Galileo is remembered by most who recall his work for the significant contributions he made in physics and applied mathematics, we should not forget that one of his more important writings dealt with the freedom to conduct scientific inquiry.

Today, scientists are beginning to acknowledge the right of society to participate in decisions directly affecting the conduct of research; at the same time there is a strong belief in certain basic rights of scientific inquiry. It is ironic that society is now seeking an opportunity to participate in scientific decision-making when during the pre-Galileo period it was science pleading for the right to contribute to society's understanding of the nature of the universe.

In my opinion, the public is properly questioning the idea of total scientific freedom. The newly-acquired capability for manipulating the most fundamental processes of life has challenged society to think about the implications of this freedom. Limitations upon experimentation on human subjects, for example, demonstrate that reasonable restrictions can be articulated and respected.

I believe that the record established in the U.S. Congress on the recombinant DNA issue during the past four years in a unique case history of the legislative process interacting with the scientific community in the determination of the need for government intervention in a scientific process. A brief of the actions of the Congress may help us estimate what may be expected, at least in the United States, during the months ahead.

The House of Representatives Subcommittee on Science, Research and Technology, because of its jurisdiction over the broad area of science policy, has a particularly strong interest in this area of research ever since Dr. James Watson focused attention on these and other developments in cell physiology in 1971. However, this committee was not alone in expressing concern about the accelerating developments in molecular biology and their potential implications for society.

In 1968, the Committee on Government Operations in the U.S. Senate held hearings on a bill which proposed the establishment of a National Commission on Health, Science and Society. One of the subjects identified in this proposal as warranting congressional attention was a class of research called "genetic engineering". This interest probably stemmed from the increasing publication of articles on such subjects as cell transformation, transduction, cell fusion, cloning and other techniques. This was also the period during which Dr. Joshua Lederberg had been testifying before congressional committees in attempts to secure increased research support for genetics in general, thus these new capabilities were somewhat familiar to some members of the U.S. Congress.

At the time that the House Science Committee was hearing from Dr. Watson, the next series of hearings was being held in the Senate on the proposal to establish the National Advisory Commission on Health Science and Society. Again, these hearings addressed the "extraordinary advances in genetics", "parallel advances in virology" and the real possibility in the near future of "genetic engineering". Questions were being asked about the ethical limits, if any, which needed to be addressed with regard to these new developments, what unique or special considerations would come into play when man became the experimental species, what sorts of criteria should be established for any genetic intervention in man, and similar matters. This period of congressional inquiry preceded the public awareness of the developing techniques involving the use of the restriction enzymes for which Drs. Daniel Nathans and Hamilton Smith recently received the Nobel prize. Perhaps the scientific leaders who testified anticipated the availability of a similar technique to that which evolved as recombinant DNA.

The issue under examination was whether to set up an advisory group to examine ethical issues in scientific research with an emphasis on human experimentation. Notwithstanding the testimony of scientists the urgency of the developments specifically in molecular biology were not perceived at that time in Congress. However, the House Committee on Science and Technology did continue to monitor the progress of developments in this field and reports on developments in "genetic engineering", including recombinant DNA, were published. Thus, a profile of concern, although at a low level, about public policy and ethical issues in genetics was present in the U.S. Congress for reaction to the accouncements emanating from the Gordon Conference in 1973 and the Asilomar Conference in 1975. The message from the scientists reflecting a concerned attitude ignited the public profile of concern and intensified the debate.

In 1975, the Senate Committee on Labor and Public Welfare held a hearing to receive a summary report on the nature of the issues discussed at the Asilomar Conference. Next was a joint hearing in September 1976 before the Subcommittee on Health and a Subcommittee of the Judiciary, both committees in the Senate. During these hearings, Senator Kennedy made it clear that his concern went beyond questions of safety and were directed toward problems dealing with the potential social consequences of the new developments in recombinant DNA. This was the first indication that legislation would be proposed, if the NIH Guidelines were perceived as not achieving a desirable level of control over industry as well as Federal research.

A number of bills to regulate DNA research were introduced during the 95th Congress in 1977 and 1978. Some of these bills proposed new Federal regulatory powers, ranging from declaring a total moratorium on the research to setting up licensing and certification procedures for researchers and laboratories and special study commissions. Four congressional committees (two in the House and two in the Senate) held extensive hearings on the DNA issue and by the beginning of the second session of the 95th Congress, the legislation introduced in 1977 had been substantially changed. Two measures received final full committee action in their respective Houses, but no legislation reached the floor for a vote during the 95th Congress.

In my own subcommittee, I had a unique opportunity to provide a neutral ground for both advocates and opponents of government regulation of recombinant DNA research. Initially, our subcommittee was not examining specific legislation. Our purpose was to examine the issue as fully as possible in the hopes that such a sequence of hearings would provide an appropriate forum for policy discussion without the atmosphere of debate necessarily associated with the examination of specific legislative proposals. I had the interesting and intellectually stimulating task of presiding over hearings which eventually consumed 12 days of the legislative calendar and involved more than 55 witnesses. My basic purpose was to examine the science policy implications of recombinant DNA molecule research. While these hearings overlapped legislative hearings held by other committees and this necessarily affected the conduct of our hearings, I believe the non-legislative orientation of my subcommittee provided a more amicable forum for all parties in the debate and prepared us to act on legislation sequentially referred to us. The record of the hearings in the House Science Subcommittee now provides 1293 pages of unique case history information on this approach to congressional oversight.

In an approach somewhat similar to our hearings, hearings were held in the Senate in November 1977. During these hearings, the Senate Science Committee examined issues such as the potential impact of legislation on scientific freedom of inquiry and the possibility of interim legislation providing greater flexibility in accommodating scientific evidence which might evolve to demonstrate a reduced risk for recombinant DNA research.

Congress devoted considerable time to the study of the issues associated with recombinant DNA research. Proposed legislation was initially focused upon safety and hazard to the environment, issues which had been identified by scientists themselves as potential problems. In my opinion, it is of great significance that the implications for the regulation of basic research and the impact on national science policies in general were emphasized, and became the focus of concern. Students of science policy will find the congressional record on this subject the source of important background data.

This change in focus followed the increasing involvement of a large number of scientists from a broader spectrum of interests than the fields of science represented by the early recombinant DNA debates. As more data were presented from experts in infectious disease, microbiology, and the plant sciences, the legislation shifted away from rigid regulatory control with severe fines toward more of a Federal oversight function with greater local involvement. This interaction between the representatives in the Congress, the interest groups opposing recombinant DNA, and the scientific researchers desiring to permit greater freedom of research, was a fascinating example of the compromise which evolves as new information and ideas are given consideration.

I think it is useful to note some of the changes in policy position which occurred during this period of congressional reassessment of the need for legislation, and to review the development of regulatory restraints as an alternative.

First of all, the initial danger was expressed as the potential for catastrophic modification of our ecosystem or the production of some new human pathogen. As the hearings were held, it was soon perceived that the hypothetical risk posed by this scenario was no greater than the hypothetical risk that termination of research might prevent the discovery of a remedy for a naturally occurring catastrophic modification of our ecosystem or new natural pathogen. The guidelines continue to provide for a conservative approach in any work with which there is still any hesitancy. In fact, the guidelines still effectively prohibit release of recombinant organisms into the environment. This is a problem which will have to receive further attention if some beneficial applications are to be examined.

Another issue related to the institutionalisation of biohazard committees and the promulgation of occupational safety guidelines for recombinant DNA research. The arrangements for public participation in the project review process and the emphasis upon laboratory safety procedures received detailed attention during the review of the original guidelines and these actions alleviated most of the congressional concern on these issues.

The concern that representatives of public interest groups were not being provided an adequate opportunity to participate in the development of research guidelines has been alleviated by the actions taken by the Secretary of Health, Education and Welfare to ensure that all interested persons are free to comment on any proposed new procedures or modifications of guidelines, and by revising the composition of NIH's special advisory committee on recombinant DNA to provide for membership by a variety of non-scientific persons.

The need for Federal preemption of state rights for regulation of individual institutions has been temporarily relieved. Proposals for using the patent process for controlling industrial research were examined and deemed inappropriate. The recent action by the Secretary of HEW in calling for the Environmental Protection Agency and the Food and Drug Administration to use existing law to regulate industrial processes involving recombinant DNA activities reduced the intensity of concern about private industrial research.

This is not to suggest that non-legislative actions have eliminated all public concern. However, the non-legislative congressional discussions, the debates within the various states, and the tremendous investment of time by the Federal research institutions in public hearings has moderated the extreme intensity of opposition to recombinant DNA research. I would like to emphasize again that the involvement of the entire scientific community in this debate contributed significantly to these policy evaluations. If the research scientists had not taken the time to collect the experimental evidence to allay some of the hypothetica fears, severely restrictive legislation might have evolved. Although the processes now at work may slow the pace of some research, it is my opinion that patience will be rewarded.

My estimate of the position of pending legislation is that there is no intent to take any further action at this time. The only public statement on the recombinant DNA issue thus far in the U.S. 96th Congress was presented by Mr. Staggers, Chairman of the House Interstate and Foreign Commerce Committee. In his statement, I believ that Mr. Staggers revealed the attitude of most members. He stated his belief that this research area can bring many benefits and increase understanding of many basic principles of cell activity. He noted that his committee would continue to follow the implementation of the NIH Guidelines and ascertain whether the scientific community is able to function effectively under these carefully constructed controls

In my short tenure with the Director of NIH's advisory committee on recombinant DNA research, I have been tremendously impressed by the talent of the professional members of the committee and the sense of responsibility for their role on the committee which is held by the public of non-scientific members. As achievements in this field move the technique along toward commercial applications, I hope that the scientific community will continue to exercise the patience, forebearance, and sense of responsibility demonstrated by compliance with the guidelines which have been established in the various countries, while providing leadership in efforts to maintain compatibility across national boundaries. Where new evidence indicates that these guidelines can be modified, these changes should occur.

From a public policy perspective, I know that many members of the scientific community became discouraged because of the tremendous amount of time required to alleviate public concern, whether real or imagined. I think that these scientists should realise that their willingness to participate in public discussions of their research will be of continuing value. The legislative bodies have gained insight from the interactions between scientists, public interest groups, and the body politic. The Federal research establishments have developed new techniques to permit a greater participation of the public in the determination of research priorities. It is my opinion that the education of the public about the uncertainties of scientific research will help secure an understanding of why progress toward the resolution of many public health problems is so unpredictable and demands continued and persistent support.

I believe much progress has been made both by government and science in resolving issues of regulation of scientific research. In my view, the responsible attitude of scientists in writing the "Berg" letter was critically important to the development of our present objective of relaxing regulation because it demonstrated that scientists can be trusted to choose responsible and careful courses of conduct.